MYELINATION AND DEMYELINATION

Implications for Multiple Sclerosis

MYELINATION AND DEMYELINATION

Implications for Multiple Sclerosis

Edited by

Seung U. Kim

University Hospital
University of British Columbia
Vancouver, British Columbia, Canada

PLENUM PRESS • NEW YORK AND LONDON

Library of Congress Cataloging in Publication Data

Satellite Symposium on Myelination and Demyelination: Implications for Multiple
Sclerosis (1987: Vancouver, B.C.)
  Myelination and demyelination.

  "Proceedings of the Twenty-Second Canadian Congress of Neurological Sciences,
Satellite Symposium on Myelination and Demyelination: Implications for Multiple
Sclerosis, held June 22, 1987, in Vancouver, British Columbia, Canada"—T.p.
verso.
  Includes bibliographies and index.
  1. Multiple sclerosis—Pathophysiology—Congresses. 2. Myelination—Congress-
es. 3. Demyelination—Congresses. I. Kim, Seung U. II. Canadian Congress of
Neurological Sciences (22nd: 1987: Vancouver,B.C.) III. Title. [DNLM: 1.
Demyelinating Diseases—congresses. 2. Myelin Sheath—physiology—congresses. 3.
Mutliple Sclerosis—etiology—congresses. WL 360 S253m 1987]
RC377.S25   1987   616.8′3407   89-3684
ISBN 0-306-43118-1

Proceedings of the Twenty-Second Canadian Congress of Neurological Sciences
Satellite Symposium on Myelination and Demyelination: Implications
for Multiple Sclerosis, held June 22, 1987, in Vancouver,
British Columbia, Canada

© 1989 Plenum Press, New York
A Division of Plenum Publishing Corporation
233 Spring Street, New York, N.Y. 10013

PREFACE

In June 1987, neurobiologists, immunologists, molecular biologists, virologists and neurologists from several countries met in Vancouver to discuss recent advances of relevance to multiple sclerosis. The symposium was a part of the 22nd Canadian Congress of Neurological Sciences meeting and was sponsored by funds from the Multiple Sclerosis Society of Canada and the Medical Research Council of Canada. The presentations covered five major topics: basic neurobiology, molecular biology, the role of viruses in demyelination, immune function and dysfunction in multiple sclerosis, and clinical magnetic resonance imaging studies. It was heartening to note that scientists from several different disciplines were working towards a common end-point: the understanding and treatment of multiple sclerosis.

In this book, speakers at the symposium have each presented a chapter of their findings and discussions. In addition, some non-participants at the symposium have been invited to submit chapters in order to give this volume a more complete scope. It is hoped that the reader will find this book a useful reference for several subjects of interest to multiple sclerosis.

In closing, I would like to thank the following for their help and support: the Multiple Sclerosis Society of Canada and the Medical Research Council of Canada for their financial support; the contributors of this book for their manuscripts; Dr. A. Eisen, Mrs. K. Eisen, Mrs. P. Bodnarchuk and Mrs. M. Kim for their efforts in planning and organizing the symposium; and Ms. Catherine Schikowski for her secretarial assistance.

Seung U. Kim, M.D. Ph.D.
Editor

CONTENTS

NEUROIMMUNOLOGY OF MULTIPLE SCLEROSIS

CLINICAL ASPECTS OF MULTIPLE SCLEROSIS

HUMAN OLIGODENDROCYTES IN CULTURE: BIOLOGY AND IMMUNOLOGY

Seung U. Kim

Division of Neurology
University of British Columbia
Vancouver, Canada

INTRODUCTION

In the central nervous system (CNS), oligodendrocytes (OL) are responsible for the formation and maintenance of myelin which facilitates saltatory conduction of nervous impulses along axons. Abnormality and loss of OL and myelin result in several neurological disorders in human, among them multiple sclerosis (MS) for which there is no effective treatment (37,40,41). To better understand the pathogenesis of these disorders, it would be helpful if an enriched OL population could be isolated, grown in culture, and the basic properties of these cells studied in a controlled environment.

During the past few years, several reports have described the isolation, enrichment and culture of OL from the brain of several mammalian species. The isolation process employs one of two basic techniques: (a) bulk isolation of OL following trypsin digestion and density gradient centrifugation, and (b) isolation of OL by differential adhesion and subsequent shaking from brain cell cultures established previously. Neonatal or adult brains from calf, pig, sheep, cat, rat, monkey and human (13,17,23,32,38,47,66) were used. These cultured OL expressed cell type specific marker such as galactocerebroside, myelin basic protein, cyclic nucleotide phosphodiesterase and myelin associated glycoprotein (23,24,32).

Large quantities of pure cultured OL can provide us with a unique and important model system to investigate the biochemical and immunological properties of the OL and their response to viral or immunological agents suspected of being causative factors of the disease.

This chapter provides a review of our attempts and achievements in the studies of oligodendrocytes and multiple sclerosis using human and bovine OL in culture.

Oligodendrocyte Isolation and Culture

For the isolation of human OL, brains obtained 3-20 hours postmortem were used. The patients had no evidence of neurological disease. Their age varied from 18-95 years. During the past 5 years (1983 -

1988), we have isolated and grown human OL from 120 individuals. Except for several series in which bacterial contamination destroyed isolated OL, we have been successful in obtaining healthy OL cultures.

White matter tissue weighing 5 - 50 grams was cut into small pieces and incubated in calcium- and magnesium-free Hanks' balanced salt solution (BSS) containing 0.25% trypsin and 20 µg/ml DNase for 40 minutes at 37 degrees C in a shaking water bath. The tissue digests were dissociated into single cells by gentle pipetting, washed 3 times in BSS, and passed through a 100 µm nylon mesh filter fashioned in a Buchner funnel. Isolated cells in 20 ml BSS were mixed with 1 ml 10X concentrated BSS and 9 ml Percoll (Pharmacia). A gradient was formed by 30 minutes centrifugation at 15,000 rpm in a Beckman high speed refrigerated centrifuge with a fixed angle rotor. OL collected between an upper dense myelin layer and lower erythrocyte layer were retrieved, diluted in BSS, and harvested by low speed centrifugation for 10 minutes at 1,500 rpm. These OL were suspended in feeding medium containing 5% fetal bovine serum, 5 mg/ml glucose, and 20 ug/ml gentamicin in Eagle's minimum essential medium, and seeded on polylysine-coated 9 mm Aclar plastic coverslips (Allied Chemical). Cell yield was 2 x 10^6 cells/gm tissue. Cell vitality checked by trypan blue exclusion immediately after the cell isolation ranged from 75-90%. After 24 hours, the coverslips were flooded with 3 ml of feeding medium. The details of the isolation and culture techniques can be found in previous reports (28,32).

Cell Type Specific Markers for Oligodendrocytes

In the past the identification of major neural cell types in culture has been dependent primarily upon morphological criteria, using phase contrast microscopy, which can be inaccurate and confusing. Recently, immunochemical methods using cell type specific markers have become a useful alternative and solution for this problem. Now several cell type specific markers are available for identifying major cell types in CNS cultures (Table I) (21,23,54). The markers considered specific for oligodendrocytes by biochemical and immunocytochemical criteria are galactocerebroside, sulfatide (sulfogalactosylceramide), ganglioside GM4 (sialosylgalanctosylerceramide), myelin basic protein, proteolipid and myelin associated glycoprotein, among others.

Since the first report of Raff et al (55) demonstrating a positive immunostaining of rat OL by rabbit anti-galactocerebroside (GC) serum, antibody directed against GC has been the most reliable and commonly applied cell type specific marker for OL in culture. However, its usefulness is less evident in tissue sections because of difficulty preserving surface antigens of OL cell body and processes in sections (46). Thus, GC immunostaining can be detected only in cell membrane of OL and Schwann cells and is not seen with other neural cell types.

Other specific cell markers for OL include several major components of myelin proteins. CNS myelin is very rich in lipid (more than 70% in dry weight) and the rest is made up of proteins. The myelin proteins consist mainly of proteolipid (PLP), 50% of total protein; myelin basic protein (MBP), 35% of total protein; 2',3'-cyclic nucleotide-3'-phosphohydrolase (CNP), 5% of total protein; myelin associated glycoprotein (MAG) 1% of total protein, and several enzymes. Antisera or antibodies specific for these major myelin proteins naturally become excellent candidates for the cell type-specific marker for OL. With immunocytochemical methods, PLP and MBP have been shown to be present in OL before myelination in the studies using tissue sections (2,18,65).

Several investigators have utilized antisera specific for MBP to identify OL in culture and studied its expression by assessing the degree of immune reaction intensity during development. Whereas these studies have been performed mostly in rat and mouse cells (3,5,11,22,44,54), we have demonstrated MBP immunoreaction in OL isolated and cultured from adult human brains (23,32). Double immunofluorescence staining for surface GC and intracellular MBP in human OL grown in serum-free medium is shown in Figure 1.

As for the immunostaining of cultured OL by PLP antibody, Dubois-Dalcq et al have reported that the PLP expression by neonatal rat OL came one week after galactocerebroside expression and 1-2 days after MBP and MAG phenotypes (11). In both adult human and bovine OL cultures, we have seen positive immunostaining of these cells by the same rabbit anti-PLP serum used by the authors cited above.

High enzyme activity of CNP has been found in central myelin and isolated OL (50,51). In cultured OL obtained from the brains of mouse (4-5 weeks), calf (2-5 months) and human (18-95 years), we have previously demonstrated a specific and intense immunofluorescence staining of OL cell bodies and processes indicating that the CNP immunostaining is primarily localized in OL and not in astrocytes or neurons (24). A similar immunostaining of cultured OL by CNP antibody was reported by a French group (59).

MAG is a minor transmembrane glycoprotein of CNS myelin that is localized on the periaxonal and non-compacted areas of the myelin sheath (53,67). The only available report on the immunocytochemical demonstration of MAG in cultured OL was that of Dubois-Dalcq in which rat OL obtained from neonatal animals was found to react to anti-MAG antibody after 1 week in culture (11). These authors reported that expression of MAG, MBP and PLP appeared in neonatal rat OL about 1 week after the immunostaining by GC antibody. We have also found that the adult human and bovine OL grown in culture could be immunostained by anti-MAG polyclonal or monoclonal antibodies. However, we were not able to obtain a positive MAG immunostaining after fixation in acid-alcohol as described by Dubois et al. Only when cultured OL was reacted to the antibody prior to fixation was positive staining obtained.

Ganglioside is the generic term for glycolipids containing sialic acids. Recently, an exclusive localization of GM4 ganglioside, a minor class of ganglioside, was reported in isolated human myelin and OL preparations (14). It is thus possible to utilize an antibody directed against GM4 ganglioside as a cell type-specific marker for cultured OL. Double immunofluorescence studies in adult human glial cell cultures have shown that all the GC-positive OL were labelled by GM4 antibody. However, it is noteworthy that about 80% of glial fibrillary acidic protein (GFAP)-positive astrocytes were also found to react to GM4 antibody (25). On the other hand, GD3 ganglioside, another minor ganglioside found in the brain, was found to be a more specific OL marker as compared to GM4 ganglioside. It was demonstrated that all of GC positive OL were labelled by GD3 antibody, while less than 5% of GFAP-positive astrocytes was immunostained by GD3 ganglioside antibody (25). From our results it is evident that GD3 ganglioside could be used as a cell type-specific marker for OL, while GM4 gangliosides considered as specific markers for both OL and astrocytes.

In addition to GM4 and GD3 gangliosides, we also investigated immunoreactions of human neuronal cells to antibodies specific for GM1 gangliosides and GQ1C ganglioside. The results are summarized in Table 2 in which specific binding of antibodies to different classes of CNS cells can clearly be recognized.

The recent development of techniques for producing hybridomas that synthesize monoclonal antibodies against single antigens has essentially revolutionized the approach to production and application of specific antibodies. By using this new hybridoma technique, a group in Heidelberg has produced 4 related monoclonal antibodies reactive to OL surface antigens (60). These antibodies are reported to be specific for galactocerebroside or sulfatide, but details have not been published.

Using bovine brain synaptosome fraction as immunogen, Ranscht et al have produced monoclonal antibody against galactocerebroside which specifically labelled the surface of cultured rat OL and Schwann cells (56). This monoclonal antibody has been used widely by many investigators in the studies of OL development in culture. More recently, monoclonal antibodies directed against various myelin proteins such as MBP and MAG have been produced and used in the studies of developmental expression of these proteins *in vivo* and *in vitro* (8,10,43).

Poduslo et al produced a rabbit antiserum directed against whole cell preparation of bovine OL which reacted positively with the surface of the isolated OL (52). Reaction could be blocked by absorbing the antiserum with isolated OL though, surprisingly, not with myelin. Similar antisera have been produced by Abramsky et al (1), Traugott et al (69) and Lisak et al (39). These antisera produced by different investigators were found to react with surface and intracellular antigens of isolated OL and brain sections. These antisera were produced in rabbits by repeated injections of bovine OL isolated by the sucrose density gradient method of Poduslo and Norton (51). Recently we have also raised a series of antisera in rabbits injected with human and bovine OL isolated by Percoll density gradient method which reacted with the surface antigens of human OL in culture. As yet the nature of the antigen(s) to which these antisera are directed have not been characterized.

Serum-Free Chemically Defined Medium for Oligodendrocytes

The growth of most cell types in culture requires the addition of serum to the nutrient medium. Serum is a complex mixture and its components are not well characterized. The role of these active serum components is to provide not only greater nutritional support but also special factors which are required for the attachment, survival, proliferation, and differentiation of the cultured cells. Several approaches have been taken to eliminate serum requirements by adapting cell lines to the medium without serum, or by the addition of active components of serum to the medium. Sato and his associates have succeeded in growing various cell lines in serum-free media supplemented with several hormones and growth factors (4). These investigators demonstrated that one of the main functions of serum is to provide a mixture of hormones which stimulate cell growth. More recently, we have applied this approach to primary neural culture in order to learn the minimal nutritional requirements of the developing nervous system. We have demonstrated that insulin, when substituted for serum, was important for the survival and differentiation of chick dorsal root ganglion neurons in culture (62,63). We have also shown that by the use of serum-free chemically defined medium, it was possible to isolate and grow astrocytes selectively in cultures of newborn rat cerebrum (27,31).

In an attempt to formulate a serum-free chemically defined medium for long-term culture of OL isolated from adult human brain, we have modified a formula originally designed for rat astrocyte cultures (DM2) (31).

After 24-48 hours in 5% fetal bovine serum containing medium,

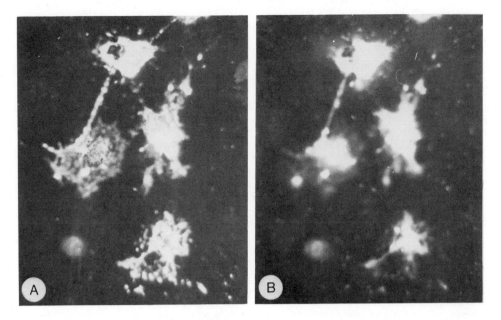

Figure 1   Double immunolabelling of cultured human oligodendrocytes
by galactocerebroside and myelin basic protein antibodies. X 700
A: Galactocerebroside.
B: Myelin basic protein.

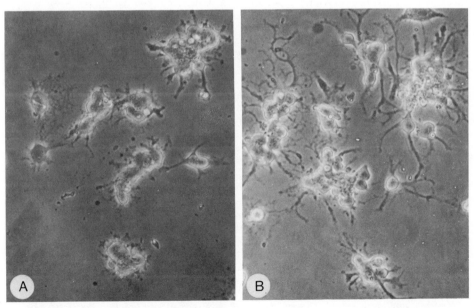

Figure 2   Phase contrast microscopy of human oligodendrocytes
isolated from the brain of 76 year old male and cultured for 68 days.
X 500
A: Oligodendrocytes grown in serum containing medium.
B: Oligodendrocytes grown in DM3 serum-free medium.

coverslips carrying human OL were washed twice in Hanks' balanced salt solution and fed with F12 plus 5% fetal calf serum or with serum-free chemically defined medium (DM3) composed of F12 supplemented by hormones, growth factors and trace metal ions. F12 synthetic medium was supplemented with 2 ml 50X MEM essential amino acid solution, 1 ml 100X MEM non-essential amino acid solution and 1 ml 100X MEM vitamin solution with the following hormones, growth factors and trace metal ions: Insulin 20 μg/ml, triiodothyronine 3×10^{-8}M, hydrocortisone 5×10^{-8}M, transferrin 10 μg/ml, glucose 255 mg/ml, human serum albumin 75 mg/ml, gentamicin 20 μg/ml, HEPES 1×10^{-2}M, manganese chloride 5×10^{-9}M, ammonium molybdate 3×10^{-9}M, nickel sulfate 3×10^{-10}M, and sodium silicate $1 \times 1-^{5}$M, sodium vanadate 5×10^{-9}M and sodium selenite 3×10^{-8}M. In addition, F12 medium contains 1×10^{-8}M copper sulfate, 3×10^{-6}M ferric sulfate and 1×10^{-6}M zinc sulfate.

Representative cultures of OL grown in serum-containing or serum-free chemically defined medium (DM3) are shown in Figure 2. The cells grown in DM3 acquired obviously longer and thicker processes than cells grown in serum-containing medium. The longest period that OL were maintained without ill effects was 4 months (30). After this period, OL became granular, atrophic, and fragmented, indicating that DM3 formula lacks essential nutrients/growth factor(s) which could sustain long-term survival of OL in vitro. Since OL grown in serum-containing medium (the concentration of serum could be as low as 1%), could survive more than 6 months, this nutrient/growth factor(s) apparently is a component of fetal bovine or horse serum. Identification of this factor in serum requires further investigation.

Modulation of Antigenic Expression by Oligodendrocytes

It has been known for some time that OL-enriched cultures may become overgrown by astrocytes and that after a month or two in vitro, GC-positive oligodendrocytes make up less than 60% of the cell population (38).

The phenotypic expression and biological modifications occurring in human OL during their maintenance in vitro has been followed in this laboratory. Following the replating procedure, most of the small round cells (8-10 μM diameter) became attached to the cover slips within 24 h and by 4-5 days they became bipolar or multipolar cells. At this stage typically 95-98% of them were OL expressing GC on their surface. Most GC-negative cells were GFAP-positive astrocytes and very few (less than 1%) were fibronectin-positive fibroblasts (Table 3). However, after 2 weeks in vitro, the OL population decreased while the number of astrocytes increased considerably. In addition, many cells (7-20% of total cell population) were immunolabelled by both GC and GFAP antibodies (26).

The concurrent occurrence of reduction in OL population and increased astrocyte population was an unexpected finding for us. We speculated that an active proliferation of astrocytes and the paucity in OL mitosis during the early phases of culture might underlie this shift in cell population dynamics. To examine this hypothesis, we exposed cultures to [^{3}H]-thymidine for 24 h, immunostained with GC antiserum and GFAP antibody, and developed the cultures for autoradiography. None of the GC-positive or GFAP-positive cells incorporated radiolabelled thymidine although a small number of fibroblasts did. This result suggested the possibility of cellular transformation/modulation of OL to astrocytes (26).

To determine whether this transformation could be impeded, we exposed cells to different concentrations of dibutyryl cyclic AMP (dbcAMP) for a short period. We found that dbcAMP at a final concentration of 10^{-3}M produced a significant increase of the GC-positive cells with a concomitant reduction in both GFAP-positive and double-positive cells. A lower concentration of dbcAMP (10^{-4}M) was less effective, while a higher concentration (10^{-2}M) was toxic.

In the second set of experiments we exposed cells for 5 days to butyric acid, dbcAMP, 8-bromo cyclic AMP, and RO-1724, an inhibitor of cyclic nucleotide phosphodiesterase (Table 4). It was found that dbcAMP and 8-bromo cyclic AMP treatment produced a marked increase in GC-positive cells in comparison with control cultures.

It has been known that cyclic AMP derivatives, or inhibitors of cyclic nucleotide phosphodiesterase capable of raising the intracellular level of cyclic AMP, affect a number of events in cultured cells. One of these effects is to promote differentiation in rodent primary astrocyte cultures (36,61). More recently it has been reported that dbcAMP induces 2',3'-cyclic nucleotide 3'-phosphohydrolase activity in rat OL. In addition, cyclic AMP derivatives have been shown to promote the expression of galactocerebroside on the surface of rat Schwann cells (64).

Dibutyryl cyclic AMP did not stimulate glial cell division but promoted a transformation from GFAP-positive cells to GC-positive cells. The interconvertibility of astrocytes and OL by external environmental influences has already been reported (54). Moreover, a recent study in the developing human fetal spinal cord of 8-10 weeks gestation has shown that GFAP-positive astrocytes are the first distinguishable neuroglial element and that cells with morphological and immunocytochemical features of OL appear later, when myelin formation begins (7). These authors also observed "transitional forms" between astrocytic and oligodendrocytic elements in which cells expressed markers for both OL and astrocytes. Thus, they have suggested that radial glia may give rise to both astrocytes and OL.

In the present cultures, most of the adult glial cells _in vitro_ became GC-positive OL within 2-7 days. After several weeks in culture, a marked expansion of the astrocytic population (20-30%) was observed, along with a significant number of "transitional form" cells expressing both GC and GFAP markers. These data suggest that glial cells might have a high degree of plasticity and under certain conditions, e.g., _in vitro_ and in absence of neurons, many lose their specificity and revert from a mature to a more undifferentiated stage. The presence of cyclic AMP derivatives might create a favorable condition similar to that _in vivo_ so that glial cells start to show the same characteristics and properties observed _in vivo_. It appears that cyclic AMP may play an important role in glial cell differentiation. It may also be involved in the process of priming OL to become associated with axons.

Major Histocompatibility Complex Antigens

The major histocompatibility complex (MHC) is a tightly linked cluster of genes located on chromosome 17 in mouse and on chromosome 6 in humans. These genes encode for a series of highly polymorphic cell surface glycoproteins which can be divided in two classes: Class I (H-2 in rodents and HLA-A,B,C in humans) and Class II (Ia in rodents and HLA-D system in humans). The MHC proteins are important in a phenomenon called MHC or T cell restriction. In the context of immune regulation, T lymphocytes cannot recognize an antigen on the cell surface if the appropriate MHC antigen is not simultaneously expressed by that cell.

Class I proteins are the usual restriction elements for cytotoxic T cells whereas Class II molecules are the restriction elements for helper T cells (33,57,71).

Class I glycoproteins have a relatively widespread distribution while Class II antigens are expressed primarily on cells involved in immune function. The latter include B lymphocytes, some cells of macrophage-monocyte lineage and certain accessory cells of the immune response such as dendritic cells and Langerhans cells (73).

Earlier studies using human brain sections have suggested that the cells of the CNS lack either Class I or Class II MHC antigens (11,15,72) and this has been partly responsible for the concept of immunological privilege of the brain. More recently, however, conflicting results have begun to emerge. Several investigators have presented negative findings of MHC antigen expression in normal human brain sections (9,35,45,68). On the other hand, other investigators have reported a limited distribution of MHC antigens in human brain sections (12a,15a). The reasons for these conflicting and controversial findings among investigators are not clear. Several factors could be considered: one is the use of brain sections, either fixed or fresh frozen. MHC antigens may be destroyed or altered during the process of section preparation or the antibody may have poor access to the antigen of interest; another is the possibility that the expression of MHC antigen levels in brain may be so low as to be below the sensitivity of the methods of detection employed. Tissue culture studies may circumvent some of the difficulties associated with tissue sections and several investigators have resorted to them. The obvious importance of MHC antigens in immune regulation, and the possibility that aberrant expression of these glycoproteins may lead to certain autoimmune disease, makes further studies on cultured cells, especially those from humans, of special importance (23,32a).

Of 14 series of human adult OL cultures analyzed in our laboratory for expression of HLA-A,B,C, 13 showed positive immunostaining. The extent of Class I expression varied, ranging from 5% of cells in two series to 90% or more in eight series. Only one sample was negative for HLA-A,B.C expression. One series was analyzed at four different time points and each time the percentage of GC-positive OL that presented with HLA-A,B,C was over 90%. Neither age of donors nor cause of death, appeared to determine the number of cells that expressed HLA-A,B,C (32a).

Of 12 series of human adult OL cultures that were double stained with antibodies specific for HAL-DR and galactocerebroside, 6 showed positive immunolabelling. The extent of HLA-DR expression varied, ranging from 4 to 16%. Six samples were found negative for HLA-DR expression. One series of OL was analyzed for HLA-DR at 3 different times in culture and never expressed HLA-DR antigen. As in the case of Class I antigens, the age of donors, their cause of death or length of time in culture did not influence the immunolabelling by HLA-DR antibody (25).

The brain has been considered to be privileged for three obvious reasons. Firstly, access of cells and macromolecules to the brain proper is restricted by the endothelial cells and astrocytes that surround the blood vessels of the brain, forming the blood-brain barrier. Secondly, the brain has been thought to lack lymphatic drainage, although more recent findings suggest otherwise. Thirdly, cells within the parenchyma of the brain are thought to be negative for MHC antigen expression (9,35,45,68). Our results have shown that a substantial population of adult human OL express both HLA-A,B,C and HLA-DR antigens. In many cases, MHC Class I antigen was detected in over 90% of OL, while Class II antigen was found in less than 15% of OL populations.

Table 1 Cell Type Specific Markers for Oligodendrocytes

Galactocerebroside
Sulfatide
Ganglioside GM_4
Ganglioside GD_3
Myelin basic protein (MBP)
Proteolipid protein (PLP)
Myelin-associated glycoprotein (MAG)
2',3'-Cyclic nucleotide 3'-phosphodiesterase
Carbonic anhydrase C
Glycerol phosphate dehydrogenase

Table 2 Immunoreavtivity of Anti-Ganglioside Antibodies with Cultured Human Neural Cells

	GM1(%)	GM4(%)	GD3(%)	GQ1C(A2B5)(%)
Oligodendrocytes (GC^+)	100	100	100	0
Astrocytes ($GFAP^+$)	80	80	5	0
Fibroblasts (FN^+)	5	0	0	0
Neuons (NF^+)	80	0	0	60-70

Table 3 Immunofluorescence Staining of Cultured Human Glial Cells for Cell Type-Specific Markers

Days in Vitro	GC^+	$GFAP^+$	$Both^+$	Neither
2	462 (97.0%)	8 (1.7%)	0	6 (1.3%)
7	388 (92.6%)	26 (6.2%)	0	5 (1.2%)
21	603 (69.1%)	146 (16.7%)	114 (13.1%)	10 (1.1%)
72	498 (51.2%)	252 (25.9%)	212 (21.8%)	12 (1.2%)

Table 4 Effects of Cyclic AMP Derivatives and RO-1724 on Human Glial Cells in Culture

Treatment	GC^+(%)	GF^+(%)	$Both^+$(%)
5 days			
Control	60.5	19.5	20.0
8-Bromo cAMP	70.6 (+10.1)	16.3 (-3.2)	13.1 (-6.9)
Dibutyryl cAMP	80.2 (+19.7)	13.8 (-5.7)	6.0 (-14.0)
RO-1724	74.8 (+14.3)	11.4 (-8.1)	13.8 (-6.2)
Butyrate	62.1 (+1.6)	18.2 (-1.3)	19.7 (-0.3)

It is evident from our study that the MHC expression, both for Class I and II antigens, by galactocerebroside-immunostaining positive human OL was detected in the absence of any inducing factors, and where the possibility of artefactual immunostaining was eliminated by careful studies of control cultures.

The widespread distribution of MHC antigens of both Class I and II on cultured human OL as presented in our studies (25,32a) is only one of the kind in the literature. For the negative results of MHC expression that are obtained from human brain sections (9,35,45,68), a possibility is that the antigen of interest might have been destroyed during the process of tissue preparation.

It is our conclusion that a large number of adult human OL express detectable levels of MHC antigens of both Class I and Class II. The presence of MHC antigens in human OL as described in our study then suggest that the brain is not as immunoprivileged as previous studies have indicated.

Myelinotoxic Factor in MS Serum and Cerebrospinal Fluid

The first evidence for a circulating myelinotoxic antibody in the sera of multiple sclerosis patients was the observation that sera from some patients in the midst of acute attack is capable of demyelinating rat central nervous system (CNS) cultures (6). Figures published by Bornstein suggest that serum from approximately 50% of patients with multiple sclerosis demyelinate rat CNS cultures. Studies of similar demyelination induced in mouse CNS cultures indicate that the active factor, probably gamma globulin, is present in cerebrospinal fluid from MS patients as well (29). One reported attempt to identify the antigen target of this circulating antibody by absorption with myelin and brain fractions showed that demyelinating activity was absorbed only by a pellet containing blood vessels and OL. An immunolabelling technique was used to demonstrate the presence of anti-oligodendrocyte antibody sera from MS patients (1) and in this study, positive binding occurred in 19 of 21 MS patients and none in normal controls or other neurological disease. Other investigators, however, did not confirm these observations (21,70). Since these studies used poorly characterized, non-viable, non-human OL as target cells, we considered it important to investigate if sera and cerebrospinal fluids obtained from MS patients can bind immunologically to human OL grown in culture. Several separate experiments performed so far have not produced any positive staining of OL by MS serum. However, it is too early to discount this lead, because the immunocytochemical technique we employed might not be sensitive enough to detect the presence of cytotoxic factor in MS cerebrospinal fluid or serum.

HTLV-I Virus Infection

In multiple sclerosis, the causes of destruction of OL and myelin sheaths are unknown. Viral infection has been suspected to play a role in the pathogenesis of MS, but no virus has been identified to link directly to the disease (19). Recently Koprowski and his associates reported a close association between a human retrovirus of HTLV family and some MS patients (34). They have found that sera and CSF from MS patients contain antibodies that cross-react with HTLV proteins. In addition, they have shown the presence of HTLV RNA in T cells cultured from CSF of such patients. The results suggest that these MS patients may have been infected by an HTLV probably related to HTLV-I.

HTLV-I was discovered first in 1980 and is the cause of adult T cell leukemia in Japan (16). Recently, Osame and his colleagues in Japan have reported the occurrence of aggregated cases of neurological patients with spinal myelopathy which is caused by HTLV-I (49). In the West Indies, in clinically similar tropical spastic/ataxic paraparesis (TSP), more than half of patients showed HTLV-I antibody in serum and more recently also in their CSF (74). Saida and his associates have reported the detection of HTLV-I antibody in sera of some Japanese MS patients by radioimmunoassay and Western blots (48), confirming the possible involvement of HTLVs in the pathogenesis of MS as reported by Koprowski and his colleagues. A number of reports are now available that suggest a possible association of HTLV-I or related viruses to the etiology of MS and other chronic CNS myelopathies (20,58).

In order to clarify the role of HTLV-I in the possible infection of CNS neural elements, it would be desirable to obtain a purified population of such cells, maintain them in vitro, and then assess the effects of HTLV-I on these cells under a controlled environment. Recently we have performed a study at our laboratory in collaboration with Dr. T. Saida of Kyoto. We infected mixed cell cultures of human OL and astrocytes with HTLV-I by co-culturing these cells with MT2, a HTLV-I producing T cell line. After 7-14 days of exposure to virus, the presence of HTLV-I antigen in OL and astrocytes was detected by double labelling immunofluorescence microscopy using rabbit anti-galactocerebrosideor anti-glial fibrillary acidic protein serum and mouse monoclonal antibody specific for HTLV-I p19 gag protein. In addition, electron microscopic examination of sister cultures demonstrated the production of virus particles inside and outside of OL and astrocytic plasma membrane.

These results indicate that HTLV-I is capable of infecting human OL and astrocytes in addition to lymphoid cells and would lend support to the contention that HTLV-I plays a primary role in the pathogenesis of HAM and TSP, chronic neurological disorders of the CNS.

CONCLUSIONS

The importance of the oligodendrocytes in the formation and maintenance of myelin in the CNS and in the understanding of the clinical problems off multiple sclerosis is self-evident. It is now possible to isolate a large number of human oligodendrocytes from autopsied brain and maintain them in culture for long periods. Using these cultures, basic biological, physiological and immunological properties and traits of oligodendrocytes can be investigated in the absence of other cell types which may interfere with phenotypic expression of the cells in question. Using these cultured cells it is also possible to study the effect of putative growth factors for oligodendrocytes such as lymphokines, epidermal growth factor, platelet derived growth factor and axonal membrane preparation which are reported to be mitogenic for oligodendrocytes.

Using analytical techniques currently available to us, much has been learned during the past 7 years since we have isolated human oligodendrocytes in quantity for the first time. The challenge of the future is to identify and understand molecular signals and mechanisms that regulate oligodendroglial structure and function during their development and maturation using the well-characterized culture system described in this chapter.

ACKNOWLEDGEMENTS

The works reported herein are supported by grants from the Medical Research Council of Canada, the Multiple Sclerosis Society of Canada, the National Multiple Sclerosis Society (U.S.) and the Jacob Cohen Funds for the Research in Multiple Sclerosis.

REFERENCES

1. Abramsky O, Lisak R, Silberberg D, Pleasure D: Antibodies to oligodendroglia in patients with multiple sclerosis. N Eng J Med 297:1207-1211, 1977

2. Agrawal HC, Hartman BK, Shearer W, Kambach S, Margolis F: Participation and immunohistochemical localization of rat brain myelin proteolipid protein. J Neurochem 28:495-508, 1977

3. Barbarese E, Pfeiffer SE: Developmental regulation of myelin basic protein in dispersed cultures. Proc Natl Acad Sci USA 78:1953-1957, 1981

4. Barnes D, Sato G: Methods for growth of cultured cells in serum free medium. Analyt Biochem 102:255-266, 1980

5. Bologa-Sandru L, Siegrist HP, Z'Graggen A, Hoffman K, Weissman U, Dahl D, Herschkowitz J: Expression of antigenic markers during the development of oligodendrocytes in mouse brain cultures. Brain Res 210:217-229, 1981

6. Bornstein MB, Appel S: Tissue culture studies of demyelination. Ann NY Acad Sci 122:280-286, 1965

7. Choi BH, Kim R: Expression of glial fibrillary acidic protein in immature oligodendroglia. Science 223:407-409, 1984

8. Chou CJ, Cox A, Fritz RB, Wood JG, Kibler RF: Monoclonal antibodies to human myelin basic protein. J Neurochem 46:47-53, 1986

9. Daar AS, Fuggle S, Fabre J, Ting A, Morris P: The detailed distribution of MHC Class II antigens in normal human origins. Transplan 38:293-298, 1984

10. Dobersen M, Gascon P, Trost S, Hammr J, Goodman S, Noronha A, O'Shannessy D, Brady R, Quarles RH: Murine monoclonal antibodies to the myelin-associated glycoprotein react with large granular lymphocyte of human blood. Proc Natl Acad Sci USA 82:552-555,

11. Dubois-Dalcq M, Behar T, Hudson L, Lazzarini RA: Emergence of three myelin proteins in oligodendrocytes cultured without neurons. J Cell Biol 102:384-392, 1986

12. Edidin M: The tissue distribution and cellular location of transplantation antigens. In: Kahan B, Reisfeld R (eds.), Transplantation antigens, Academic Press, New York, pp 125-240, 1972

12a. Frank E, Pulver M, Tribolet N: Expression of Class II MHC antigens on reactive astrocytes and endothelial cells within gliosis surrounding metastases and abscesses. J Neuroimmunol 12:29-36, 1986

13. Gebick-Harter PJ, Althaus H, Schwartz P, Neuhoff V: Bulk separation and long-term culture of oligodendrocytes from adult pig brain. J Neurochem 42:357-368, 1981

14. Hamberger A, Svennerholm L: Composition of gangliosides and phospholipids of neuronal and glial cell enriched fractions. J Neurochem 18:1821-1829, 1971

15. Hammering G: Tissue distribution of Ia antigens and their expression on lymphocyte subpopulations. Transplant Rev 30:64-82, 1976

15a. Hauser S, Bhan A, Gilles F, Hoban C, Reinherz E, Schlossman S, Weiner H: Immunohistochemical staining of human brain with monoclonal antibodies that identify lymphocytes, monocytes and the Ia antigen. J Neuroimmunol 5:197-205, 1983

16. Hinuma Y, Nagata K, Hanaoka M, Nakai M, Matsumoto T, Kinoshita K, Shirakawa S, Miyosh I: Adult T-cell leukemia: Antigen in an ATL cell

line and detection of antibodies to the antigen in human sera. Pro Natl Acad Sci USA 78:6476-6480, 1981

17. Hirayama M, Silberberg D, Lisak R,Pleasure D: Long-term culture of oligodendrocytes isolated from rat corpus callosum by Percoll density gradient. J Neuropath Exp Neurol 42:16-28, 1983

18. Itoyama Y, Sternberger NH, Kies MW, Cohen S, Richardson EP, Webster H: Immunocytochemical method to identify myelin basic protein in oligodendroglia and myelin sheaths of the human nervous system. Ann Neurol 7:157-166, 1980

19. Johnson RT: Viral infections of the nervous system. Raven, New York, 1982

20. Johnson RT: Myelopathies and retroviral infections. Ann Neurol 21:113-116, 1987

21. Kennedy PGE: Neural cell markers and their applications to neurology. J Neuroimmunol 2:35-53, 1982

22. Kennedy PGE, Lisak RP: Astrocytes and oligodendrocytes in dissociated cell culture of adult rat optic nerve. Neurosci Lett 16:229-233, 1980

23. Kim SU: Antigen expression of glial cells grown in culture. J Neuroimmunol 8:255-282, 1985

24. Kim SU, McMorris A, Sprinkle T: Immunofluorescence demonstration of 2',3'-cyclic nucleotide 3'-phosphodiesterase in cultured oligodendrocytes of mouse, rat, calf and human. Brain Res 300:195-199, 1984

25. Kim SU, Moretto G, Shin D: Expression of Ia antigens on the surface of human oligodendrocytes and astrocytes in culture. J Neuroimmunol 10:141-149, 1985

26. Kim SU, Moretto G, Shin D, Lee V: Modulation of antigenic expression in cultured adult human oligodendrocytes by derivatives of cyclic AMP. J Neurol Sci 69:81-91, 1985

27. Kim SU, Moretto G, Lee V, Yu RK: Neuroimmunology of gangliosides in human neurons and glial cells in culture. J Neurosci Res 15:303-321, 1986

28. Kim SU, Moretto G, Ruff B, Shin D: Culture and cryopreservation of adult human oligodendrocytes and astrocytes. Acta Neuropath 64:172-175, 1984

29. Kim SU, Murray MR, Tourtellotte W, Parker J: Demonstration in tissue culture of myelinotoxicity in CSF and brain extract from multiple sclerosis patients. J Neuropath Exp Neurol 29:420-431, 1970

30. Kim SU, Shin DH, Paty DW: Long term culture of human oligodendrocytes in serum-free chemically defined medium. In: Experimental Allergic Encephalitis: A Useful Model for Multiple Sclerosis, Ed. by E. Alvord, A Liss, New York, pp 207-214, 1984

31. Kim SU, Stern J, Kim M, Pleasure D: Culture of purified rat astrocytes in serum-free medium supplemented with mitogen. Brain Res 274:79-86, 1983

32. Kim SU, Sato Y, Silberberg DH, Pleasure DE, Rorke LB: Long term culture of human oligodendrocytes. J Neurol Sci 62:295-301, 1983

32a. Kim SU, Yong VW, Takei F: Expression of MHC Class I antigens (HLA-ABC) by cultured human oligodendrocytes and astrocytes. J Neuroimmunol (In press)

33. Klein J, Juretic A, Baxevanis C, Nagy Z: The traditional and a new version of the mouse H-2 complex. Nature 291:455-460, 1981

34. Koprowski H, DeFreitas EC, Harper ME: Multiple sclerosis and human T-cell lymphotropic retroviruses. Nature 318:154-160, 1985

35. Lamson LA, Hickey W: Monoclonal antibody analysis of MHC expression in human brain biopsies: Tissue ranging from histologically normal to that showing different levels of glial tumor involvement. J Immunol 136:4054-4062, 1986

36. Lim R, Mitsunobu K, Li W: Maturation stimulating activity of brain extract and dibutyrul cyclic AMP on dissociated embryonic brain cells in culture. Exp Cell Res 79:243-246, 1973

37. Lisak R: Multiple sclerosis: Evidence for immunopathogenesis. Neurology (Minn.) 30:99-105, 1980

38. Lisak R, Pleasure D, Silberberg D, Manning M, Saida T: Long-term culture of bovine oligodendrocytes isolated with a Percoll gradient. Brain Res 213:165-170, 1981

39. Lisak R, Pruss P, Kennedy D, Abramsky O, Pleasure D, Silberberg D: Antisera to bovine oligodendroglia raised in guinea pig bind to surface of rat oligodendroglia and Schwann cells. Neurosci Lett 17:119-124, 1980

40. McAlpine D, Lumsden CE, Acheson ED: Multiple sclerosis: A reappraisal. Churchill Livingston, Edinburgh, 1972

41. McFarlin DE, McFarland H: Multiple sclerosis. New Eng J Med 307:1183-1188, 1982

42. McMorris FA: Cyclic AMP induction of the myelin enzyme cyclic nucleotide phosphodiesterase in oligodendrocytes. J Neurochem 41:506-515, 1983

43. Miller SL, Pleasure D, Herlyn M, Atkinson B, Ernst C, Tachovsky T, Baird L: Production and characterization of monoclonal antibodies to peripheral and central nervous tissue myelin. J Neurochem 43:394-400, 1984

44. Mirsky R, Winter J, Abney ER, Pruss RM, Gautilovic S, Raff MC: Myelin specific proteins and glycolipids in rat Schwann cells and oligodendrocytes in culture. J Cell Biol 84:483-494, 1980

45. Natal PG, Bigotti A, Nicotta M, Viora M, Manfredi D, Ferrone S: Distribution of human Class I histocompatibility antigens in normal and malignant tissues of nonlymphoid origin. Cancer Res 44:4679-4687, 1984

46. Norton W, Abe I, Poduslo S, DeVries G: The lipid composition of isolated brain cells and axons. J Neurosci Res 1:57-75, 1975

47. Norton W, Farouq M, Fields K, Raine C: The long-term culture of bulk isolated bovine oligodendroglia from adult brain. Brain Res 270:295-310, 1983

48. Ohta M, Ohta K, Mori F, Saida K, Saida T: Sera from patients with multiple sclerosis react with human T-cell lymphotrophic virus-I gag proteins but not env proteins: western blotting analysis. J Immunol 137:3440-3443, 1986

49. Osame M, Matsumoto M, Usuku K, Izumo S, Ijichi N, Amitani H, Tara M, Igata A: Chronic progressive myelopathy associate with elevated antibodies to HTLV-I and adult T-cell leukemia-like cells. Ann Neurol 21:117-122, 1987

50. Pleasure DE, Abramsky O, Silberberg D, Parris T, Saida T: Lipid synthesis by an oligodendroglial fraction in suspension culture. Brain Res 134:377-382, 1977

51. Poduslo S, Norton W: Isolation and some chemical properties of oligodendroglia from calf brain. J Neurochem 19:727-736, 1972

52. Poduslo S, McFarland H, McKhann G: Antiserum to neurons and to oligodendroglia from mammalian brain. Science 197:270-272, 1977

53. Quarles RH: Myelin-associated glycoprotein in development and disease. Dev Neurosci 6:285-303, 1985

54. Raff MC, Fields K, Hakomori S, Mirsky R, Pruss R, Winter J: Cell type-specific markers for distinguishing and studying neurons and the major classes of glial cells in culture. Brain Res 174:283-308, 1979

55. Raff MC, Mirsky R, Fields K, Lisak R, Dorman S, Silberberg D, Gregson N, Leibowitz S, Kennedy M: Galactocerebroside is a specific cell surface antigenic marker for oligodendrocytes in culture. Nature 274:813-816, 1978

56. Ranscht B, Clapshaw P, Price J, Noble M, Seifert W: Development of oligodendrocytes and Schwann cells studied with a monoclonal antibody against galactocerebroside. Proc Natl Acad Sci USA 79:2709-2713, 1982

57. Roitt IM: Essential immunology. Blackwell Scientific, Oxford, 1980

58. Romain GC: Retrovirus-associated myelopathies. Arch Neurol 44:659-663, 1987

59. Roussel G, Labourdette G, Nussbaum JL: Characterization of oligodendrocytes in primary cultures from brain hemispheres of newborn rat. Dev Biol 81:372-378, 1981

60. Schachner M, Kim SU, Zehnle R: Developmental expression in central and peripheral nervous system of oligodendrocyte cell surface antigens (0 antigens) recognized by monoclonal antibodies. Dev Biol 83:328-338, 1981

61. Shapiro DL: Morphological and biochemical alterations in fetal rat brain cell cultures in the presence of monobutyryl cyclic AMP. Nature 241:203-204, 1973

62. Snyder EY, Kim SU: Hormonal requirements for neurons in culture. Neurosci Lett 13:225-230, 1979

63. Snyder EY, Kim SU: Insulin: Is it a nerve survival factor. Brain Res 196:565-571, 1980

64. Sobue G, Pleasure D: Schwann cell galactocerebroside induced by derivatives of adenosine 3',5'-monophosphate. Science 224:72-74, 1984

65. Sternberger HH, Itoyama Y, Kiew MW, Webster H: Myelin basic protein demonstrated immunocytochemically in oligodendroglia prior to myelin sheath formation. Proc Nat Acad Sci USA 75:2521-2524, 1978

66. Szucht S, Stefansson K, Wollman R, Dawson G, Arnason B: Maintenance of isolated oligodendrocytes in long-term culture. Brain Res 200:151-164

67. Trapp BD, Quarles RH: Immunocytochemical localization of the myelin-associated glycoprotein. Factor artifact: J Neuroimmunol 6:231-249, 1984

68. Traugott U: Multiple sclerosis: relevance of Class I and Class II MHC expressing cells to lesion development. J Neuroimmunology 16:283-302, 1987

69. Traugott U, Snyder S, Norton W, Raine CS: Characterization of antioligodendrocyte serum. Ann Neurol 4:431-439, 1978

70. Traugott U, Snyder S, Raine CS: Oligodendrocyte staining by multiple sclerosis serum is non-specific. Ann Neurol 6:13-20, 1979

71. Vitella E, Capra J: The protein products of the murine 17th chromosome: genetics and structure. Adv Immunol 26:147-193, 1978

72. Williams K, Hart D, Fabre J, Morris P: Distribution and quantitation of HLA-ABC and DR antigens on human kidney and other tissues. Transplan 29:274-279, 1980

73. Winchester R, Kunkel H: The human Ia system. Adv Immunol 28:221-298, 1980

74. Vernant J, Maurs L, Gessain A, Barin F, Gout O, Delaporte J, Sanhadji K, Buisson G, de-The G: Endemic tropical spastic paraparesis associated with HTLV-I: A clinical and seroepidemiological study of 25 cases. Ann Neurol 21:123-130, 1987

75. Zeller NK, Behar T, Dubois-Dalcq M, Lazzarini RA: The timely expression of myelin basic protein gene in cultured rat brain oligodendrocytes is independent of continuous neuronal influences. J Neurosci 5:2955-2962, 1985

2

NEUROLOGICAL LINEAGES AND NEUROLOGICAL DISEASES

Kenji Mokuno, Pierluigi Baron, Judy Grinspan,
Gen Sobue, Barbara Kreider, and David Pleasure

Children's Hospital of Philadelphia and
Department of Neurology, University of Pennsylvania
Philadelphia, PA

INTRODUCTION

While neuroglial cells have been regarded as "support cells" for neurons since the beginning of this century, it has been only in recent years that the nature of such "support" has begun to be appreciated. It is now clear that neuroglial cells provide neurons with essential substrates such as glutamine and remove or inactivate such toxic metabolic products as NH_{4+} (46,47,97,98), regulate activities of potassium and other ions in the extracellular space (56,57), and permit saltatory conduction of nerve impulses by forming myelin. In addition to these metabolic support functions, Schwann cells in the peripheral nervous system (PNS) synthesize many proteins necessary for neuronal development and survival; these include extracellular matrix constituents such as type IV collagen, fibronectin and laminin (5,6,14), cell adhesion proteins such as N-CAM and myelin associated glycoprotein (MAG) (51), and soluble proteins such as nerve growth factor (NGF) (2,29,30). Similar protein synthetic trophic functions are performed in central nervous system (CNS) by astroglia; these include the synthesis of extracellular matrix components and growth factors, for example insulin-like growth factor (1,13,20,21,58,65). It is now thought likely, as well, that neuroglia regulate the properties of the blood-brain barrier. CNS neuroglia, probably astroglia, induce brain capillaries to express the tight junction phenotype that is required for a competent blood-brain barrier (89) and it is possible that astroglia participate in the regulation of regional cerebral blood flow by modulating perivascular potassium concentrations (62).

This chapter will not focus on these "support" roles of neuroglia, but will address a closely related topic: how are the proliferation, differentiation, and metabolism of the various classes of neuroglia regulated under normal circumstances, and what are the consequences of perturbation of these regulatory mechanisms? We will review current information on the lineages of neuroglia of the peripheral and central nervous systems and their regulation and then briefly consider what is known about the functions and roles of these lineages in various human diseases.

NEUROGLIAL LINEAGES

Peripheral Nervous System

The work of Le Douarin in chick-quail chimeras has been most valuable in documenting the differentiation of neural crest precursor cells into Schwann cells, autonomic and sensory neurons, neuroendocrine cells and melanocytes (42). Webster and coworkers employed classical electron microscopic observations to provide a clear picture of the subsequent segregation of Schwann cells into myelin-forming and non-myelinating adult phenotypes (93). Nerve cross-anastomosis experiments by several laboratories demonstrated plasticity of these Schwann cell phenotypes, with non-myelin forming Schwann cells capable of myelination if provided with an axon of appropriately large caliber. More recent immunohistological and molecular biological studies have provided further details on the effects of contact with axons of various caliber on the differentiation of Schwann cells. Schwann cells are induced by both large and small axons to express surface galactocerebroside and cytosolic proteolipid and to down-regulate surface expression of low affinity NGF receptors (86,88,95) but only by small axons to express a cytoskeletal protein resembling glial fibrillary acidic protein (18,35) and only by large axons to synthesize myelin (94).

The glial cells of the enteric nervous system were largely unstudied till the work of Jessen and Mirsky (34,35). Employing immunohistological methods, they demonstrated that these glia are irregular or multiprocess-bearing cells that express GFAP-like and glutamine synthetase immunoreactivity, in these respects resembling CNS astroglia.

Central Nervous System

Classical neuroanatomic techniques permitted identification of oligodendroglia, astrocytes, ependymal cells, and microglia in adult CNS and Muller cells in the retina. Astroglial foot processes were known to abut the basal lamina of CNS capillaries, and it was therefore presumed that astroglia regulate metabolite fluxes between bloodstream and neuropil. Ependymal cells were noted to be ciliated (50) and were thought to participate in absorption or secretion of CSF. Radial glia were suspected to be precursors of astroglia, and perhaps of oligodendroglia (8).

Immunohistological approaches to the identification of CNS neuroglia began with the work of Bignami, Dahl and Eng (3,15), who established GFAP as a "marker" for astroglia. Two types of astroglia were described: "protoplasmic", relatively poor in GFAP and located mainly in gray matter; and "fibrous", with more glial filaments and located mainly in white matter. It was initially unclear whether these were representatives of distinct astroglial lineages, different stages in differentiation along a single lineage, or simply different astroglial phenotypes dictated by their locale.

A decade ago, information on neuroglial lineages in CNS was very limited. It was assumed that divergence of oligodendroglial from astroglial lineages occurred early in development and in an irreversible fashion. Three successive stages in oligodendroglial maturation were delineated by combined microscopy and tritiated thymidine radioautography--immature light, transitional, and post-mitotic dark oligodendroglial. The relationship between small resting microglia and microglia "activated" by infection or other disease processes was appreciated. However, the ontogeny of microglia and the significance of microglial activation were unclear.

The development of cell type-specific antibodies in addition to anti-GFAP (eg. anti-galactocerebroside [anti-galC]) to identify mature oligodendroglia; the anti-ganglioside monoclonal antibody A2B5, which, in the rat, binds to the plasma membrane surface of precursor cells in the oligodendroglial lineage (10,59,60,71,74); and anti-glutamine synthetase, which binds to the interior of mature astroglia (61), permits identification of various neuroglial types and description of their stages of differentiation. These immunohistologic reagents, when combined with advances in tissue culture that allow serial observation on neuroglia in the presence or absence of neurons and growth factors (38), selective killing of single classes of neuroglia by complement and antibody dependent cytolysis, tritiated thymidine radioautography to identify cells undergoing mitosis, and retrovirus-mediated gene transfer techniques to identify all the descendants of single infected cells (69), have permitted more critical analysis of neuroglial lineages.

Application of these new techniques has permitted a number of important advances in our understanding of CNS glial lineage relationships. Miller et al were able to demonstrate the existence of at least two non-interconvertable classes of astroglia. "Type 1" astrocytes are abundant in gray matter, appear in rat brain before birth, maintain processes that abut brain capillaries, and are responsible for the astroglial reaction to various forms of brain injury (52). The immunohistological phenotype of these cells in the rat is GFAP + A2B5-. The lineage relationships of type 1 astroglia are not understood, but it is clear that type 1 astroglial progenitors diverge from progenitors within the oligodendroglial lineage very early in CNS development. "Type 2" astroglia, which are abundant in optic nerve and white matter, appear only postnatally in the rat, contribute to the glial limiting membrane of optic nerve and send processes to abut nodes of Ranvier. The immunohistological phenotype of these cells in the rat is GFAP + A2B5+. At the same time, Raff's group (59,91) and others (23) described the characteristics of an immediate precursor for both the type 2 astrocyte and the oligodendrocyte: the oligodendrocyte-type 2 astrocyte ("O2A") cell. In the rat, this O2A cell has the immunophenotype A2B5 + galC-GFAP-. Cells of this phenotype are small, round, bi- or multi-polar, and are both actively motile and actively mitotic. Their migration through optic nerve, and possibly other regions of immature CNS, serves to seed the neuropil with precursors for mature oligodendroglia and type 2 astroglia (81). Unfortunately, the monoclonal antibody A2B5 fails to mark such common precursor cells in species other than the rat, and there is a great need for a more generally applicable histological means for identification of this branch point in the oligodendroglial lineage.

Herndon et al (28) and Ludwin (49) clearly documented that oligodendroglia in mature brain of experimental animals are capable of mitosis following demyelination induced by viral or other diseases, thus indicating that the persistent nature of demyelination. These observations suggest that the irreversible demyelination observed, for example, in multiple sclerosis, cannot be ascribed simply to an irreversibly post-mitotic state for oligodendroglia in adult brain.

Price and Cepko showed that lineage relationships in brain and retina can be worked out by retrovirus-mediated transfer of the bacterial gene for beta-galactosidase, a convenient marker for the descendants of cells infected with this retroviral construct (68,69). This technology has shown that precursor cells common to glial and neuronal lineages persist in brain and retina to a developmental stage much later than previously suspected.

Several laboratories have demonstrated that CNS microglia derive from the systemic monocyte lineage and have many of the properties of macrophages in non-neural tissues. These include phagocytosis, display of MHC class II antigens and production of various monokines (22,27,63,100). At least two conclusions can be drawn from these observations. First, brain microglia have the capacity to act both as antigen-processing cells and effector cells during the evolution of immune disorders of the CNS. Second, it should be possible to repopulate brains of patients with inherited lysosomal or peroxisomal disorders with cells containing the missing or inactive protein by marrow transplantation, though one cannot necessarily assume that expression of such a protein in microglia would be of therapeutic value.

Linser and Moscona showed that the activity of glutamine synthetase, an enzyme vital for ammonia detoxification, is induced in Muller cells by axonal contact and repressed by axotomy (46,47). This is one of the better documented examples of an effect of neuronal contact on neuroglial phenotype.

REGULATION OF NEUROGLIAL LINEAGES

To what extent is the program for proliferation and differentiation of neuroglia controlled by mechanisms intrinsic to the differentiating precursor cells themselves, and to what extent by exogenous signs (eg. hormones, growth factors, cell contact-dependent phenomena)? That which is known can be summarized in a few sentences. Survival and initial proliferation of the O2A precursors of mature oligodendroglia and type 2 astroglia is dependent upon intimate contact with neurons (9), but later development of oligodendroglia, including the timely appearance of the various myelin-specific proteins can proceed in the absence of neurons (39,99). A protein in serum, not yet fully characterized, induces O2A cells to differentiate toward type 2 astroglia; in its absence, the oligodendroglial phenotype is favored (60). Proliferation of cells of the O2A lineage is enhanced by type 1 astroglia (59), and this appears to be accomplished by means of the secretion by these astrocytes of platelet-derived growth factor and insulin-like growth factor (1,10,71,74). Basic fibroblast growth factor also stimulates proliferation of cells of the oligodendroglial lineage (12). Proliferation of type 1 astrocytes appears to be inhibited by neuronal contact (26,84), and phenotypic and metabolic maturation of type 1 astroglia to be enhanced by contact with endothelial cell basal lamina (25).

In contrast to the behaviour of type 1 astroglia in CNS, proliferation of Schwann cells in PNS is enhanced by axonal contact (76,84). Schwann cell mitosis is also unstimulated by agents which increase Schwann cell intracellular adenosine 3',5'-monophosphate (85) and by fibroblast growth factor, glial growth factor, and PDGF (67) (Hardy and Pleasure, submitted for publication). Axonal contact induces Schwann cells to down-regulate surface expression of NGF receptors and neural cell adhesion molecules (N-CAM), but to upregulate synthesis of galactolipids such as galC and glycoproteins such as P_o (36,37,43,51,53,88). One of the signals involved in this neuronal modulation of Schwann cell phenotype appears to be an elevation in Schwann cell cyclic AMP content (43,53,80,86).

The recent application of methods for in situ hybridization to localize NGF mRNA has demonstrated that Schwann cell levels of this mRNA are high in the immature animal, diminish in the adult, and rise early during Wallerian degeneration; this rise in Schwann cell synthetic

capacity for this protein is due to the action of macrophages which
invade the degenerating nerve and release interleukin-1 (2,29,30,45).

NEUROGLIA IN DISEASE

This is a very brief summary of a large body of data. Only selected
references are provided, and the reader is urged to consult standard
texts (eg. Dyck et al, Peripheral Neuropathy, 1984) for further details.

Inherited Disorders of Oligodendroglia and Schwann Cells

A number of genetic defects affecting the synthesis of proteins
required for normal function of myelin-forming cells have been recognized
(87). Mutation of the X chromosome affecting the gene for proteolipid,
the principal structural protein of CNS myelin, causes sex-linked
dysmyelination in the mouse (the "jimpy" strain), in "myelin-deficient"
rats, and Pelizaeus-Merzbacher disease in man (54). An autosomal
mutation affecting the myelin basic protein gene also causes CNS
demyelination in the mouse (66), but no human analogue has yet been
recognized.

Autosomal mutations affecting either lyosomal arylsulfatase A or
galactocerebrosidase interfere with metabolism by oligodendroglia and
Schwann cells of myelin galactosphingolipids, causing dysmyelinative
encephalopathy and neuropathy. A mouse analogue of human
galactocerebrosidase deficiency has also been recognized (the "twitcher"
strain). Metabolism of myelin lipids is also impaired and the integrity
of myelin is compromised in human inherited peroxisomal disorders
affecting the PNS (phytanic acid oxidase deficiency or Refsum's disease)
or both CNS and PNS (adrenoleukodystrophy).

Neurofibromatosis (NF, von Recklinghausen's disease) is a dominantly
inherited predisposition to tumors containing Schwann-like cells (64).
NF is one of the most frequent inherited disorders affecting the nervous
system. The existence of two distinct forms of NF, long suspected on
clinical grounds, has now been verified by genetic analysis. Systemic
NF, characterized by cafe au lait spots and axillary freckles, Lisch
nodules of the iris, malformations of the sphenoid, vertebrae and tibia
in addition to subcutaneous and plexiform Schwann cell tumors, is due to
a mutation on chromosome 17, in the neighborhood of but not in the gene
for NGF receptor (79). The much rare central form of NF, characterized
by bilateral acoustic neurinomas and a paucity of cutaneous
manifestations, is carried on chromosome 22 (78), and appears to result
from inactivation of an anti-oncogene (40,78). The genetic defects in
these two forms of neurofibromatosis are expressed primarily in neural
crest-derived cells, especially in Schwann cells, but there is an
increased incidence of tumours in neural tube-derived cells as well, for
example optic gliomas.

Acquired Disorders of Oligodendroglia and Schwann Cells

Immune mechanisms compromising the survival or function of myelin
forming cells play a role in the pathogenesis of multiple sclerosis (MS)
and the Guillain-Barre syndrome (GBS) (31,32,92), and both cell-mediated
and serologically mediated immune processes involving sensitization to
such myelin components as myelin basic protein, P_2 basic protein, and
galactocerebroside participate in the pathogenesis of experimental models
that simulate some of the clinical and pathological features of MS and
GBS. For example, rabbits immunized repeatedly with galC develop
antibody-mediated PNS demyelination, and Lewis rats sensitized to P_2 PNS

myelin protein develop T-lymphocyte-mediated PNS demyelination (75). Both microglia and astroglia may play roles in the pathogenesis of these immune-mediated disorders by presenting antigens and secreting monokines and thereby facilitating lymphocyte-mediated damage to myelin and to myelin synthesizing cells (4,22,27,63,100).

Viral infections of neuroglia are increasingly recognized (7,11,90). Ependymal cells express surface binding sites for herpes simplex, mumps, measles and other viruses, which may participate in the pathogenesis of viral encephalitides. In some instances (eg. coronavirus in mice, papova virus in humans), oligodendroglial infection leads to CNS demyelination. Both direct cytolytic effects of the infection on myelin-forming cells and the triggering by the viral infection of an inflammatory process consequent to a delayed hypersensitivity mechanism play roles in these disorders. Neuroglia can serve as reservoirs for latent virus (eg. herpes simplex within gasserian ganglion Schwann cells) or for chronic viral infections (eg. HIV) (7,19). In some instances (eg. simian sarcoma viral infection), growth factors encoded by the viral genome (eg. v-sis) induce tumors (eg. glioblastoma).

Bacterial infections of neuroglia also occur. The most common example is the accumulation of Hansen's bacilli within the Schwann cells of subcutaneous nerve twigs in lepromatous leprosy, resulting in the loss of pain and temperature perception that is a characteristic feature of this disease.

Neuroglia are affected by various endocrine diseases. Myelination is delayed in infantile hypothyroidism, and some adults with hypothyroidism develop a demyelinative neuropathy. Diabetes mellitus also predisposes to PNS demyelination, perhaps due in part to the presence of aldose reductase within Schwann cells, but axonal degeneration is more common and more severe.

Toxins known to selectively affect neuroglia are far less numerous than those injuring neurons. Perhaps the best example is the protein exotoxin secreted by C. diphtheriae, which inhibits Schwann cell protein synthesis and induces a delayed demyelinative polyneuropathy in patients with pharyngeal or wound diphtheria.

Examples of the intermediary metabolism by neuroglia of toxins affecting primarily neurons are also now recognized. Neuroglia play a protective role in some instances, for example the detoxification of ammonia and of excitatory neurotoxins by astroglia (97,98), while in other cases, a non-toxic precursor is converted to a toxic product by neuroglia (eg. MPTP to MPP+).

Secondary Responses of Neuroglial Cells

Astrogliosis occurs in CNS in response to many types of injury, and appears to be primarily a response of type 1 astrocytes (52). In most instances, the relative contributions of astroglial proliferation, astroglial hypertrophy, and accumulation of GFAP+ fibrils to this astroglial reaction have not been worked out. An even more glaring lacune in our knowledge is the lack of information on the functional significance of astrogliosis--is it a reparative process, or instead, an impediment to regeneration (17,48,77). Perhaps the answer is that either is true, depending on age and species (82).

Retinal Muller cells also respond to injury by accumulation of GFAP (16). In addition, when maintained in culture in the absence of neurons, these cells can assume a lenslike phenotype, including the expression of antigens characteristic of lens cells (55).

Schwann cells respond to axotomy caused by Wallerian degeneration by upregulation of expression of NGF receptor and N-CAM (51,80,88) and increased synthesis of NGF--all events likely to enhance subsequent axonal regeneration. Schwann cell mitosis is transiently, but dramatically augmented during the early stages of Wallerian degeneration (76) and the Schwann cells form columns ("bands of Bungner") through which the regenerating axonal sprouts then propagate. Whether this Schwann cell proliferation is induced by the exposure to axonal fragments (84), to myelinic fragments (96), or both, remains to be established. Schwann cells also evince a mitogenic response to demyelination (24), but in this instance do not up-regulate NGF receptor expression (86).

Microglia are activated by CNS infections and immune disorders to a macrophage-like phenotype, becoming actively phagocytic, expressing MHC components necessary for antigen presentation, and secreting monokines (22,27,100).

Oligodendroglial proliferation also occurs in mature brain in response to many types of injury (28,49) and is followed by accelerated synthesis of myelin constituents (41). The relative contributions of mature oligodendroglial and residual precursor cells to this proliferation are unclear. Invasion of CNS by nerve root Schwann cells also occurs in response to CNS demyelination, and can lead to considerable remyelination, for example of MS spinal cord lesions (33). Such aberrant remyelination is recognizable by the presence of basal lamina encircling the myelinating cell (which occurs normally in PNS but not CNS) and by immunohistological techniques which detect myelin constituents normally restricted to PNS (eg. P_2, P_0). The functional consequences of such Schwann cell remyelination of CNS have not been established.

REFERENCES

1.   Ballotti R, Nielsen F, Pringle N, Kowalski A, Richardson W, Van Obberghen E, Gammeltoft S: Insulin-like growth factor 1 in cultured rat astrocytes: expression of the gene, and receptor tyrosine kinase. Embo J 6:3633, 1987
2.   Bandtlow C, Heumann R, Schwab M, Thoenen H: Cellular localization of nerve growth factor synthesis by in situ hybridization. Embo J 6:891, 1987
3.   Bignami A, Dahl D: Differentiation of astrocytes in the cerebellar cortex and the pyramidal tracts of the newborn rat. An immunofluorescence study with antibodies to a protein specific to astrocytes. Brain Res 49:393, 1973
4.   Birnbaum G, Clinchy B, Widner M: Recognition of major histocompatibility complex antigens on murine glial cells. J Neuroimmunol 12:225, 1986
5.   Bunge M, Williams A, Wood P: Neuron-Schwann cell interaction in basal lamina formation. Dev Biol 92:449, 1982
6.   Carey D, Eldridge C, Cornbrooks C, Timpl R, Bunge R: Biosynthesis of type IV collagen by cultured rat Schwann cells. J Cell Biol 97:473, 1983
7.   Cheng-Mayer C, Rutka J, Rosenblum M, McHugh T, Stites D, Levy J: Human immunodeficiency virus can productively infect cultured human glial cells. Proc Natl Acad Sci 84:3526, 1987
8.   Choi B, Kim RC, Lapham LW: Do radial glia give rise to both astroglial and oligodendroglial cells? Dev Brain Res 8:119, 1983
9.   David S, Miller R, Patel R, Raff M: Effects of neonatal transection on glial cell development in the rat optic nerve: evidence

that the oligodendrocyte-type 2 astrocyte cell lineage depends on axons for its survival. J Neurocytol 13:961, 1984

10. Dubois-Dalcq M: Characterization of a slowly proliferative cell along the oligodendrocyte differentiation pathway. EMBO J 6:2587, 1987

11. Dubois-Dalcq M, Rentier B, Hooghe-Peters E, Haspel M, Knobler R, Holmes K: Acute and persistent viral infections of differentiated nerve cells. Reviews of Infectious Diseases 4:999, 1982

12. Eccleston PA, Silberberg DH: Fibroblast growth factor is a mitogen for oligodendrocytes in vitro. Dev Brain Res 21:315, 1985

13. Edmondson J, Hatten M: Glial-guided granule neuron migration in vitro: a high-resolution time-lapse video microscopic study. J Neurosci 7:1928, 1987

14. Eldridge C, Bunge M, Bunge R, Wood P: Differentiation of axon-related Schwann cells in vitro. I. Ascorbic acid regulates basal lamina assembly and myelin formation. J Cell Biol 105:1023, 1987

15. Eng L, Vanderhaeghen J, Bignami A, Gerst I: An acidic protein isolated from fibrous astrocytes. Brain Res 28:351, 1971

16. Erickson P, Fisher S, Guerin C, Anderson D, Kaska D: Glial fibrillary acidic protein increases in Muller cells after retinal detachment. Exp Eye Res 44:37, 1987

17. Fallon J: Preferential outgrowth of central nervous system neurites on astrocytes and Schwann cells as compared with nonglial cells in vitro. J Cell Biol 100:198, 1985

18. Fields K, McMenamin P: Schwann cells cultured from adult rats contain a cytoskeletal protein related to astrocyte filaments. Dev Brain Res 20:259, 1985

19. Funke I, Hahn A, Rieber E, Weiss E, Riethmuller G: The cellular receptor (CD4) of the human immunodeficiency virus is expressed on neurons and glial cells in human brain. J Exp Med 165:1230, 1987

20. Giulian D, Allen R, Baker T, Tomozawa Y: Brain peptides and glial growth. I. Glia-prompting factors as regulators of gliogenesis in the developing and injured central nervous system. J Cell Biol 102:803, 1986

21. Giulian D, Young D: Brain peptides and glial growth. II. Identification of cells that secrete glia-promoting factors. J Cell Biol 102:812, 1986

22. Giulian D, Baker T, Shih L, Lachman L: Interleukin-1 of the central nervous system is produced by ameboid microglia. J Exp Med 164:594, 1986

23. Goldman J, Geier S, Hirano M: Differentiation of astrocytes and oligodendrocytes from germinal matrix cells in primary culture. J Neurosci 6:52, 1986

24. Griffin JW, Drucker W, Gold BG, Rosenfeld J, Benzaquen M, Charnas LP, Fahnestock KE, Stocks EA: Schwann cell proliferation and migration during paranodal demyelination. J Neurosci 7:682, 1987

25. Grinspan JB, Lieb M, Stern J, Rupnick M, Williams S, Pleasure D: Rat brain microvessel extracellular matrix modulates the phenotype of cultured rat type 1 astroglia. Dev Brain Res 33:291, 1987

26. Hatten M: Neuronal regulation of astroglial morphology and proliferation in vitro. J Cell Biol 100:384, 1985

27. Hayes G, Woodroofe M, Cuzner M: Microglia are the major cell type expressing MHC class II in human white matter. J Neurol Sci 80:25, 1987

28. Herndon R, Price D, Weiner L: Regeneration of oligodendroglia during recovery from demyelinating disease. Science 195:693, 1977

29. Heumann R, Kordsching S, Bandtlow C, Thoenen H: Changes of nerve growth factor synthesis in non-neuronal cells in response to sciatic nerve transection. J Cell Biol 104:1623, 1987

30. Heumann R, Lindholm D, Bandtlow C, Meyer M, Radeke M, Misko T, Shooter E, Thoenen H: Differential regulation of mRNA encoding nerve growth factor and its receptor in rat sciatic nerve during development,

degeneration and regeneration: role of macrophages. Proc Natl Acad Sci 84:8733, 1987

31. Hirayama M, Yokochi T, Shimokata K, Iida M, Fujiki N: Induction of human leukocyte antigen-A,B,C and -DR on cultured human oligodendrocytes and astrocytes by human gamma-interferon. Neurosci Let 72:369, 1986

32. Hofman F, von Hanwehr R, Dinarello C, Mizel S, Hinton D, Merrill J: Immunoregulatory molecules and IL-2 receptors identified in multiple sclerosis brain. J Immunol 136:3239, 1986

33. Itoyama Y, Webster H, Richardson E Jr, Trapp B: Schwann cell remyelination of demyelinated axons in spinal cord multiple sclerosis lesions. Ann Neurol 14:339,1979

34. Jessen K, Mirsky R: Astrocyte-like glia in the peripheral nervous system: an immunohistochemical study of enteric glia. J Neurosci 3:2206, 1983

35. Jessen K, Mirsky R: Glial fibrillary acidic polypeptides in peripheral glia. Molecular weight, heterogeneity and distribution. J Neuroimmunol 8:377, 1985

36. Jessen K, Mirsky R, Morgan L: Myelinated, but not unmyelinated axons, reversibly down-regulate N-CAM in Schwann cells. J Neurocytol 16:681, 1987

37. Jessen K, Mirsky R, Morgan L: Axonal signals regulate differentiation of non-myelin-forming Schwann cells: an immunohistochemical study of galactocerebroside in transected and regenerating nerves. J Neurosci 7:3362, 1987

38. Kim SU: Antigen expression by glial cells in culture. J Neuroimmunol 8:255, 1985

39. Knapp P, Bartlett W, Skoff R: Cultured oligodendrocytes mimic in vivo phenotypic characteristics: cell shape, expression of myelin-specific antigens, and membrane production. Dev Biol 120:356, 1987

40. Knudson A Jr: Hereditary cancer, oncogene, and antioncogenes. Cancer Res 45:1437, 1985

41. Kristensson K, Holmes K, Duchala C, Zeller N, Lazzarini R, Dubois-Dalcq M: Increased levels of myelin basic protein transcripts gene in virus-induced demyelination. Nature 322:544, 1986

42. Le Douarin N: Cell line segregation during peripheral nervous system ontogeny. Science 231:1515, 1986

43. Lemke G, Chao M: Axons regulate Schwann cell expression of the major myelin and NGF receptor genes. Development 102:499, 1988

44. Liesi P, Kaakkola S, Dahl D, Vaheri A: Laminin is induced in astrocytes of adult brain by injury. EMBO J 3:683, 1984

45. Lindholm D, Heumann R, Meyer M, Thoenen H: Interleukin-1 regulates synthesis of nerve growth factor in non-neuronal cells of rat sciatic nerve. Nature 330:658, 1987

46. Linser P, Moscona A: Induction of glutamine synthetase in embryonic neural retina: localization in Muller fibers and dependence on cell interactions. Proc Natl Acad Sci 76:6476, 1979

47. Linser P, Moscona A: Hormonal induction of glutamine synthetase in cultures of embryonic retinal cells: requirements for neuron-glia contact interactions. Dev Biol 96:529, 1983

48. Liuzzi F, Lasek R: Astrocytes block axonal regeneration in mammals by activating the physiological stop pathway. Science 237:642, 1987

49. Ludwin S: Reaction of oligodendrocytes and astrocytes to trauma and implantation. A combined autoradiographic and immunohistochemical study. Lab Invest 52:20, 1985

50. Manthorpe C, Wilkin G, Wilsin J: Purification of viable ciliated cuboidal ependymal cells from rat brain. Brain Res 134:407, 1977

51. Martini R, Schachner M: Immunoelectron microscopic localization of neural cell adhesion molecules (L1, N-CAM, and MAG) and their shared

carbohydrate epitope and myelin basic protein in developing sciatic nerve. J Cell Biol 103:2439, 1986

52. Miller R, Abney E, David S, ffrench-Constant C, Lindsay R, Patel R, Stone J, Raff M: Is reactive gliosis a property of a distinct subpopulation of astrocytes? J Neurosci 6:22, 1986

53. Mokuno K, Sobue G, Reddy U, Wurzer J, Kreider B, Hotta H, Baron P, Ross A, Pleasure D: Regulation of Schwann cell nerve growth factor receptor by cyclic adenosine 3',5'-monophosphate. J Neurosci Res, in press

54. Morello D, Dautigny A, Plum-Dinh D, Jollies P: Myelin proteolipid protein (PLP and DM-20) transcripts are deleted in jimpy mutant mice. EMBO J 5:3489, 1986

55. Moscona A: Conversion of retina glia cells into lenslike phenotype following disruption of normal cell contacts. Cur Top Dev Biol 20:1, 1986

56. Newman E: Regional specialization of retinal glial cell membrane. Nature 309:155, 1984

57. Newman E, Frambach D, Odette L: Control of extracellular potassium levels by retinal glial cell K^+ siphoning. Science 225:1174, 1984

58. Noble M, Fok-Seang J, Cohen J: Glia are a unique substrate for the in vitro growth of central nervous system neurons. J Neurosci 4:1892, 1984

59. Noble M, Murray K: Purified astrocytes promote the in vitro division of a bipotential glial progenitor cell. EMBO J 3:2433, 1984

60. Noble M, Murray K, Stroobant P, Waterfield MD, Riddle D: Platelet-derived growth factor promotes division and motility and inhibits premature differentiation of the oligodendrocyte-type 2 astrocyte progenitor cell. Nature 333:560, 1988

61. Norenberg MD, Martinez-Hernandez A: Fine structural localization in astrocytes of rat brain. Brain Res 161:303, 1979

62. Paulson O, Newman E: Does the release of potassium from astrocyte endfeet regulate cerebral blood flow? Science 237:896, 1987

63. Perry V, Hume D, Gordon S: Immunohistochemical localization of macrophages and microglia in the adult and developing mouse brain. Neurosci 15:313, 1985

64. Pleasure D, Kreider B, Sobue G, Ross A, Koprowski H, Sonnenfeld K, Rubenstein A: Schwann-like cells cultured from human dermal neurofibromas: immunohistological identification and response to Schwann cell mitogens. Ann NY Acad Sci 486:227, 1986

65. Poltorak M, Sadoul R, Keilhauer G, Landa C, Fahrig T, Schachner M: Myelin-associated glycoprotein, a member of the L2/HNK-1 family of neural cell adhesion molecules, is involved in neuron-oligodendrocyte and oligodendrocyte-oligodendrocyte interaction. J Cell Biol 105:1893, 1987

66. Popko B, Puckett C, Lai E, Shine H, Readhead C, Takahashi N, Hunt S III, Sidman R, Hood L: Myelin deficient mice: expression of myelin basic protein and generation of mice with varying levels of myelin. Cell 48:713, 1987

67. Porter S, Clark MB, Bunge RP: Schwann cells stimulated to proliferate in the absence of neurons retain full functional capacity. J Neurosci 6:3070, 1986

68. Price J: Retroviruses and the study of cell lineage. Development 101:409, 1987

69. Price J, Turner D, Cepko C: Lineage analysis in the vertebrate nervous system by retrovirus-mediated gene transfer. Proc Natl Acad Sci 84:156, 1987

70. Raff M, Abney E, Cohen J, Lindsay R, Noble M: Two types of astrocytes in cultures of developing rat white matter: differences in morphology, surface gangliosides and growth characteristics. J Neurosci 3:1280, 1983

71. Raf MC, Lillien LE, Richardson WD, Burne JF, Noble MD: Platelet derived growth factor from astrocytes drives the clock that times oligodendrocyte development in culture. Nature 333:562, 1988

72. Ratner N, Bunge RP, Glaser L: A neuronal cell surface heparin sulfate proteoglycan is required for dorsal root ganglion neuron stimulation of Schwann cell proliferation. J Cell Biol 101:744, 1985

73. Reichenbach A, Wohlrab F: Morphometric parameters of Muller (glial) cells dependent on their topographic localization in the nonmyelinated part of the rabbit retina. A consideration of functional aspects of radial glia. J Neurocytol 15:451, 1986

74. Richardson WD, Pringle N, Mosley MJ, Westermark B, Dubois-Dalcq M: A role for platelet derived growth factor in normal gliogenesis in the central nervous system. Cell 53:309, 1988

75. Rostami A, Burns J, Brown M, Rosen J, Zweiman B, Lisak R, Pleasure D: Transfer of experimental allergic neuritis with P_2-reactive T-cell lines. Cellular Immunol 91:354, 1985

76. Salzer J,Bunge R: Studies of Schwann cell proliferation. I. An analysis in tissue culture of proliferation during development, Wallerian degeneration, and direct injury. J Cell Biol 84:739, 1980

77. Schwab M, Thoenen H: Dissociated neurons regenerate into sciatic but not optic nerve explants in culture irrespective of neurotrophic factors. J Neurosci 5:2415, 1985

78. Seizinger B, Martuza R, Gusella J: Loss of genes on chromosome 22 in tumorigenesis of human acoustic neuroma. Nature 322:644, 1986

79. Seizinger B, and others: Genetic linkage of von Recklinghausen neurofibromatosis to the nerve growth factor receptor gene. Cell 49:589, 1987

80. Shuman S, Hardy M, Sobue G, Pleasure D: A cyclic adenosine 3′,5′-monophosphate (cAMP) analogue induces synthesis of a myelin-specific glycoprotein by cultured Schwann cells. J Neurochem 50:190, 1988

81. Small RK, Riddle P, Noble M: Evidence for migration of oligodendrocyte-type 2 astrocyte progenitor cells into the developing rat optic nerve. Nature 328:155, 1987

82. Smith G, Miller R, Silver J: Changing role of forebrain astrocytes during development, regenerative failure, and induced regeneration upon transplantation. J Comp Neurol 251:23, 1986

83. Sobue G, Pleasure D: Adhesion of axolemmal fragments to Schwann cells: a signal- and target-specific process closely linked to axolemmal stimulation of Schwann cell mitosis. J Neurosci 5:379, 1985

84. Sobue G, Pleasure D: Astroglial proliferation and phenotype are modulated by neuronal plasma membrane. Brain Res 324:175, 1986

85. Sobue G, Shuman S, Pleasure D: Schwann cell responses to cyclic AMP: proliferation, change in shape, and appearance of surface galactocerebroside. Brain Res 362, 23, 1986

86. Sobue G, Yasuda T, Mitsuma T, Ross A, Pleasure D: Expression of nerve growth factor receptor in human peripheral neuropathies. Ann Neurol 24:64, 1988

87. Sorg B, Agrawal D, Agrawal H, Campagnoni A: Expression of myelin proteolipid protein and basic protein in normal and dysmyelinating mutant mice. J Neurochem 46:379, 1986

88. Taniuchi M, Clark H, Johnson E Jr: Induction of nerve growth factor receptor in Schwann cells after axotomy. Proc Natl Acad Sci 83:4094, 1986

89. Tao-Cheng J, Nagy Z, Brightman M: Tight junctions of brain endothelium in vitro are enhanced by astroglia. J Neurosci 7:3293, 1987

90. Tardieu M, Weiner H: Viral receptors on isolated murine and human ependymal cells. Science 215:419, 1982

91. Temple S, Raff M: Clonal analysis of oligodendrocyte development in culture: evidence for a developmental clock that counts cell divisions. Cell 44:773, 1986

92. Traugott U, Scheinberg L, Raine C: On the presence of Ia-positive endothelial cells and astrocytes in multiple sclerosis lesions and its relevance to antigen presentation. J Neuroimmunol 8:1, 1985

93. Webster H, Martin J, O'Connell M: The relationships between interphase Schwann cells and axons before myelination: a quantitative electron microscopic study. Dev Biol 32:401, 1973

94. Windebank AJ, Wood P, Bunge RP, Dyck PJ: Myelination determines the caliber of dorsal root ganglion neurons. J Neurosci 5:1563, 1985

95. Yan Q, Johnson E Jr: A quantitative study of the developmental expression of nerve growth factor (NGF) receptor in rats. Dev Biol 121:139, 1987

96. Yoshino JE, Mason PW, DeVries JH: Developmental changes in myelin-induced proliferation of cultured Schwann cells. J Cell Biol 104:655, 1987

97. Yudkoff M, Nissim I, Pleasure D: [^{15}N]-Aspartic acid metabolism in cultured astrocytes: studies with gas chromatography-mass spectrometry. Biochem J 241:193, 1987

98. Yudkoff M, Nissim I, Pleasure D: Astrocyte metabolism of ^{15}N- and ^{13}C-glutamine: implications for the glutamine-glutamate cycle. J Neurochem, in press

99. Zeller N, Behar T, Dubois-Dalcq M, Lazzarini R: The timely expression of myelin basic protein gene in cultured rat brain oligodendrocytes is independent of continuous neuronal influences. J Neurosci 5:2955, 1985

100. Zucker-Franklin D, Warfel A, Grusky G, Frangione B, Teitel D: Novel monocyte-like properties of microglial/astroglial cells. Lab Invest 57:176, 1987

3

HUMAN GLIAL CELLS AND GROWTH FACTORS

Voon Wee Yong, Myong W. Kim, and Seung U. Kim

Division of Neurology, Department of Medicine
The University of British Columbia
Vancouver, Canada

INTRODUCTION

In the central nervous system (CNS), macroglial cells are classified as astrocytes or oligodendrocytes, while in the peripheral nervous system, Schwann cells, the equivalent of CNS oligodendrocytes, form the only glial cell type. During the normal development of the nervous system, glial cells or their precursors undergo proliferation at certain time points, possibly influenced by the schedule of an internal 'time clock' and just before the onset of myelination. The macroglial cells eventually become post-mitotic and normally do not divide in the adult nervous system except in response to injury. The latter is well documented for astrocytes during insults of various kinds (10,35,68) and for Schwann cells during Wallerian degeneration (1,5,64) and internodal (segmental) or paranodal demyelination (21). In the case of oligodendrocytes, however, the literature has been controversial. While some reports have shown that oligodendrocytes cannot proliferate after injury (10,68), others have suggested that mitosis of oligodendrocytes can occcur after physical brain trauma (40,41,42), following recovery from demyelinating viruses (26,62), and in an experimental allergic encephalomyelitis (EAE) model of multiple sclerosis (61).

Given that macroglial cells may undergo cell division under certain conditions, it is beneficial to identify and understand the factors that control their proliferation, be these endogenous growth factors, contact with axons or exogenous agents. In disease states where cells are lost, such as for oligodendrocytes in multiple sclerosis, induction of mitosis may achieve a critical number of the same cell type to enable restoration of function. This aim is not without clinical support. In multiple sclerosis, proliferation of oligodendrocytes, albeit limited in its extent, has already been reported in the relatively normal white matter adjacent to disease plaques (60). What is therefore required are stimuli to enhance this limited regenerative capacity already observed.

In vivo studies of glial cell proliferation are intrinsically difficult. Dividing cells may not be reliably identified as having undergone mitosis and morphological characterization of the cell type may also present problems. The difficulties existing in the identification of dividing oligodendrocytes were underlined by Cavanagh (10) who remarked that even when he could not find definite evidence for oligodendrocyte proliferation, it was conceivable that they had divided, but that in so doing, they had become morphologically indistinguishable from a normal oligodendrocyte.

For such reasons, many studies of glial cell proliferation have resorted to tissue culture techniques. Enriched population of a particular cell type can be attained (29,30,34,43,72) and cells can be also readily identified by immunocytochemistry for the presence of their cell type specific markers, such as galactocerebroside for oligodendrocytes and glial fibrillary acidic protein (GFAP) for astrocytes (4,33,55,57).

Using cultured cells, many mitogens have been reported for glial cells and the results are summarized on Tables I and II. For astrocytes, these include fibroblast growth factor (FGF) (54), platelet-derived growth factor (PDGF) (25), epidermal growth factor (EGF) (37,66,71), glial growth factor from the bovine pituitary (GGF-BP) (6,29.54), myelin basic protein (7,65) and interleukin-1 (17). For Schwann cells, among the reported mitogens are axolemma fragments derived from either the peripheral or central nervous system (11,52,64,69), a myelin-enriched fraction (78), GGF-BP (6,52,53,56), cAMP (56,70), and forskolin, an adenyl cyclase activator that results in cellular accumulation of cAMP (53). Mitogens for Schwann cells have recently been comprehensively summarized by Ratner et al (59). In the case of oligodendrocytes, proliferative factors are reported to be 'glia promoting factors' (16,18,19), contact with neuronal axons (73,74), PDGF (3), FGF (13), T-cell derived factors (44) and interleukin-2 (2).

To identify factors that can induce cell proliferation, it is necessary to have a technique that is reproducible and unambiguous. Three methods are commonly used, but each has its own limitations. The first relies on the counting of cell numbers in a given area (16,18,22,48). The second utilizes the incorporation of ^{3}H-thymidine or ^{125}I-iododeoxyuridine into the DNA of dividing cells followed by quantification of the amount of radioactivity in a scintillation counter (2,44,51,56,66). For these two methods, unless the cell population is homogeneous, results can be misleading. Type I astrocytes have a fibroblast-like morphology (58) and it is conceivable that when a positive result for astrocytes is reported by these techniques, the cells that were proliferating were mainly fibroblasts, which normally thrive in culture. Another potential source of error is that the original density of cells may vary from one sample to the next such that comparisons of test samples from controls become flawed.

The third method also employs ^{3}H-thymidine incorporation into the DNA of dividing cells, but then uses autoradiography to determine the number of cells with silver grains on their nuclei (3,29,71,78). The cell type undergoing mitosis can be positively identified if such cells have been additionally immunostained with their cell type specific markers. Although this method is commonly used, the disadvantages include the high level of background that is frequently encountered with autoradiography (68) and the long period (days or even weeks) that is required for the development of the radiolabelling. Subjectivity associated with dis-crimination between unlabelled and weakly labelled cells can also constitute a problem.

Due to such difficulties, we have developed a double labelling immunofluorescence technique to assess cell proliferation (75). The method uses nuclear incorporation of bromodeoxyuridine (BrdU), an analog of thymidine, as an index of DNA replication (12,20,49). By using a specific antibody directed against BrdU and another antibody directed against the cell type specific marker (for example, GFAP for astrocytes), the proliferating cell type can be readily identified. This technique is simple, rapid, reproducible and unambiguous.

TABLE I.   REPORTED MITOGENS FOR ASTROCYTES AND OLIGODENDROCYTES IN CULTURE

Reported mitogen	Sources of cells	Method*	Ref.
ASTROCYTES			
Glial growth factor from bovine pituitary	Neonatal rat brain Neonatal rat brain	C D	6 29.54
Fibroblast growth factor	Neonatal rat brain	D	54
Epidermal growth factor	Neonatal mouse brain Neonatal rat brain Human glial cell line	D B C	37 66 71
Platelet-derived growth factor	Human glial cell line	C	25
Myelin basic protein	Mouse embryonic brain Adult mouse brain or C6TK cell line	D B	7 65
Interleukin-1	Neonatal rat brain	B	17
OLIGODENDROCYTES			
'Glia promoting factors'	Neonatal rat brain	A	16,18,19
Fibroblast growth factor	Neonatal rat brain	D	13
Platelet-derived factor	Neonatal rat brain	C	3
Axonal contact	Adult rat spinal cord	D	73,74
T-cell derived factors	Neonatal rat brain	B	44
Interleukin-2	Neonatal rat brain	B	2

*Refers to the method of assessing cell proliferation: A=Counting of cells not identified by immunostainings, B=^{3}H-thymidine or ^{125}I-iododeoxyuridine uptake followed by scintillation counting, C=^{3}H-thymidine incorporation followed by autoradiography with cell-type specific markers not used, D=^{3}H-thymidine uptake followed by autoradiography, cell-type specific markers used.

As shown on Tables I and II, most studies of mitogens for glial cells have relied on cells isolated from rodent nervous system. Results using human glial cells are scarce and it cannot be assumed that mitogens for rodent cells will necessarily have similar effects on human cells. For this reason, we have embarked on an extensive study on factors that might induce proliferation of cultured human glial cells. This chapter, using the BrdU method, summarizes our findings with fetal astrocytes, fetal Schwann cells, adult astrocytes and adult oligodendrocytes. Part of the results have been published elsewhere (76,77).

TABLE II.   REPORTED MITOGENS FOR SCHWANN CELLS IN CULTURE

Reported mitogen	Sources of cells	Method*	Ref.
Glial growth factor	Neonatal rat sciatic	B	56
from bovine	Neonatal rat sciatic	C	52,53
pituitary	Adult human sural	C	52
	Adult human trigeminal	C	47
Axolemma	Neonatal rat sciatic	C	52,64,69
	Adult human sural	C	52
	Neonatal rat dorsal root ganglion	C	11
cAMP	Neonatal rat sciatic	B	56
	Neonatal rat sciatic	C	70
Myelin	Neonatal rat sciatic	C	78
Forskolin	Neonatal rat sciatic	C	53

*Refer Table I for legend

MATERIALS AND METHODS

Cell Culture

Human adult astrocytes and oligodendrocytes were obtained by previously described methods (30.31) from the autopsied corpus callosum of subjects of ages between 60 and 90 years old. In brief, the corpus callosum was cut into fragments of 3^3 mm and incubated with 0.25% trypsin and 0.002% DNAse in calcium- and magnesium-free Hanks' balanced salt solution (CMF-HBSS) for 1 h at 37 degrees C. Dissociated cells were passaged through a nylon mesh of 100 um pore size, mixed with Percoll (Pharmacia) (final concentration of 30% Percoll) and centrifuged at 15000 rpm for 25 minutes. Oligodendrocytes and astrocytes, floating between an upper myelin layer and a lower erythrocyte layer, were diluted with 3 volumes of HBSS and harvested by centrifugation at 1500 rpm for 10 minutes. The cells were washed twice in HBSS, suspended in feeding medium, and plated onto 9 mm Aclar plastic coverslips coated with 10 ug/ml polylysine (Sigma, approximate molecular weight of 400,000) at a density of 10^4 cells/coverslip. Feeding medium consisted of 5% fetal calf serum, 5 mg/ml glucose and 20 ug/ml gentamicin in Eagle's minimum essential medium. The cells were used for the present proliferation studies after 7 weeks in culture.

Fetal astrocytes were isolated from human fetuses, legally and therapeutically aborted, of 8-10 weeks gestation. Whole brain was cut into fragments of 2^3 mm and incubated with 0.25% trypsin and 0.002% DNase in CMF-HBSS for 30 min at 37 degrees C. The fragments were then dissociated into single cells by gentle pipetting. The suspension was centrifuged at 1500 rpm to collect the dissociated cells. Two washings in HBSS ensued and the cells were finally suspended in feeding medium identical to that for adult astrocyte/oligodendrocyte. The cells were seeded onto polylysine-coated Aclar coverslips at a density of 10^4 cells/coverslip. Cell types include neurons, fibroblasts and astrocytes. Oligodendrocytes could only be detected occasionally using immunostainings of antibodies directed against galactocerebroside. The proliferation experiments were performed after the cells have been in culture for at least 11 days.

For the isolation of Schwann cells, dorsal root ganglia were obtained from fetuses of 8-10 weeks gestation. Dissociated cultures were established as previously described (32) in a manner similar to that for fetal astrocytes above. Feeding medium was Eagle's minimum essential medium supplemented with 10% fetal calf serum, 5 mg/ml glucose and 20 µg/ml gentamicin. No nerve growth factor was added and this resulted in depletion of neuronal populations (38,39,46) such that within 2 weeks in vitro, neurons comprised less than 1% of the total cell population; Schwann cells (30%) and fibroblasts (70%) being predominant. These neuron-deficient cultures were used for the present proliferation studies.

Incubation of cells with growth factors and BrdU

For fetal astrocytes, the test growth factors were added to the culture medium for 3 days prior to immunofluorescence studies while BrdU (Sigma, 10 uM) was introduced for the last 2.5 h. Preliminary experiments had indicated that for control fetal astrocytes, 2.5 h of incubation with BrdU-containing medium was adequate to achieve BrdU nuclear labelling of 10-30% of cells. For adult astrocytes and oligodendrocytes, test growth factors and BrdU were added for 3 days before immunostainings were performed. BrdU was added for 3 days because preliminary studies had shown that these adult cells were slowly dividing cells, if at all. In addition, the concentration (10 µM) of BrdU added for 3 days was found not to be toxic to the cells as assessed by cell morphology and survival. In the case of fetal Schwann cells, test factors and 10 uM BrdU were applied to the culture medium simultaneously for 2 days prior to immunostainings.

Test growth factors were recombinant Interleukin-2 (Genzyme Corporation, 5 and 50 U/ml), EGF and FGF (Bethesda Research Laboratories, 0.5 and 5 µg/ml), PDGF (Collaborative Research, 5 and 50 mU/ml), dibutyryl cAMP (DBcAMP, Sigma, 100 µM and 1 mM), nerve growth factor (NGF, Sigma, 100 ng/ml and 1 µg/ml), forskolin (Sigma, 10 and 100 µM), a tumor-promoting agent 4B-phorbol 12, 13-dibutyrate (PDB, Sigma, 10 and 100 nM) (8,50), and GGF-BP (1 µg/ml). GGF-BP was prepared from adult bovine pituitary homogenates by ammonium sulfate fractionation, carboxy-methylcellulose and phosphocellulose column chromatographies (6) and was a kind of gift of Dr. D. Pleasure (29). In addition, a combination of GGF-BP (1 µg/ml) and forskolin (10 µM) was tested.

Immunocytochemistry

The general process is outlined in Figure 1 and has been described in detail for astrocytes (75). After removal of BrdU-containing medium and washing the cells thoroughly with phosphate-buffered saline (PBS) to

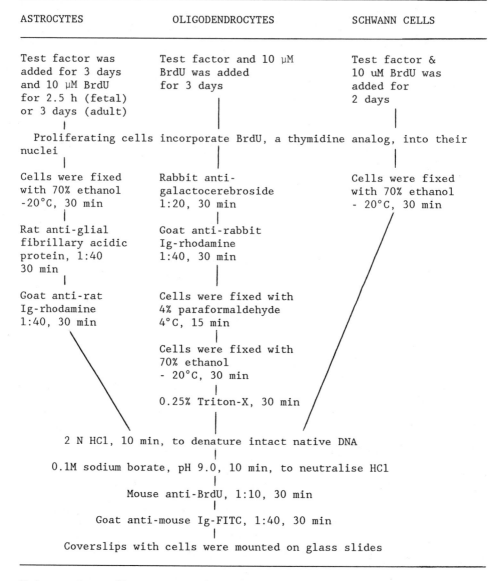

ASTROCYTES	OLIGODENDROCYTES	SCHWANN CELLS

Test factor was added for 3 days and 10 μM BrdU for 2.5 h (fetal) or 3 days (adult)

Test factor and 10 μM BrdU was added for 3 days

Test factor & 10 uM BrdU was added for 2 days

Proliferating cells incorporate BrdU, a thymidine analog, into their nuclei

Cells were fixed with 70% ethanol -20°C, 30 min

Rabbit anti-galactocerebroside 1:20, 30 min

Cells were fixed with 70% ethanol - 20°C, 30 min

Rat anti-glial fibrillary acidic protein, 1:40 30 min

Goat anti-rabbit Ig-rhodamine 1:40, 30 min

Goat anti-rat Ig-rhodamine 1:40, 30 min

Cells were fixed with 4% paraformaldehyde 4°C, 15 min

Cells were fixed with 70% ethanol - 20°C, 30 min

0.25% Triton-X, 30 min

2 N HCl, 10 min, to denature intact native DNA

0.1M sodium borate, pH 9.0, 10 min, to neutralise HCl

Mouse anti-BrdU, 1:10, 30 min

Goat anti-mouse Ig-FITC, 1:40, 30 min

Coverslips with cells were mounted on glass slides

Under an immunofluorescence microscope, the % of astrocytes, oligodendrocytes or Schwann cells with BrdU labelling in their nuclei was tabulated. These results of test coverslips were divided by similar results obtained from control coverslips to give the proliferation index (PI) of the test factor.

Figure 1.   METHOD FOR DETERMINING PROLIFERATION
             Cells were cultured on 9 mm Aclar coverslips and maintained
             in serum containing medium.

ensure complete removal of unbound BrdU, astrocytes and Schwann cells on coverslips were fixed with 70% ethanol at -20 degrees C for 30 min. Rat monoclonal antibody to glial fibrillary acidic protein (GFAP) (1:2) or polyclonal rabbit anti-GFAP was applied to astrocytes for 30 min, followed by goat anti-rat immunoglobulin conjugated to rhodamine (1:40). This enabled subsequent identification of astrocytes. For oligodendrocytes, prior to the ethanol fixation process, cells were incubated with rabbit antiserum against galactocerebroside (1:20, 30 min) followed by goat anti-rabbit immunoglobulin conjugated to rhodamine (1:40) for another 30 min. Oligodendrocytes were then fixed with 4% paraformaldehyde at 4 degrees C for 15 min, followed by ethanol fixation. Without the paraformaldehyde pretreatment, galactocerebroside immunostainings tended to be obliterated by the acid treatment described below. After the ethanol fixation step, cells were incubated with 0.25% Triton-X in PBS for 30 minutes. This improvisation was necessary because the paraformaldehyde treatment resultedd in the nuclear membrane being less permeable to the BrdU antibody.

Hydrochloric acid at 2 M concentration was introduced to the cells for 10 min to denature DNA. This step is essential because the antibody to BrdU cannot bind to BrdU of intact DNA. Following a wash in PBS, sodium borate at pH 9 and 0.1 M was added for 10 min to neutralize the HCl. After rinsing off the borate with PBS, mouse monoclonal antibody to BrdU was applied for 30 min at 1:10 dilution. This was followed by goat anti-mouse immunoglobulin conjugated to fluorescein (1:40) for 30 min. The coverslips were then mounted on glass slides with glycerol-PBS and examined on a Zeiss Universal fluorescence microscope equipped with phase contrast, fluorescein and rhodamine optics. Except for the cell fixation processes, the entire staining and acid denaturation procedure was carried out at room temperature.

Thus, proliferating cells could be detected by their nuclear incorporation of BrdU. Astrocytes could be simultaneously recognized by the presence of GFAP while oligodendrocytes were identified by galactocerebroside. No double stainings were performed on Schwann cells but these cells could be readily identified by their bipolar, spindle-shaped morphology.

Sources of antibodies were as follows: mouse anti-BrdU monoclonal from Becton Dickinson, rat anti-GFAP monoclonal from Dr. V. Lee (36), polyclonal rabbit anti-GFAP from Dako Corporation, and all secondary antibodies (e.g. goat anti-mouse immunoglobulin-fluorescein) were from Cappel Laboratories. Antiserum to galactocerebroside purified from bovine brain was raised in rabbits in this laboratory (15). Dilutions of antibodies were made in PBS.

Tabulation of data

Under the immunofluorescence microscope, GFAP or galactocerebroside positive cells (astrocytes and oligodendrocytes respectively) were identified and counted. Of these, the number of identified cells with BrdU labelling in their nuclei was assessed to obtain the percentage of astrocytes or oligodendrocytes that were proliferating on the particular coverslip. Cell countings were done at x20 magnification in 10-20 fields such that between 100-400 cells were tabulated. The result was divided by the average of similar results obtained from control coverslips to give the proliferation index (PI) of the test factor. Thus, a PI value of 1 represents no mitotic capability while a PI value of 2 indicates a two-fold mitotic response. Value of 1.5 or greater was taken as a positive proliferative result.

For Schwann cells, the percentage of bipolar, spindle shaped cells with BrdU-nuclear labelling was similarly tabulated and the PI value obtained.

In all cases, controls were sister cultures that were exposed to BrdU but not to any of the test growth factor.

RESULTS

For the convenience of the reader this section has been sub-divided into the three types of glial cells.

Astrocytes

Figure 2 shows positive immunolabellings of a control fetal astrocyte culture with the antibodies directed against GFAP and BrdU. The GFAP staining was cytoplasmic while BrdU labelling was nuclear. Different degrees of nuclear labelling by BrdU could be seen, presumably representing different stages of the mitotic cycle of the cell. Although there was occasional cross-reactivity between the two secondary antibodies when mouse anti-BdrU and rat anti-GFAP were used simultaneously, both being directed against rodent immunoglobulins, the distinct spatial labellings of cytoplasmic GFAP on the one hand and nuclear BrdU on the other, makes the results clear, interpretable and valid. However, to avoid confusion, we have recently begun to use polyclonal rabbit anti-GFAP and mouse anti-BdrU antibodies.

As shown on Table III, four mitogens were identified for fetal astrocytes. These were GGF-BP at 1 µg/ml, PDGF at 5 and 50 mU/ml, FGF at the higher concentration tested (5 µg/ml), and PDB at 100 nM but not 10 nM. All other agents tested produced a PI value of about 1 and thus were ineffective as mitogens. Figure 3 shows a higher density of BrdU labelling for fetal astrocytes exposed to GGF-BP when compared to controls.

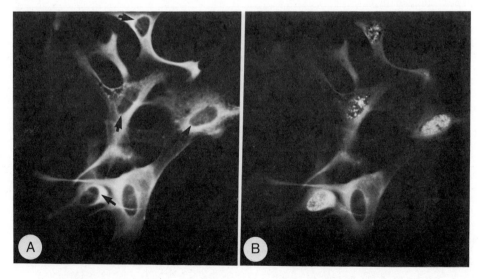

Figure 2. Control fetal astrocyte culture showing GFAP (A) and bromodeoxyuridine (B) incorporation. Astrocytes undergoing proliferation are shown by arrows in A. X 1600.

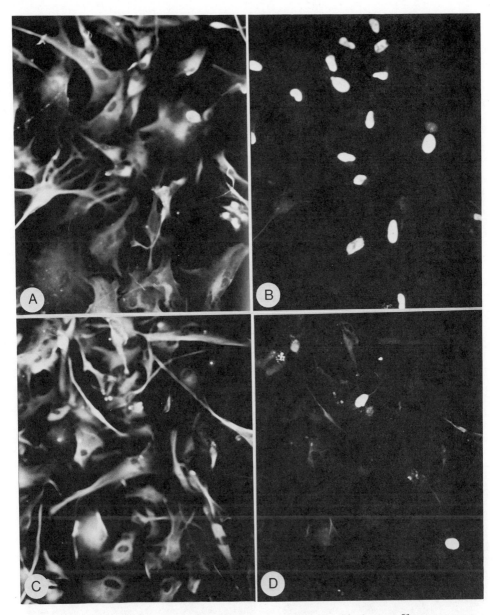

Figure 3. GFAP (A, C) and bromodeoxyuridine (B, D) immunofluorescence of control fetal astrocytes (A, B) or astrocytes exposed to GGF-BP (C, D). Note the higher density of bromodeoxyuridine labelling in culture exposed to GGF-BP (D). X 800.

The proliferative ability of fetal astrocytes appeared to be dependent on the period of time in culture. Figure 4 represents the percent of control astrocytes (those that were not treated with growth factors) that incorporated BrdU as a function of time in vitro. At 11 days of culture, 57% of astrocytes had positive labelling of BrdU in their nuclei. This gradually declined such that after 112 days in vitro, only 11% of fetal astrocytes incorporated BrdU.

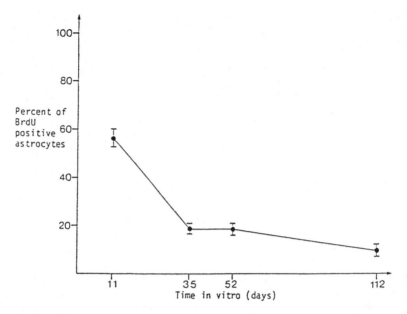

Figure 4.   Percent of control human fetal astrocytes undergoing
proliferation as a function of time in culture.   Each point is the mean $\pm$
SEM of 3-6 determinations.

In the case of adult astrocytes, no BrdU labelling could be observed
for control astrocytes or those exposed to the various growth factors.
It appeared that these adult cells had lost the ability to undergo
mitosis.   Figure 5 shows the absence of BrdU incorporation in the nucleus
of adult astrocytes.

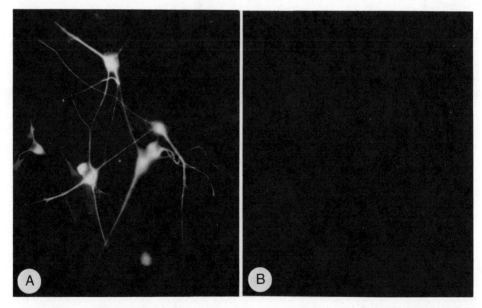

Figure 5.   Human adult astrocytes do not incorporate bromodeoxyuridine
into their nuclei. A: GFAP immunofluorescence, B: bromodeoxyuridine
labelling. X 1600.

It should be noted that most fetal astrocytes were flat and fibroblast-like (Figures 2 and 3) while most of their adult counterparts were protoplasmic and process-bearing (Figure 5). The fibroblast-like astrocytic morphology has been designated Type I while protoplasmic and process-bearing astrocytes have been described as Type II (55). The failure of adult astrocytes to incorporate BrdU did not seem to be due to their Type II morphology since similar cells in fetal astrocyte cultures were found to incorporate BrdU.

TABLE III.   PROLIFERATION INDEX (PI) OF VARIOUS GROWTH FACTORS ON HUMAN FETAL ASTROCYTES AND SCHWANN CELLS IN CULTURE

Growth factor	Concentration		PI astrocyte	PI Schwann
Interleukin-2	5	U/ml	1.2 ± 0.2	1.2 ± 0.1
	50	U/ml	0.9 ± 0.1	1.0 ± 0.1
Glial growth factor from bovine pituitary	1	µg/ml	3.6 ± 0.1*	1.7 ± 0.1*
Platelet-derived growth factor	5	mU/ml	2.1 ± 0.1*	1.0 ± 0.1
	50	mU/ml	1.8 ± 0.2*	1.7 ± 0.2*
Fibroblast growth factor	0.5	µg/ml	1.0 ± 0.2	1.0 ± 0.1
	5	µg/ml	1.5 ± 0.1*	1.2 ± 0.2
Epidermal growth factor	0.5	µg/ml	0.5 ± 0.1	0.9 ± 0.1
	5	µg/ml	1.2 ± 0.1	1.2 ± 0.1
Nerve growth factor	100	ng/ml	1.1 ± 0.1	1.5 ± 0.1*
	1	µg/ml	1.4 ± 0.2	1.7 ± 0.1*
Phorbol-12, 13- dibutyrate	10	nM	1.1 ± 0.1	1.7 ± 0.2*
	100	nM	3.1 ± 0.4*	1.6 ± 0.1*
Dibutyryl cAMP	100	µM	1.1 ± 0.0	1.1 ± 0.1
	1	mM	0.9 ± 0.0	0.8 ± 0.0
Forskolin	10	µM	not tested	0.9 ± 0.2
	100	µM	not tested	1.1 ± 0.1
Glial growth factor and forskolin	1	µg/ml	not tested	1.8 ± 0.1*
	10	µM		

PI values are mean ± SEM, of between 3 and 6 coverslips each. PI for control is 1. *Denotes effective mitogens. SEM of less than 0.05 is shown as 0.0. The average % of control fetal astrocytes that incorporated BrdU was 16 ± 1.2 (n=12) while that of control fetal Schwann cells was 34 ± 2.1 (n=13) (mean ± SEM).

Oligodendrocytes

As was the case with adult astrocytes, adult oligodendrocytes did not incorporate BrdU into their nuclei even in the presence of any of the test growth factors. This inability was not due to the stage of differentiation of oligodendrocytes in culture since undifferentiated oligodendrocytes (Figure 6), as well as their differentiated counterparts (Figure 7), did not show positive labellings with BrdU. That the BrdU method works for oligodendrocytes is shown by galactocerebroside and BrdU stainings of rat oligodendroccyte isolated on postnatal day 1 and incubated with BrdU for 48 hours (Figure 8).

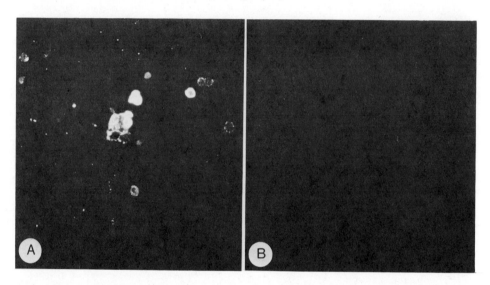

Figure 6.   Undifferentiated human adult oligodendrocytes do not incorporate bromodeoxyuridine into their nuclei. A: galactocerebroside immunostaining, B: bromodeoxyuridine labelling. X 800.

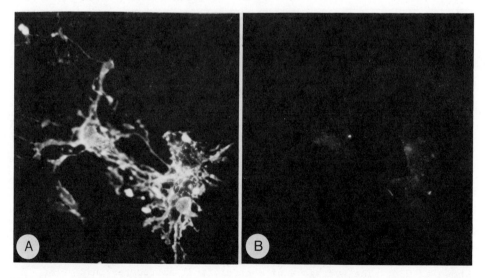

Figure 7.   Differentiated human adult oligodendrocytes do not incorporate bromodeoxyuridine into their nuclei. A: galactocerebroside immunostaining, B: bromodeoxyuridine labelling. X 800.

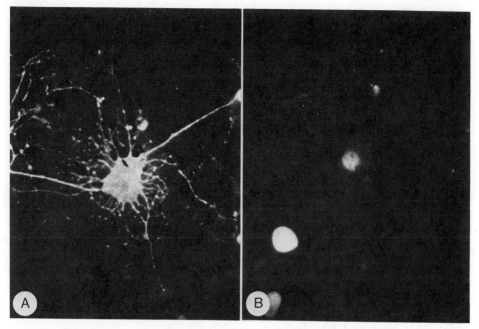

Figure 8. Postnatal day 1 rat brain at 3 days in vitro was incubated with 10 um BrdU for 48 hours. A: galactocerebroside immunostaining, B: BrddU labelling. X 800. The positive BrdU labelling is evidence that the method can detect mitosis of oligodendrocytes.

Schwann cells

In Table III, the PI values of various growth factors on human fetal Schwann cells are presented. Mitogenic factors were GGF-BP (1 ug/ml), NGF (100 ng/ml and 1 ug/ml), PDB (10 and 100 nM), and the combination of GGF (1 ug/ml) and forskolin (10 uM). PDGF produced a positive proliferative response only at the higher concentration tested (50 mU/ml). On the average (n of 13 cultures containing between 100-200 Schwann cells each), 34% of control Schwann cells incorporated BrdU into their nuclei over a 48 hour period. Figure 9 compares BrdU immunolabelling of control Schwann cells with those exposed to GGF-BP. As noted, more Schwann cells were labelled in cultures treated with GGF-BP.

DISCUSSION

The search for factors that can promote proliferation of glial cells has generated many studies, some with contradictory results. For instance, while EGF was found to be mitogenic for neonatal rodent astrocytes in the studies of Simpson et al (66) and Leutz and Schachner (37), negative result was obtained by Pruss et al (54). A probable reason for the discrepancies in the literature lies in the methods commonly used to detect cell proliferation, namely ^{3}H-thymidine uptake followed by scintillation counting or autoradiography (Tables I and II). As aforementioned, these techniques can have potential flaws such as cell identification, the lack of a homogeneous cell population in the case of scintillation counting, and the subjectivity that may become important in autoradiography when differentiating between weakly labelled and unlabelled cells.

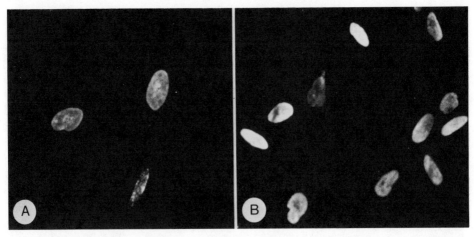

Figure 9.   Bromodeoxyuridine immunofluorescence of control human fetal Schwann cells (A) or those exposed to GGF-BP (B).   X 800.

In the present series of experiments, a less ambiguous method (immunofluorescence of BrdU) was used to assess whether or not several soluble factors have the potential to produce proliferation of glial cells derived from human fetuses or adults.   In addition, since astrocytes and oligodendrocytes were simultaneously labelled with their cell type specific markers, positive identification of the proliferating cell type becomes possible.

Our results for fetal astrocytes show that four agents were capable of producing a mitogenic response: GGF-BP, PDGF, FGF and phorbol ester. The effects of FGF and phorbol ester were dose-dependent in that only the higher concentrations tested resulted in proliferation.   The findings with GGF-BP confirm the reports of others for rat astrocytes (6,29,54). On the other hand, the negative actions of EGF contrast the findings of Simpson et al (66) and Leutz and Schachner (37).   Besides the difference in the technique used to assess cell proliferation, a possible explanation for the discrepancy is that the astrocytes in the present study are derived from humans and not rodents.   Another possibility is that the present experiment was conducted in the present of serum-containing medium while many previous studies were performed in serum-free medium.   Indeed, Morrison et al (48) showed that the proliferation of astrocytes in 10% serum-containing medium was about 3 times greater than those maintained in serum-free medium supplemented with a mitogen, FGF.   Thus, the presence of serum in the present experiment, itself capable of promoting mitosis, might have masked the proliferative ability of EGF.   However, serum-containing medium presents a more physiological environment to the cells than that of serum-free medium and the present series of experiments was performed in serum-containing medium for that reason.   The capability of phorbol ester to stimulate division of fetal astrocytes, as well as fetal Schwann cells (see below), has never been reported and may represent a widespread tumor-inducing property of phorbol esters (for review, see reference 8).

Figure 4 shows that the potential of control astrocytes to undergo proliferation decreases with the length of time that they have been maintained in culture.   Many cell types lose their proliferative ability after repeated mitosis and passaging (14,23,24) and in the present experiment, human fetal astrocytes appear to fall into this category.

That adult astrocytes in culture were incapable of mitosis confirms our earlier observations (31) using ^{3}H-thymidine radiolabelling to detect cell division. This failure did not appear to be due to their Type II morphology (58) since some of the Type II astrocytes in the fetal cultures incorporated BrdU. In addition, Miller et al (45) showed that both Type I and Type II astrocytes could undergo mitosis in response to injury. From the results in Figure 4, where fetal astrocytes lose their proliferative capability with aging in vitro, it appears that the most likely reason for the failure of adult astrocytes to undergo mitosis is due to the advanced age of the adult astrocytes, being derived from humans of between ages 60-90. It is important to elucidate the mechanism by which proliferation of astrocytes (gliosis) is initiated. Gliosis is a central problem in CNS regeneration in which uncontrolled astrocytic multiplication blocks paths of regrowing nerves.

As shown on Table I, factors such as 'glial promoting factors', FGF, PDGF, axonal contact and Interleukin-2 have been reported to produce mitosis of oligodendrocytes isolated mostly from neonatal rat brain. It is important to confirm these findings since such results have important implications for diseases such as multiple sclerosis where oligodendrocytes are lost. Promoting remaining oligodendrocytes in such diseases to proliferate, and then to differentiate to produce remyelination, may eventually result in restoration of lost functions. Since multiple sclerosis is a disease of adults, it is especially necessary to test such agents on human adult oligodendrocytes. We attempted such an experiment and found that adult human oligodendrocytes in culture do not undergo mitosis. In addition, none of the agents tested, such as interleukin-2, could promote proliferation. The search for effective agents continues.

Several studies have examined the factors that might induce the mitosis of neonatal rat Schwann cells in vitro (Table II). The results have led Raff et al (56) to propose that there might exist two pathways that led to the proliferation of Schwann cells: one involved cAMP and the other did not. This was supported by the observation that the combination of forskolin (an adenyl cyclase activator with resultant accumulation of cAMP) and GGF-BP (which did not increase cAMP content) produced a greater proliferation index on rat Schwann cells than those obtained when either was used individually (53). We repeated these experiments using human fetal Schwann cells. In agreement with the literature (6,52,53,56), GGF-BP was found to increase the proliferation rate of human fetal Schwann cells (Table III). Contrasting the results of others (54,56,64), NGF (at 100 ng/ml and 1 ug/ml) and PDGF (at the higher concentration of 50 mU/ml) were mitogenic for human fetal Schwann cells. The mitogenic property of NGF is interesting in view of the presence of nerve growth factor receptors on these cells (Kim et al, J Neuroscience Research, in press).

Unlike the hypothesis of Raff et al (56), we found no evidence for the requirement of a cAMP-dependent mechanism in increasing the mitotic response of human fetal Schwann cells. This was shown by the lack of effects of two concentrations of dibutyryl cAMP or forskolin (Table II). The combination of forskolin and GGF-BP did not result in a higher response than that of GGF-BP alone, leading to the conclusion that the positive effect of the combination was due to GGF only.

In conclusion, we have assessed the ability of various agents to promote proliferation of human glial cells in culture. We began with the hypothesis that growth factors described for rodents cannot be assumed to be similarly effective in humans and indeed, this may be the main reason why some of the results presented here contradict those in the literature. The previous observation that interleukin-2 promoted proliferation of oligodendrocytes (44) was of special interest because interactions between the immune system and the nervous system are increasingly being recognized to be important in the pathogenesis of several nervous disorders. However, our results with human glial cells could not replicate that finding. The observations that phorbol ester, thought to activate directly protein kinase C (9,28,50), stimulated the mitosis of human fetal astrocytes and Schwann cells, suggest that activation of this important ubiquitous enzyme may trigger key events that lead to cell proliferation. Human adult astrocytes and oligodendrocytes could not be induced to proliferate in the present series of experiments and it is important to continue the search for factors that might induce their mitosis in culture. As stated, such experiments have strong implications for diseases such as multiple sclerosis and CNS injury.

SUMMARY

Although several mitogens for glial cells isolated from the rodent nervous system have been described, few studies have examined the factors that can promote proliferation of human glial cells. Using a new double immunofluorescence technique, we assessed various growth factors on the proliferation of cultured human glial cells. Cells studied were fetal astrocytes, fetal Schwann cells, adult astrocytes and adult oligodendrocytes. Mitogens effective for fetal astrocytes were glial growth factor from the bovine pituitary, platelet-derived growth factor, fibroblast growth factor and 4B-phorbol 12, 13-dibutyrate. The proliferative capability of fetal astrocytes decreases with aging in vitro. For fetal Schwann cells, effective agents were glial growth factor from the bovine pituitary, platelet-derived growth factor, nerve growth factor, and 4B-phorbol 12, 13-dibutyrate. Adult astrocytes and oligodendrocytes did not normally divide in culture and none of the agents tested were effective in inducing their proliferation. The report that interleukin-2 was a mitogen for oligodendrocytes could not be replicated in the present study on any of the glial cell types.

ACKNOWLEDGEMENTS

This study was supported by grants from the Medical Research Council of Canada, the Multiple Sclerosis Society of Canada, and Jacob Cohen Fund for Research into Multiple Sclerosis. The skilled technical assistance of David Osborne is acknowledged. V.W. Yong was a postdoctoral fellow of the Multiple Sclerosis Society of Canada.

REFERENCES

1.   Abercrombie M, Johnson ML: Quantitative histology of Wallerian degeneration. I. Nuclear population in rabbit sciatic nerve. J Anat 80:37-50, 1946

2.   Benveniste EN, Merrill JE: Stimulation of oligodendroglial proliferation and maturation by interleukin-2. Nature 321:610-613, 1986

3.   Besnard F, Perraud F, Sensenbrenner M, Labourdette G: Platelet-derived growth factor is a mitogen for glial but not for neuronal rat brain cells in vitro. Neurosci Letts 73:287-292, 1987

4.   Bignami A, Eng L, Dahl D, Uyeda C: Localization of the glia fibrillary acidic protein in astrocytes by immunofluorescence. Brain Res 43:429-435, 1972

5.   Bradley WG, Asbury AK: Duration of synthesis phase in neurilemma cells in mouse sciatic nerve during degeneration. Exp Neurol 26:275-282, 1970

6.   Brockes JP, Lemke GE, Balzer DR Jr: Purification and preliminary characterization of a glial growth factor from the bovine pituitary.   J Biol Chem 255:8374-8377, 1980

7.   Bologa L, Deugnier MA, Joubert R, Bisconte JC: Myelin basic protein stimulates the proliferation of astrocytes: Possible explanation for multiple sclerosis plaque formation.   Brain Res 346:199-203, 1985

8.   Boutwell RK: The function and mechanism of promoters of carcinogenesis.   Crit Rev Toxicol 2:419-443, 1974

9.   Burgess SK, Sahyoun N, Blanchard SG, LeVine H III, Chang KJ, Cuatrecasas P: Phorbol ester receptors and protein kinase C in primary neuronal cultures: Development and stimulation of endogenous phosphorylation.   J Cell Biol 102:312-319, 1986

10.   Cavanagh JB: The proliferation of astrocytes around a needle wound in the rat brain.   J Anat 106:471-487, 1970

11.   DeVries GH, Salzer JL, Bunge RP: Axolemma-enriched fractions isolated from PNS and CNS are mitogenic for cultured Schwann cells.   Dev Brain Res 3:295-299, 1982

12.   Dolbeare F, Gratzner H, Pallavicini MG, Gray JW: Flow cytometric measurement of total DNA content and incorporated bromodeoxyuridine.   Proc Natl Acad Sci (USA) 80:5573-5577, 1983

13.   Eccleston PA, Silberberg DH: Fibroblast growth factor is a mitogen for oligodendrocytes in vitro.   Dev Brain Res 21:315-318, 1985

14.   Freshney RI: Culture of animal cells: A manual of basic technique.   Alan R Liss, Inc, New York, 1983

15.   Fry J, Lisak R, Manning M, Silberberg D: Serological techniques for detection of antibody to galactocerebroside.   J Immuno Methods 11:185-193, 1976

16.   Giulian D, Tomozawa Y, Hindman H, Allen RL: Peptides from regenerating central nervous system promote specific populations of macroglia.   Proc Natl Acad Sci (USA) 82:4287-4290, 1985

17.   Giulian D, Lachman LB: Interleukin-1 stimulation of astroglial proliferation after brain injury.   Science 228:497-499, 1985

18.   Giulian D, Allen RL, Baker TJ, Tomozawa Y: Brain peptides and glial growth.   I. Glia-promoting factors as regulators of gliogenesis in the developing and injured central nervous system.   J Cell Biol 102:803-811, 1986

19.   Giulian D, Young DG: Brain peptides and glial growth.   II. Identification of cells that secrete glia-promoting factors.   J Cell Biol 102:812-820, 1986

20.   Gratzner HG: Monoclonal antibody to 5-bromo- and 5-iododeoxyuridine: A new reagent for detection of DNA replication.   Science 218:474-475, 1982

21.   Griffin JW, Drucker N, Gold BG, Rosenfeld J, Benzaquen M, Charnas LR, Fahnestock KE, Stocks EA: Schwann cell proliferation and migration during paranodal demyelination.   J Neurosci 7:682-699, 1987

22.   Hatten ME: Neuronal regulation of astroglial morphology and proliferation in vitro.   J Cell Biol 100:384-396, 1985

23.   Hay RJ, Strechler BL: The limited growth span of cell strains isolated from the chick embryo.   Exp Gerontol 2:123-135, 1967

24.   Hayflick L, Moorhead PS: The serial cultivation of human diploid cell strains.   Exp Cell Res 25:585-621, 1961

25.   Heldin CH, Wasteson A, Westermark B: Partial purification and characterization of platelet factors stimulating the multiplication of normal human glial cells.   Exp Cell Res 109:429-437, 1977

26.   Herndon RM, Pricce DL, Weiner LP: Regeneration of oligodendroglia during recovery from demyelinating disease.   Science 195:693-694, 1977

27.   Immamoto K, Paterson J, Leblond CP: Radioautographic investigation of gliogenesis in the corpus callosum of young rats.   I. Sequential changes in oligodendrocytes.   J Comp Neurol 180:115-138

28. Kikkawa U, Takai Y, Tanaka Y, Miyake R, Nishizuka Y: Protein kinase C as a possible receptor protein of tumor-promoting phorbol esters. J Bil Chem 258:11442-11445, 1983

29. Kim SU, Stern J, Kim MW, Pleasure DE: Culture of purified rat astrocytes in serum-free medium supplemented with mitogen. Brian Res 274:79-81, 1983

30. Kim SU, Sato Y, Silberberg DH, Pleasure DE, Rorke L: Long-term culture of human oligodendrocytes. Isolation, growth and identification. J Neuro Sci 62:295-301, 1983

31. Kim SU, Moretto G, Shin DH, Lee VM: Modulation of antigenic expression in cultured adult human oligodendrocytes by derivatives of adenosine 3', 5'-cyclic monophosphate. J Neurol Sci 69:81-91, 1985

32. Kim SU, Kim KM, Moretto G, Kim JH: The growth of fetal human sensory ganglion neurons in culture: A scanning electron microscopic study. Scan Electron Microscopy II:843-848, 1985

33. Kim SU: Antigen expression by glial cells grown in culture. J Neuroimmunol 8:255-282, 1985

34. Kreider BQ, Messing A, Doan H, Kim SU, Lisak RP, Pleasure DE: Enrichment of Schwann cell cultures from neonatal rat sciatic nerve by differential adhesion. Brain Res 207:433-444, 1981

35. Latov N, Nilayer G, Zimmerman EA, Johnson WG, Silverman JA, Detenionini R, Cote L: Fibrillary astrocytes proliferate in response to brain injury. Dev Biol 72:381-384, 1979

36. Lee V, Page K, Wu H, Schlaepfer WW: Monoclonal antibodies to gel excised glial filament protein and their reactivity with other intermediate filament proteins. J Neurochem 42:25-32, 1984

37. Leutz A, Schachner M: Epidermal growth factor stimulates DNA-synthesis of astrocytes in primary cerebellar cultures. Exp Tissue Res 220:393-404, 1981

38. Levi-Montalcini R, Meyer H, Hamburger V: In vitro experiments as the effects of mouse sarcomas 180 and 37 on spinal and sympathetic ganglia of the chick embryo. Cancer Res 14:49-57, 1954

39. Levi-Montalcini R, Angeletti PU: Nerve growth factor. Physiol Rev 48:534-569, 1968.

40. Ludwin SK: Proliferation of mature oligodendrocytes after trauma to the central nervous system. Nature 308:274-275, 1984

41. Ludwin SK: Reaction of oligodendrocytes and astrocytes to trauma and implantation: A combined autoradiographic and ummunohistochemical study. Lab Invest 52:20-30, 1985

42. Ludwin SK, Bakker DA: Can oligodendrocytes attached to myelin proliferate? J Neurosci 8:1239-1244, 1988

43. McCarthy KD, de Vellis J: Preparation of separate astroglial and oligodendroglial cell cultures from rat cerebral tissue. J Cell Biol 85:890-902, 1980

44. Merrill JE, Kutsunai S, Mohlstrom C, Hofman F, Groopman J, Colde DW: Proliferation of astroglia and oligodendroglia in response to human T cell-derived factors. Science 224:1428-1430, 1984

45. Miller RH, Abney ER, David S, French-Constant C, Lindsay R, Patel R, Stone J, Raff MC: Is reactive gliosis a property of a distinct subpopulation of astrocytes? J Neurosci 6:22-29, 1986

46. Mobley WC, Server AC, Ishii DN, Riopelle RP, Shooter EM: Nerve growth factor. New Engl J Med 297:1096-1104, 1977

47. Moretto G,Kim SU, Shin DH, Pleasure DE, Rizzuro N: Long-term cultures of human adult Schwann cells isolated from autopsied materials. Acta Neruopathol 64:15-21, 1984

48. Morrison RS, Saneto RP, de Vellis J: Developmental expression of rat brain mitogens for cultured astrocytes. J Neurosci Res 8:435-442, 1982

49. Morstyn G, Hsu SM, Kinsella T, Gratzner H, Russo A, Mitchell JB: Bromodeoxyuridine in tumors and chromosomes detected with a monoclonal antibody. J Clin Invest 72:1844-1850, 1983

50. Nishizuka Y: Studies and perspectives of protein kinase C. Science 233:305-311, 1986

51. Pettman B, Weibel M, Daune M, Sensenbrenner M, Labourdette G: Stimulation of proliferation and maturation of rat astroblasts in serum-free culture by an astroglial growth factor. J Neurosci Res 8:463-476, 1982

52. Pleasure DE, Kreider B, Shuman S, Sobue G: Tissue culture studies of Schwann cell proliferation and differentiation. Dev Neurosci 7:364-373, 1985

53. Porter S, Clark MB, Glaser L,. Bunge RP: Schwann cells stimulated to proliferate in the absence of neurons retain full functional capability. J Neurosci 6:3070-3078, 1986

54. Pruss RM, Bartlett PF, Gavrilovic J, Lisak RP, Rattray S: Mitogens for glial cells: A comparison of the response of cultured astrocytes, oligodendrocytes and Schwann cells. Dev Brain Res 2:19-35, 1982.

55. Raff MC, Mirsky R, Fields KL, Lisak RP, Dorfman SH, Silberberg DH, Gregson NA, Leibowitz S, Kennedy MC: Galactocerebroside is a specific cell-surface antigenic marker for oligodendrocytes in culture. Nature 274:813-816, 1978

56. Raff MC, Abney E, Brockes JP, Hornby-Smith A: Schwann cell growth factors. Cell 15:813-822, 1978

57. Raff MC, Fields KL, Hakomori SI, Mirsky R, Pruss RM, Winter J: Cell type-specific markers for distinguishing and studying neurons and the major classes of glial cells in culture. Brain Res 174:283-308, 1979

58. Raff MC, Abney ER, Cohen J, Lindsay R, Noble M: Two types of astrocytes in cultures of developing rat white matter: differences in morphology, surface gangliosides, and growth characteristics. J Neurosci 3:1289-1300, 1983

59. Ratner N, Bunge RP, Glaser L: Schwann cell proliferation in vitro: An overview. In Rubenstein AE, Bunge RP, Houseman DE (eds): Neurofibromatosis, Ann NY Acad Sci 486:170-181, 1986

60. Raine CS, Scheinberg L, Waltz JM: Multiple sclerosis: Oligodendrocyte survival and proliferation in an active established lesion. Lab Invest 45:534-546, 1981

61. Raine CS, Traugott U: Chronic relapsing experimental autoimmune encephalomyelitis: Ultrastructure of the central nervous system of animals treated with combinations of myelin components. Lab Invest 48:275-284, 1983

62. Rodriguez M, Lennon VA, Benveniste EN, Merrill JE: Remyelination of oligodendrocytes stimulated by antiserum to spinal cord. J Neuropathol Experimental Neurol 46:84-95, 1987

63. Rogers AW: Techniques of autoradiography. Elsevier, Amsterdam, 1979

64. Salzer JL, Bunge RP: Studies of Schwann cell proliferation. I. An analysis in tissue culture of proliferation during development, Wallerian degeneration, and direct injury. J Cell Biol 84:739-752, 1980

65. Sheffield WD, Kim SU: Myelin basic protein causes proliferation of lymphocytes and astrocytes in vitro. Brain Res 132:580-584, 1977

66. Simpson DL, Morrison R, de Vellis J, Herschman HR: Epidermal growth factor binding and mitogenic activity on purified populations of cells from the central nervous system. J Neurosci Res 8:453-462, 1982

67. Skoff RP, Vaughn JE: An autoradiographic study of cellular proliferation in degenerating rat optic nerve. J Comp Neurol 141:133-156, 1971

68. Skoff RP: The fine structure of pulse labelled (^{3}H-thymidine) cells in degenerating rat optic nerve. J Comp Neurol 161:595-612, 1975

69. Sobue G, Brown MJ, Kim SU, Pleasure D: Axolemma is a mitogen for human Schwann cells. Ann Neurol 15:449-452, 1984

70. Sobue G, Shuman S, Pleasure D: Schwann cell responses to cyclic AMP: Proliferation, change in shape, and appearance of surface

galactocerebroside. Brain Res 362:233-32, 1986

71. Westermark B: Density dependent proliferation of human glia cells stimulated by epidermal growth factor. Biochem Biophy Res Comm 69:304-310, 1976

72. Wood PM: Separation of functional Schwann cells and neurons from normal peripheral nerve tissue. Brain Res 115:361-375, 1976

73. Wood PM, Williams AK: Oligodendrocyte proliferation and CNS myelination in cultures containing dissociated embryonic neuroglia and dorsal root ganglion neurons. Dev Brain Res 12:225-241, 1984

74. Wood PM, Bunge RP: Evidence that axons are mitogenic for oligodendrocytes isolated from adult animals. Nature 320:756-758, 1986

75. Yong VW, Kim SU: A new double labelling immunofluorescence technique for the determination of proliferation of human astrocytes in culture. J Neurosci Methods 21:9-16, 1987

76. Yong VW, Kim SU, Pleasure DE: Growth factorss for fetal and adult human astrocytes in culture. Brain Res 444:59-66, 1988

77. Yong VW, Kim SU, Kim MW, Shin D H: Growth factors for human glial cells in culture. Glia 1:113-123, 1988

78. Yoshino JE, Mason PW, DeVries GH: Developmental changes in myelin-induced proliferation of cultured Schwann cells. J Cell Biol 104:655-660, 1987

4

MOLECULAR GENETICS OF MYELIN BASIC PROTEIN IN MOUSE AND HUMANS

John Kamholz

Department of Neurology
University of Pennsylvania
Philadelphia, PA

INTRODUCTION

Myelin is a multi-lamellar membrane structure which ensheathes the axon and acts to increase nerve conduction velocity without a significant increase in axonal diameter. Myelin occurs mainly in vertebrates (1), although some invertebrates have been found to have a myelin like structure as well. The myelin sheath is synthesized by oligodendrocytes in the central nervous system (2) (CNS) and Schwann cells in the peripheral nervous system (3) (PNS). The myelin sheath synthesized by these two cell types is morphologically similar and consists of a series of concentrically wrapped membrane bilayers which are extensions of the oligodendrocyte or Schwann cell plasma membrane (4). During myelination, these membrane extensions are repeatedly wrapped about the axon and are then compacted at both apposed cytoplasmic and extracellular membrane surfaces to form the intraperiod and major dense lines seen by electron microscopy in mature myelin (5) [see Figure 1]. Maturation of the myelinating cell and deposition of the myelin sheath is, thus, accompanied by a number of drastic morphological and physiological changes which also include the synthesis of a set of new proteins found only in myelin (6). Although Schwann cells and oligodendrocytes synthesize morphologically similar membrane structures, they have different embryological origins and distinct biological properties (2,3). This chapter concerns mainly oligodendrocyte gene expression, thus the following discussion will focus on the oligodendrocyte and CNS myelination.

In humans, CNS myelination begins with the appearance of the medial longitudinal fasciculus at the beginning of the second trimester and is virtually complete by the end of the second year (7), although the association cortex may continue to myelinate until the tenth decade (8). During this critical period, brain weight increases dramatically and this increase is predominantly due to the synthesis of myelin (9). The multiple fiber tracts of the brain each myelinate at different times and at different rates (7). In general, however, more caudal structures myelinate earlier than more rostral ones, creating a gradient of myelination from caudal to rostral within the neuraxis (7, 8, 10). The factors involved in regulating this complex pattern of myelin biogenesis are currently unknown. Many studies, however, have revealed a sudden increase in the glial cell population in the brain prior to the onset of myelination (11, 12, 13). This cellular proliferation, called "myelination gliosis", is accompanied by changes in glial cell morphology (14) as well as the accumulation of the lipid components of myelin (15, 16). This pattern suggests that the production of substances which promote oligodendrocyte precursor migration, proliferation and subsequent maturation occurs locally around specific axon tracts to be myelinated. Current evidence from studies of rat optic nerve cultures suggests that type I or protoplasmic astrocytes secrete a factor(s) which promotes oligodendrocyte precursor proliferation (17, 18, 19). Whether these substances are produced by neurons, astrocytes or some other cell type in vivo, however, is still an open question. It is clear from this data, however, that regulation of the overall timing of myelination is complex and may involve cellular interactions between oligodendrocytes as well as several other cell types.

After this initial phase of cellular proliferation, the developing oligodendrocyte sends out processes which make contact with and wrap around axons. Stainable myelin then begins to accumulate and myelin compaction occurs. During this process, each oligodendrocyte will myelinate many different axons, in some instances 35 or more (20, 21), unlike the Schwann cell in the PNS which myelinates only one. At this stage, local events in the area of the myelinating saxon appear to determine the extent of myelination or the number of wraps the oligodendrocyte process makes about the axon. Usually the larger the axon, the thicker its myelin sheath (22). Electron microscopic studies of mouse brain (23) and optic nerve (24) myelination have shown single oligodendrocytes myelinating axons of different sizes. In each instance, the larger axon had more wraps of myelin, suggesting that myelin thickness is specified independently for each axon in an area far from the oligodendrocyte cell body. The mechanisms involved in regulating this process are not known, but again point out the importance of cellular interactions in the control of myelin biogenesis.

The myelin sheath laid down during this complex series of events is composed mainly of lipid, 75% by weight (25), and also contains a number of structural proteins (6). In the CNS, the major structural proteins in myelin are proteolipid protein (PLP) which makes up about 50% of the myelin protein, myelin basic protein (MBP) which makes up about 30% of the protein and myelin associated glycoprotein or MAG which makes up about 1% of the myelin protein. Cyclic nucleotide phosphohydrolase (CNP) and the myelin lipid biosynthetic enzymes make up the rest of the myelin protein. With the myelinating oligodendrocyte itself, a unique temporal sequence of expression of these myelin proteins has been demonstrated both in vitro (26) and in vivo (27) during the period prior to sheath formation. Galactocerebroside, a specific lipid, is synthesized first, then MBP, MAG and finally PLP are produced and myelination commences. For MBP, PLP and MAG, control of new protein synthesis occurs at the level of transcription (28, 29, 30, 21) and this is likely to be the case for the lipid biosynthetic enzymes as well. Myelination is thus accompanied by the expression of a unique set of myelin specific genes, although the signals which control the onset of synthesis of these myelin specific proteins are not understood. Unlike Schwann cells in the PNS, oligodendrocytes do not require the presence of axons to carry out this program once it has been set in motion (26, 29). In summary then, the process of myelination involves regulation of the timing of oligodendrocyte proliferation, migration and maturation, temporal regulation of the expression of a set of myelin specific genes, and finally, control of the extent of myelination.

Recent application of the techniques of molecular biology to the study of myelination have significantly increased our knowledge of the myelin structural proteins, including MBP (32, 33), PLP (30, 34, 35) and MAG (31, 36). These techniques have been used to determine the primary structures of each of these proteins, to analyze the structure of the genes encoding them, and to describe their patterns of gene expression during myelination. The genes for each of these proteins have been mapped to specific chromosomal locations (37, 38, 39), and in the cases of PLP (40, 41, 42) and MBP (43, 44), mutations in these genes have been shown to result in profound disruption of the process of myelination. In the following sections, I will discuss the molecular genetics of MBP. These data, although interesting in themselves, should be viewed in the context of myelination as described above, as well as a prelude for future work aimed at understanding the cellular and molecular mechanisms involved in the overall regulation of myelination.

Human and Rodent MBP mRNA and Gene Structure: Alternative Splicing of the Primary MBP Transcript Produces a Family of MBPs

In rodent myelin there is a family of four MBPs of 21.5, 18.5, 17.0 and 14.0 kd. The 18.5 and 14.0 kd forms make up the bulk of the MBP, while the 17.0 and 21.5 kd forms are present in much smaller amounts (45). These four proteins share a common amino acid sequence and differ by the presence of an extra sequence in either the N terminus, the C terminus, or both. The 21.5 kd protein has both extra sequences, the 18.5 kd protein has the C terminal sequence, the 17.0 kd form has the N terminal sequence and the 14 kd form has neither. In vitro translation of rodent brain mRNA can produce all four proteins, suggesting that they are not formed from a single MBP by post-translational modification (46). Several laboratories have subsequently isolated cDNAs for each of the four mouse MBPs, and have determined the mouse MBP gene structure as well (32, 33). These workers found that there was a single MBP gene of 32 kb, consisting of seven exons which contained all of the coding information in the largest MBP. Thus, the four forms of MBP could be accounted for by alternative splicing of a primary MBP transcript from a single MBP gene. A summary of these results is shown on the left hand side of Figure 2. As can be seen in the figure, alternative splicing of either exon 2 or exon 6 produces the four rodent MBP isoforms. Neither the mechanism(s) producing these alternatively spliced transcripts nor their functional significance for myelination are known.

Using a similar molecular biological approach, Kamholz et al (47) isolated cDNAs encoding three separate human MBPs of 21.5, 18.5 and 17.2 kd. These cDNAs shared a common sequence and differed by the presence of extra sequences in either the 5' or 3' end of their coding regions. The human MBP gene structure was also determined, and was found to be approximately 45 kb in length. The gene structure was also determined, and was found to be approximately 45 kb in length. The gene was similar in structure to its rodent counterpart, consisting of seven exons, and contained all of the coding information of the largest MBP (J. Kamholz, unpublished observations). Alternative splicing of a single human MBP transcript can thus also account for the three human MBP isoforms. These results are summarized on the right hand side of Figure 2. Alternative use of either exon 2 or exon 5 will produce the three human MBPs. Exon 6, one of the alternatively spliced mouse exons, does not participate in alternative splicing in the human case. Although both the human and mouse MBP transcripts undergo alternative splicing, the splicing patterns in the two species are thus slightly different. Neither the significance of this difference, nor the mechanism which produces it, is currently understood.

Alternative splicing of a single transcript producing a family of

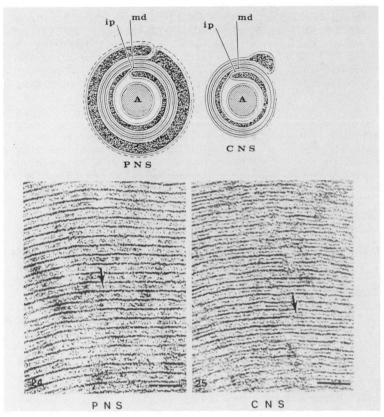

Figure 1   Schematic representation of the major features of PNS and
CNS myelin (above) and electron micrographs of PNS and CNS myelin
(below).   ip- intraperiod line; md- major dense line; A- axon. The
arrows in the electronmicrographs indicate the intraperiod line.

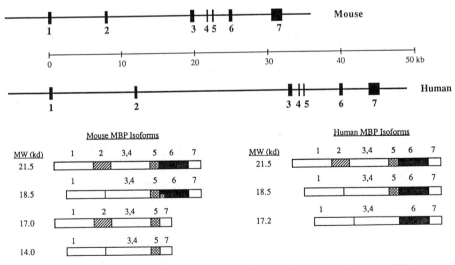

Figure 2   Schematic representation of the human and mouse MBP genes
(above) and the corresponding mRNAs and proteins produced from them
(below).   The numbering scheme for both the genes and mRNAs refers to the
MBP exon sequences. kd- kilodaltons; kb- kilobase pairs.

proteins from a single transcription unit, has also been found for a large number of genes other than MBP (see 48, 49 for recent reviews). In fact, both proteolipid protein (PLP), the major CNS myelin protein, as well as myelin associated glycoprotein (MAG), are found in more than one isoform produced by alternative splicing (36, 30). The phenomena of alternative splicing is thus ubiquitous in brain as well as other tissues. Although it is quite common, the function of alternative splicing within the cell remains elusive (50). For the MBPs, however, several lines of indirect evidence suggest that alternative splicing may play an important role in myelination.

Alternative Splicing of MBP is Conserved in Mammalian Evolution and Regulated During Myelination

MBP is conserved in myelin throughout vertebrate evolution, and in a number of mammalian species, several minor isoforms of the protein are also found, along with a major protein of 18.5 kd. A minor 21.5 kd MBP has been found in man, mouse, chimpanzee, guinea pig, cow, rabbit and sheep (51, 52). Using antibodies raised to a peptide encoded by MBP exon 2, which is alternatively spliced in both mouse and human, we have specifically detected the guinea pig 21.5 kd MBP isoform (F de Ferra and J Kamholz, unpublished). Thus, the guinea pig MBP must have an exon 2 like sequence in its largest MBP, and it is likely that this protein is formed by alternative splicing. By analogy, the cow, rabbit and sheep 21.5 kd MBPs may also contain exon 2-like sequences, which can participate in alternative splicing. Myelin from several mammalian species has also been found to contain a 17.6 kd MBP variant (51, 52, 53), which is also present in chicken myelin. No detailed data are currently available for these small MBP variants, although it is likely that they are also formed by alternative splicing, as in the case of the human 17.2 kd MBP. Thus, alternative splicing is likely to be conserved throughout mammalian evolution, and may thus be important for myelination in these species.

In rodent brain, the relative concentrations of the four MBPs have been shown to vary somewhat during myelination. Early in the process, there is relatively more of the 21.5 and 17.0 kd isoforms, both of which contain exon 2 sequences, that can be found later, in more mature myelin (46). We have recently investigated myelination in human brain, and have found that there is relatively more of the mRNA encoding the 21.5 kd MBP isoform in the developing brain than can be detected in adult brain material (J Kamholz, unpublished). Since each oligodendrocyte makes all of the MBP isoforms, (F de Ferra, unpublished), regulation of alternative splicing must occur during the myelination process in order to modulate the amounts of these isoforms. These data also suggest that sequences contained within MBP exon 2 may be necessary for normal myelin development to occur.

Although there is no direct evidence to implicate minor MBP isoforms and alternative splicing as important for the myelination process, the above indirect data strongly suggest that this is the case. If this is so, what is the role of the minor MBP isoforms? Figure 3 is a cartoon version of the myelin proteins situated within the myelin membrane. MBP is located on the inner cytoplasmic surfaces, while PLP is a transmembrane protein (54). Direct interaction of these proteins within the myelin membrane, although suggested in the figure, has not been demonstrated directly. Minor MBP isoforms containing exon 2 sequences could interact with other cellular constituents- either PLP or other membrane proteins and/or lipids in the myelin membrane- and may be necessary to direct MBP to the correct position in the cell for myelin assembly to occur. Minor isoforms may also be necessary for the myelin

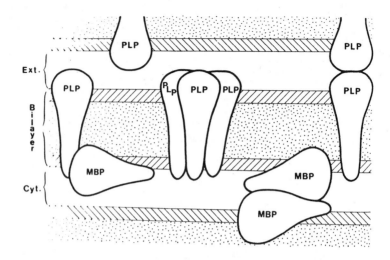

Figure 3   Diagrammatic representation of the molecular organization
of the CNS myelin bilayer. Cyt- cytoplasmic apposition; Ext-
extracellular apposition; MBP- myelin basic protein; PLP- proteolipid
protein.

```
                                                       -400
                5'...............CCAG AGCCTTCTGA AACACAGAGC
            -380                      -360
TGCAATAAGG CTGCTCCATC CAGGTTAGCT CCATCCTAGG CCAAGGGCTT
    -340                      -320                      -300
TATGAGGACT GCACATATTC TGTGGGTTTT ATAGGAGACA GCTAGGTCAA
                -280                      -260
GACCCCTCAG AGAAAGCTGC TTTGTCCGGT GCTCAGCTTT GCACAGGCCC
    -240                      -220                      -200
TGATTCATAT CTCATTGTTG TTTGCAGGAG AGGCAGATGC GAACCAGAAC
                -180                      -160
AATGGGACCT CCTCTCAGGA CACAGCGGTG ACTGGACTCC AAGCGCACAG
    -140                      -120                      -100
CGGACCCGAA GAATGCCTGG CAGGATGCCC ACCCAGCTGA CCCAGGGAGC
                -80                       -60
CGCCCCCCAC TTGATCCGCC TCTTTTCCCG AGATGCCCCG GGGAGGGAGG
    -40                       -20                       +1
ACAACACCTT CAAAGACAGG CCCTCTGAGT CCGACGCAGC TCCAGACCAT
                +20                       +40
CCAAGAAGAC AGTGCAGCCA CCTCCGAGAG CCTGGATGTG ATGGCGTCA
    +60                       +80
CAGAAGAGAC CCTCCCAGAG GCACGGA..........3'
```

Figure 4   Human MBP promoter sequence data. Numbering of the MBP
promoter sequence begins with the mRNA cap site as determined by primer
extension analysis. The ATG at which translation is initiated is
underlined, and is 47 base pairs downstream from the cap site. Also
underlined are the TATA and CAT like sequences at -30 and -80, and the
possible myelin specific promoter element at -50.

assembly process itself. If this were the case, more of these proteins would be necessary during the early stages of myelination than later, in more mature myelin, and exon 2 like sequences should be conserved, in minor MBP isoforms, throughout evolution. As seen above, both of these consequences have been found to be correct. A direct test of this hypothesis, however, will be to construct transgenic shiverer mice which express only a single MBP isoform. Shiverer mice are missing most of the MBP gene, are neurologically impaired, and make quite abnormal myelin (43, 44). If multiple MBP isoforms are necessary for myelination, these transgenic animals will remain abnormal, since myelin assembly should not occur. This type of experimental system can also be used to investigate the function of the individual myelin proteins <u>in vivo</u>. In any case, alternative splicing of a single MBP transcript, producing a family of related proteins, may prove more than a molecular curiosity, and may be crucial for the process of myelination itself.

<u>Regulation of MBP Gene Expression is Likely to Involve
Interaction of Specific DNA Sequences Upstream from the Gene
(Promoter/Enhancer Elements) with a Number of Regulatory Proteins</u>

MBP is expressed in only two cell types, oligodendrocytes and Schwann cells, and is under strict temporal control as part of a program of myelin specific gene expression. The factor(s) which control this expression, however, are currently unknown. Analysis of the expression of a large number of viral and cellular genes has shown that, in most instances, DNA sequences upstream from the gene, promoters and enhancers, are directly involved in regulating gene expression (55). These sequences interact with specific regulatory proteins, which either stimulate or inhibit transcription of the gene. Thus, in order to understand the regulation of the MBP gene, it will be necessary to identify its promoter and enhancer elements, as well as those proteins with which these sequence elements interact.

450 bp of DNA sequence upstream of the first exon of the human MBP gene is shown in Figure 4. The mRNA cap site, which delineates the transcriptional start site of the gene, is marked in bold, and is 47 bp from the start of translation (data not shown), and each of the alternatively spliced MBP isoforms begins transcription at this same A residue. Thus, there is a unique 5' end for the MBP mRNA precursor prior to splicing. No currently known regulatory sequence element can be found within this sequence. Comparison of this sequence with the published mouse promoter (33), however, reveals a significant sequence homology over a region of several hundred base pairs upstream from the cap sites. No TATA or CAT boxes, sequence elements commonly found in the first 80 bp upstream from the cap sites of many eukariotic promoters (55), are present within either promoter sequence. Both, however, contain a TATA like sequence 30 bp upstream of the cap site and a CAT like sequence 80 bp from this site. These are underlined in the figure. A 12 bp sequence, GGGGAGGGAGGA, which lies between the TATA and CAT like sequences, is also found in the JC virus 98 bp tandem repeat region (56). JC virus, which causes the human dysmyelinating disease progressive multifocal leukoencephalopathy (PML)_, is a small DNA virus which replicates mainly in oligodendrocytes. This 12 bp sequence is thus a good candidate for an oligodendrocyte specific regulatory element. The physiological significance of this sequence, as well as the rest of the putative promoter sequence, is currently under investigation.

SUMMARY AND CONCLUSIONS

As shown in the above experiments, we have used the techniques of molecular biology to analyze the structure of both the human and mouse MBP genes, and have demonstrated that a family of related MBPs is produced in each species by alternative splicing of a single MBP transcript. We have also demonstrated that this splicing pattern is regulated, producing more of exon 2 containing minor isoforms during the early stages of myelination. These data suggest that minor MBP isoforms, and thus alternative splicing, may be necessary for normal myelination to occur. And finally, we have determined a portion of the human MBP promoter and have identified a 12 bp sequence which is likely to act as a myelin specific transcription element.

Future analysis of the regulation of myelin gene expression will likely define both the regulatory sequences, promoters and enhancers, for the major myelin structural genes, as well as the proteins with which they interact. These data will describe the intercellular molecular machinery necessary for the cell to carry out the program of myelin specific gene expression. they will not, however, give a full picture of the overall process of myelination. In order to do this, the details of the intracellular events which occur during oligodendrocyte development must be more fully defined, and the axonal signals, which are likely to accompany specific oligodendrocyte-axon interactions, identified. It will also be important to understand how these extracellular events are transduced into the intercellular signals which ultimately activate the program of myelin specific gene expression. the molecular details of myelin assembly must also be worked out, as well as the function of the various myelin structural proteins in maintaining the myelin membrane structure. These data are not only necessary for understanding myelination itself, but are critical for analysis of demyelination as well as remyelination. future advances in all these areas will thus depend on a thorough integration of both molecular and cellular approaches to myelination, creating a truly molecular neurobiology.

REFERENCES

1.   Raine CS: Morphology of myelin and myelination. In: Myelin, second edition, (p. Morrell, ed), Plenum, pp 1-45, 1984
2.   Wood P, Bunge RP: The biology of the oligodendrocyte. In: Oligodendroglia, Advances in Neurochemistry, Vol. 5, pp 1-46, 1984
3.   Asbury AK: The biology of Schwann cells. In: Peripheral Neuropathy, Vol I (P.J. Dyck, P.K. Thomas, and E.H. Lambert, eds), W.B. Saunders, pp 201-212, 1975
4.   Geren BB: The formation from the Schwann cell surface of myelin in the peripheral nerves of chick myelin. Exp Cell Res 7:558, 1954
5.   Bunge MB, Bunge RP, Pappas GD: Electronmicroscopic demonstration of connections between glia and myelin sheaths in the developing mammalian central nervous system. J Cell Biol 12: 448, 1962
6.   Lees MB, Brostoff SW: Proteins of myelin. In: Myelin, second editions (P. Morrell, ed), Plenum, pp 197-224, 1984
7.   Gilles FH, Shankle W, Dooling EC: Myelinated tracts: growth patterns. In: The Developing Human Brain: Growth and Epidemiologic Pathology (FH Gilles, A Leviton and EC Dooling, eds.), John Wright, pp 117-183, 1983
8.   Yakovlev PI, Lecours A: The myelogenetic cycles of regional maturation of the brain. In: Regional Development of the Brain in Early Life (A. Minkowski, ed), Oxford, pp 3-70, 1967
9.   Norton WT: Formation, structure and biochemistry of myelin. In: Basic Neurochemistry (GJ Siegel, RW Albers, BW Agranoff, RC Katzman, eds), Little Brown, Boston, pp 63-92, 1981
10. Rorke LB, Riggs HE: Myelination of the brain in the newborn. JB Lipincott, 1969

11. Roback HN, Scherer HJ: Uber die feinere Morphologie des Frühkindlichen Gehirns unter besonderer Berücksichtigung der Gliaentwicklung. Virchows Arch.f.path.Anat 294, 365, 1935

12. Davison AN, Peters A: Myelination, Charles C. Thomas, 1970

13. Richardson EP: Myelination in the human central nervous system. Histology and Histopathology of the Nervous System (W. Haymaker and R.D. Adams, eds), Charles C. Thomas pp 146-173, 1981

14. Mickel HS, Gilles FH: Changes in glial cells during human telencephalic myelinogenesis. Brain 93:337, 1970

15. Brante G: Studies on lipids in the nervous system. Acta Physiol Scand 18:1, 1949

16. Chi J, Gilles F, Kerr C, Hare C: Sudanophilic material in the developing nervous system. J Neuropathol Exp Neurol 35:119, 1976

17. Noble M, Murray K: Purified astrocytes promote the in vitro division of a bipotential glial progenitor cell. Embo J 3:2243, 1984

18. Raff MC, Abney ER, Fok-Seang J: Reconstitution of a developmental clock in vitro: a critical role for astrocytes in the timing of oligodendrocyte differentiation. Cell 42:61, 1985

19. Temple S, Raff MC: Clonal analysis of oligodendrocyte development in culture: evidence for a developmental clock that counts cell divisions. Cell 44:773, 1986

20. Hortega P del Rio: Tercera aportacion al conocimiento morphologica e interpretacion functional de la oligodenroglia. Mem Real Soc Exp His Nat 14:5, 1928

21. Matthews MA, Duncan D: A quantitative study of morphological changes accompanying the initiation and progress of myelin production in the dorsal funiculus of the rat spinal cord. J Comp Neurol 142:1, 1971

22. Bishop GH, Clare MH, Landau WM: The relation of axon sheath thickness to fiber size in the central nervous system of vertebrates. Int J Neurosci 2:69, 1971

23. Waxman SG, Sims TJ: Specificity in central myelination: evidence for regulation of myelin thickness. Brain Res 292:179, 1984

24. Friedrich VL, Mugnaini E: Myelin sheath thickness in the central nervous system is regulated near the axon. Brain Res 274:329, 1983

25. Norton WT, Cammer W: Isolation and characterization of myelin. In: Myelin, second edition (P. Morrel, ed), Plenuem pp 147-180, 1984

26. Dubois-Dalcq M, Behar T, Hudson L, Lazzarini RA: Emergence of three myelin proteins in oligodendrocytes cultured in the absence of neurons. J Cell Biol 102:384, 1986

27. Rancht B, Clapshaw PA, Price J, Noble M, Seifert W: Development of oligodendrocytes and Schwann cells studied with a monoclonal antibody against galactocerebroside. Proc Natl Acad Sci 79:2709, 1982

28. Zeller NK, Hunkeler MJH, Campagnoni AT, Sprague J, Lazzarini RA: Characterization of mouse myelin basic protein messenger RNAs with a myelin basic protein cDNA clone. Proc Natl Acad Sci 81:18, 1984

29. Zeller N, Behar T, Dubois-Dalcq M, Lazzarini RA: The timely expression of myelin basic protein gene in cultured rat brain oligodendrocytes is independent of continuous neuronal influences. J Neurosci 5:247, 1985

30. Milner RJ, Lai C, Nave KA, Lenoir D, Ogata J, Sutcliffe JG: Nucleotide sequence of two mRNAs for rat brain myelin proteolipid protein. Cell 42:931, 1985

31. Salzer JL, Holmes WP, Colman DR: The amino acid sequence of the myelin associated glycoproteins: homology to the immunoglobulin gene superfamily. J Cell Biol 104:957, 1987

32. de Ferra F, Engh H, Hudson L, Kamholz J, Pucketrt C, Molineaux, Lazzarini RA: Alternative splicing accounts for the four forms of myelin basic protein. Cell 43:721, 1985

33. Takahashi N, Roach A, Teplow DB, Prusiner S, Hood L: Cloning and characterization of the myelin basic protein gene from mouse: one

gene can encode both 14 kd myelin basic proteins by alternative use of exons. Cell 42:149, 1985

34. Diehl HJ, Schaich M, Budzinski RM, Stoffel W: Individual exons encode the integral membrane domains of human myelin proteolipid protein. Proc Natl Acad Sci 83:9807, 1986

35. Macklin WB, Campagnoni CW, Denninger PL, Gardiner MV: Structure and expression of the mouse myelin proteolipid gene. J Neurosci Res 18:383, 1987

36. Sutcliffe JG: The genes for myelin. Trends in Genetics 3:73, 1987

37. Kamholz J, Spielman R, Gogolin K, Modi W, O'Brien S, Lazzarini R: The human myelin basic protein gene: chromosomal localization and RFLP analysis. Am J Human Genet 40:365, 1987

38. Willard HF, Riordan JR: Assignment of the gene for myelin proteolipid protein to the X-chromosome: implications for X linked myelin disorders. Science 230: 940, 1985

39. Barton DE, Arquint M, Roder J, Dunn R, Francke U: The myelin associated glycoprotein gene: mapping to human chromosome 19 and mouse chromosome 7 and expression in quivering mice. Genomics 1:107, 1987

40. Nave KA, Lai C, Bloom F, Milner R: Jimpy mutant mouse: a 74 base deletion in the mRNA for myelin proteolipid protein and evidence for a primary defect in RNA splicing. Proc Natl Acad Sci 83:9264, 1986

41. Hudson LD, Berndt JA, Puckett C, Kozak CA, Lazzarini RA: Aberrant splicing of proteolipid protein mRNA in the dysmyelinating jimpy mutant mouse. Proc Natl Acad Sci 84:1454, 1987

42. Macklin WB, Gardiner MV, King KD, Kampf K: An AG-GG transition at a splice site in the myelin proteolipid protein gene in jimpy mice results in removal of an exon. FEBS Letters 223:417, 1987

43. Molineaux S, Engh H, de Ferra F, Hudson L, Lazzarini RA: Recombination within the myelin basic protein gene created the dysmyelinating shiverer mouse mutation. Proc Natl Acad Sci 83:7542, 1986

44. Roach A, Pravphava D, Ruddle F, Hood L: Chromosomal mapping of mouse myelin protein gene and structure and transcription of the partially deleted gene in shiverer mutant mice. Cell 42:149, 1985

45. Barbarese E, Braun PE, Carson JH: Identification of prelarge and presmall basic proteins in mouse myelin and their structural relationship to large and small basic proteins. Proc Natl Acad Sci 74:3360, 1977

46. Carson JH, Nielson ML, Barbarese E: Developmental regulation of myelin basic protein expression in mouse brain. Dev Biol 96:485, 1983

47. Kamholz J, de Ferra F, Puckett C, Lazzarini RA: Identification of three forms of human myelin basic protein by cDNA cloning. Proc Natl Acad Sci 83:4962, 1986

48. Breibart RE, Andreadis A, Nadal-Ginard B: Alternative splicing: a ubiquitous mechanism for generation of multiple protein isoforms from single genes. Ann Rev Biochem 56:476, 1987

49. Andreadis A, Gallego M, Nadal-Ginard B: Generation of protein isoform diversity by alternative splicing: mechanistic and biological implications. Ann Rev Cell Biol 4:207, 1987

50. Wieczorek DF, Smith CWJ, Nadal-Ginard B: The rat ā-tropomyosin gene generates a minimum of six different mRNAs coding for striated, smooth, and monmuscle isoforms by alternative splicing. Mol Cell Biol 8:679, 1988

51. de Rosbo NK, Carnegie PR, Bernard CCA, Linthicum DS: Detection of various forms of brain myelin basic protein in vertebrates by electroimmunoblotting. Neurochem Res 9:1359, 1984

52. Waehneldt TV, Malotka J, Karin NJ, Matthieu JM: Phy7logenetic examination of vertebrate central nervous system myelin proteins by electro-immunoblotting. Neurosci Letters 57:97, 1985

53. Deibler GE, Krutzsch HC, Kies MW: A new form of myelin basic protein found in human brain. J Neurochem 47:1219, 1986

54. Braun PE: Molecular organization of myelin. In: Myelin, second edition (P. Morrel, ed), Plenum, pp 97-116, 1984

55. Serfling E, Jasin M, Schaffner W: Enhancers and eukaryotic gene expression. Trends Genet 1:224, 1985

56. Frisque RJ, Bream GL, Cannela MT: Human polyomavirus JC virus genome. J Virol 51:443, 1984

5

THE PRIMARY STRUCTURE OF MYELIN ASSOCIATED GLYCOPROTEIN

SUGGESTS A ROLE IN MYELINATION

Monique Arquint[1], Michael B. Tropak [1,2], Paul W. Johnson[1],
Robert J. Dunn[2], and John C. Roder[1,2,3]

[1]Division of Molecular Immunology and Neurobiology
Mount Sinai Hospital Research Institute
[2]Department of Medical Genetics, Univ. of Toronto
[3]Department of Immunology, University of Toronto
Toronto, Canada

INTRODUCTION

The molecular events that control the interaction of Schwann cells or oligodendrocytes with neurons and finally lead to myelination are at present not understood. A detailed knowledge of the process of myelination is needed in order to reach the long-term goal of improving the prognosis of human demyelinating disorders, possibly by promoting efficient remyelination. A body of evidence has accumulated indicating that major morphological changes during neural development are caused by the interaction of cells through adhesion molecules expressed on their surface. It appears likely that specific cell adhesion is also crucial to the process of myelination. In this review we summarize the structural characteristics of myelin associated glycoprotein (MAG) and discuss how it may be involved in cell interactions leading to myelination.

Structure of Myelin

Myelin is a multilamellar membrane structure surrounding many axons in the central (CNS) and the peripheral nervous system (PNS) (reviewed in 79). It is produced by oligodendrocytes in the CNS and Schwann cells in the PNS and typically consists of 30-50 closely apposed double membrane leaflets surrounding each axon (10,11).

Mature myelin sheaths in the central and the peripheral nervous system are morphologically related, despite major differences in their protein composition (79). The innermost myelin membrane is separated from the axon by a constant distance of 12-14 nm, the periaxonal space. The central myelin leaflet surrounding the periaxonal space, designated inner mesaxon, as well as the outermost membrane leaflet of myelin, the outer mesaxon, retains its cytoplasm. The cytoplasm in all the other myelin leaflets is extruded during the process of compaction, resulting in the tight apposition of the cytoplasmic faces of the two membranes of

one leaflet (the major dense line), and the close contact of the external
faces of two adjacent leaflets (the intraperiod line). Other areas in
mature myelin which retain their cytoplasm are the paranodal loops
flanking the nodes of Ranvier and the Schmidt-Lanterman incisures,
pockets of cytoplasm found within the compacted myelin of the peripheral
nervous system. The non-compacted areas of myelin form a continuous
channel of cytoplasm which retains contact with the cell body of the
myelinating cell (41).

The Process of Myelination

The development of techniques for separately culturing pure neurons
and Schwann cells has allowed a detailed in vitro study of the steps
leading to myelination. Our understanding of myelination in the PNS is
more advanced because Schwann cells in culture can be induced to
proliferate and myelinate more readily than oligodendrocytes. Recently,
however, successful in vitro myelination of dorsal root ganglion (DRG)
neurons by pure oligodendrocytes has been achieved and the newly
developed cell culture techniques have opened the door for the in-depth
study of myelination in the CNS (98,99).

The stepwise events that follow the initial contact of Schwann cells
with pure neurons can be distinguished experimentally (12). First the
Schwann cells are induced to proliferate by the direct contact with
neuronal processes. Schwann cell proliferation in the absence of neurons
can be achieved in vitro by the addition of glial growth factor and
forskolin, an activator of adenyl cyclase (73). After the initial
mitotic phase, the Schwann cells migrate and line up at regular intervals
along the length of axons; they bind to axons that will be myelinated at
a later stage and those that will merely be ensheathed. The binding of
Schwann cells, which may be mediated by cell adhesion molecules like N-
CAM, is therefore not a decisive factor determining the future
myelination state of axons. In the next step, termed ensheathment, the
Schwann cell completely surrounds the axon shaft once. Several smaller
axons (<1 um diameter) can be ensheathed by a single Schwann cell. These
small axons remain arrested at the ensheathment stage and are not
myelinated. Axons larger than 1 um are enveloped by a single Schwann
cell. In vitro, the myelination process is halted at this stage, unless
ascorbate and fetal calf serum are present in the medium (27). Early in
myelin synthesis, the growing number of lamellae surrounding the axon are
uncompacted; at a later stage, the myelin lamellae become flattened and
the cytoplasm is extruded. Parallel with the process of myelination, the
Schwann cell secretes extracellular matrix components (collagen type I,
III, IV and V, laminin, entactin and heparin sulfate proteoglycan) which
form the basal lamina surrounding the Schwann cell (13). Upon
maturation, the basal laminae of adjacent Schwann cells form a continuous
tube which encloses the myelinated axon along its whole length (79).
Compact myelin, morphologically indistinguishable from in vivo myelin,
can be observed in culture four weeks after the ensheathment stage.

Neural Cell Adhesion Molecules

The best characterized cell interaction molecule is the neural cell
adhesion molecule N-CAM which plays a key role during very early
development in the segregation of epithelia, the attachment of tissue
sheets, the conversion of epithelia to mesenchyme and the control of
cellular migrations (reviewed in 24, 25, 26). In the nervous system, N-
CAM mediates the fasciculation of axons, the guidance of axonal growth
cones along glial pathways and the formation of neuromuscular junctions.
The effects of N-CAM are mediated by the homophilic interaction of two N-
CAM molecules expressed on the surface of adjacent cells (15).

In the adult, N-CAM exists as three protein species of 180, 140 and 120 kD apparent molecular weight. These three forms of N-CAM share a common external domain and were recently shown to arise from a single gene which maps to chromosome 9 in the mouse (17) and to chromosome 11 in the human (66). The homophilic binding domain is situated near the N-terminus of the common external portion of N-CAM (15), and is part of a region containing five tandem repeat units. The repeats are approximately 90 amino acids in length and exhibit significant homology with the variable and constant domains of immunoglobulins (40). Each tandem repeat in N-CAM is coded for by two exons (70), in contrast to the homology units of all other members of the immunoglobulin supergene family, which are specified by a single exon.

The 180 kD species (ld chain) differs from the 140 kD form (sd chain) by a single insert of 261 amino acids in the cytoplasmic domain which is encoded by a single exon (70). The 120 kD N-CAM (ssd chain) lacks the entire cytoplasmic domain present in the 140 and 180 kD forms and contains at its C-terminal a unique stretch of 24 uncharged amino acid residues coded for by a separate exon (5). This region serves as the attachment site for phosphatidylinositol which helps to anchor the N-CAM_{120} protein in the membrane (39).

The 140 kD protein is the early form of N-CAM and is involved in embryonic pattern formation and promotes neuromuscular synapse formation in striated muscle (82). The 180 kD form of N-CAM is only found in the nervous system and is restricted to postmitotic neurons (72). Its temporal and spatial expression parallels that of another neural cell adhesion molecule, Ng-CAM. There is some evidence suggesting that N-CAM_{180} and Ng-CAM may interact on the surface of the same cell (24,92).

The two proteins together with myelin associated glycoprotein (MAG) may play a joint role at the onset of myelination (60,81).

Myelin-Associated Glycoprotein (MAG)

Several lines of evidence point to a possible cell adhesive function of MAG during myelination. MAG is present in myelin of both CNS and PNS and is an integral membrane glycoprotein like the known cell adhesion molecules studied so far (N-CAM, Ng-CAM, and J1) (26). The importance of its function is underlined by the fact that, together with MBP, it is evolutionary the most highly conserved protein in myelin (62).

Structure and Expression of MAG

MAG is a glycoprotein of 100 kD apparent molecular weight and contains a high proportion of acidic (20%) and hydrophobic (23%) residues (77). The carbohydrate portion represents 30% of the glycoprotein and was estimated to consist of up to nine asparagine-linked chains (30). The carbohydrate chains are sulfated (61) and contain a high percentage of sialic acid (18% by weight of sugar residues). The isoelectric point of the protein is unusually low (3.0 to 4.5) as a result of the sialation, sulfation, and the high ratio of acidic to basic amino acids (77).

The MAG protein consists of two forms (p67MAG and p72MAG) which vary in the size of their peptide portion, as first revealed by immunoprecipitation of _in vitro_ translation products of mouse and rat brain RNA (30). Peptide mapping of the immunoprecipitated _in vitro_ translation products indicated a great degree of similarity between the p67 and p72 polypeptides. Frail and Braun (30) first suggested that the synthesis of the two MAG species may be differentially regulated during development. The larger form, p72MAG, appears first at the onset of myelination in brain, at 10 days after birth; its level reaches a peak during myelination and declines as myelination is completed. In contrast, the small MAG, p67MAG, is first detectable during the course of myelination and steadily accumulates to become the major MAG species in the mature animal. The message for the p67 form of MAG predominates in the PNS, in particular the sciatic nerve, both during myelination (7 days after birth) and after completion of myelin synthesis (12 days after birth) (33).

It was recently determined independently by Salzer et al (84), as well as by our lab (M. Tropak, P. Johnson, M. Arquint, J. Roder, R. Dunn, submitted), that the two MAG polypeptide species arise from two differentially spliced messages originating from the same gene. The mRNAs differ only by the presence of a 45 bp insert near the 3' end of the coding sequence in the transcript specifying the smaller form of MAG, p67 (Fig. 1A). The presence of an in-frame termination codon (TGA) within the insert causes termination of translation and produces the alternative form of MAG which is 44 amino acids shorter than the polypeptide specified by the MAG transcript lacking the insert (Fig. 1B). Since the difference in size corresponds to 5 kD, the discovery of two alternately spliced mRNAs provides an elegant explanation for the origin of the two MAG polypeptide species of 67 and 72 kD molecular weight described previously (30).

The differential splicing of MAG transcripts is developmentally regulated, resulting in the prevalence of the message encoding p72MAG during early myelination and the predominance of the message specific for p67MAG in adult brain (30, 56, Tropak et al, submitted). This pattern of differential expression suggests strongly that the two forms of MAG play different roles. It is plausible that p72MAG is involved in some aspect of ensheathment, whereas p67MAG is associated with active myelination and the maintenance of the myelin sheath. The peptide segment exclusively found in the larger species, p72MAG, contains potential sites for phosphorylation by protein kinase C around residues 575-582 and 604-608 and by tyrosine kinases at Tyr-620 (Fig. 2). This suggests that the function of the larger MAG species may be modulated to a greater extent by the action of protein kinases. Such an observation has been made for $N\text{-CAM}_{180}$ where the additional cytoplasmic insert appears to contribute to a higher phosphorylation state when compared to $N\text{-CAM}_{140}$ (87).

Evidence for a Role of MAG in Myelination

Several lines of evidence point to a possible role of MAG in myelination or in the maintenance of the myelin sheath.

1. Periaxonal localization. Together with CNP (2',3'-cyclic nucleotide 3'-phosphohydrolase), MAG is the only myelin protein found localized to the innermost myelin membrane surrounding the axolemma. This periaxonal localization has been demonstrated by subfractionation and biochemical characterization of myelin and myelin-related membranes (reviewed in 78) and by immunocytochemistry at the light microscopical (88) and electron microscopical level (93,94). Two controversial reports have challenged the accepted notion of the absence of MAG from compact myelin (96,28). This finding is in contradiction to the previous immunocytochemical studies and has not been confirmed by other laboratories (94).

Recently, the absence of MAG from compact myelin and its precise localization to the innermost myelin membrane has been confirmed beyond doubt by Schachner and coworkers by electron microscopy, using polyclonal

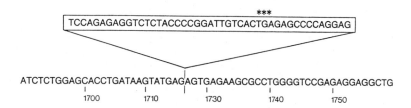

Figure 1A   Two alternately spliced MAG mRNA species differing by the presence of a 45 bp insert. Part of the sequence coding for the cytoplasmic domain of MAG is shown. The position of the 45 bp insert which is present in a differentially spliced MAG mRNA species, is shown. The translation termination codon within the insert (TGA) is indicated with an asterisk.

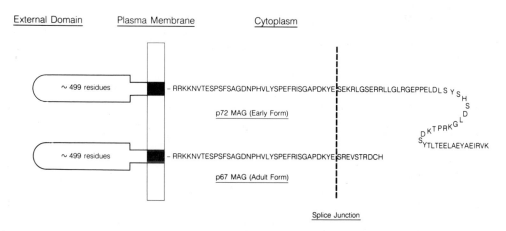

Figure 1B   MAG exists as two peptide species. The amino acid sequence of the cytoplasmic domains of the two MAG species p72MAG and p67MAG is shown. The splice junction is indicated at which the two polypeptide sequences diverge due to a 45 bp insert in the cDNA coding for p67MAG. The identical extracellular domains of p67MAG and p72MAG are not represented to scale.

anti-MAG antibodies coupled to gold particles (60). MAG was observed to be present in all non-compacted leaflets of immature myelin, but was missing upon compaction. In mature myelin, MAG was found in all non-compacted areas, including the periaxonal membrane, paranodal loops, the outer mesaxon, and Schmidt-Lanterman incisures, in confirmation of the previous findings of Trapp and Quarles (93).

The significance of the localization of MAG to the periaxonal region can be interpreted in several ways. It may reflect a role for MAG in establishing the initial contact between processes of the myelin-forming cell and the axon and/or in maintaining the established contact. Trapp and Quarles (93) have also proposed that MAG, with its highly negatively charged external domain projecting into the periaxonal space, helps maintain the constant distance of 12-14 nm between the axolemma and the myelin. The localization of MAG to other non-compacted regions furthermore suggests that it may prevent compaction of the cytoplasmic collar, the outer mesaxon, and the paranodal loops (93,94). It is conceivable that MAG interacts with cytoplasmic components of the myelinating cell. Alternatively, the bulky protein may simply be extruded from myelin during compaction and as a consequence appear exclusively in the non-compacted areas.

2. Role in the maintenance of axon/glial cell contact. Evidence for a possible role of MAG in the maintenance of the contact between the membrane of the myelin-forming cell and the axon has come from the study of the neurological mouse mutant, quaking (95). This mutant is deficient in myelin assembly and contains immature myelin, characterized by a restricted number of wraps which remain uncompacted. When adjacent cross-sections of neurons of the L_4 ventral root in the PNS were examined during myelination, it was found that areas characterized by a loss of the apposition of the Schwann cell membrane and the axon appeared to lack

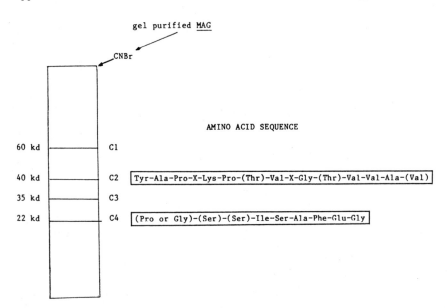

Figure 2   Gel purification and sequencing of CNBr fragments of MAG. MAG purified by electrophoresis and gel elution was cleaved with cyanogen bromide (CNBr) and the resulting peptide fragments (C_1, C_2, C_3, C_4) were separated by gel electrophoresis and sequenced by automated Edman degradation. The amino acid sequence obtained from two of the four peptide fragments (C_2, and C_4) is shown.

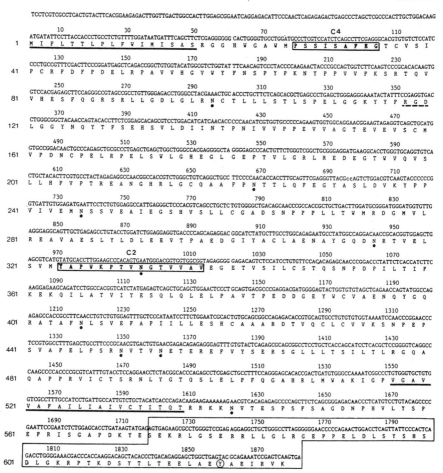

Figure 3   Nucleotide sequence of the coding region for MAG. The
sequence shown is a composite of clones MAG 1.2, MAG 6.2 and clone 1B236.
The sequence of MAG 1.2 spans from nucleotides -132 to 1062, and the
sequence of MAG 6.2 from nucleotides -68 to 1716. The boxed-in area
represents part of the published sequence of the random cDNA clone 1B236
(89), which overlaps with the isolated MAG clones between positions 921
and 1716. The numbering of nucleotides starts at the putative translation
start site. Amino acids are shown in the standard one-lettercode and
numbered in the left-hand margin. The amino acid sequences corresponding
to the sequenced CNBr peptides C2 and C4 are boxed-in. The putative
signal peptide and transmembrane regions are underlined. Potential
asparagine-linked glycosylation site is circled and the fibronectin
recognition sequence Arg-Gly-Asp is underlined with dashes.

MAG, as judged by immunocytochemistry. Other regions where the Schwann
cell membrane remained in close contact with the axolemma did express
MAG. The correlation between the lack of MAG and the absence of the
constant 12-14 nm periaxonal space was confirmed by examination of many
adjacent 1 um cross-sections. Furthermore, wherever MAG was missing from
the periaxonal membrane, the adjacent cytoplasmic collar was absent due
to the fusion of the innermost myelin leaflet, suggesting an inhibitory
effect of MAG on compaction.

Consistent with its putative cell adhesion function, MAG has recently been reported to mediate the contact between neurons and oligodendrocytes, as well as amongst oligodendrocytes (60).

3. <u>Time of expression</u>. If MAG is important for the initial contact of the myelinating cell with the axon, its expression on the cell surface should precede that of other myelin proteins. Lazzarini and coworkers have reported that, <u>in vitro</u>, immunostaining for MAG can be detected several days before MBP is observed first in the cytoplasm of cultured rat oligodendrocytes (23). <u>In vivo</u>, the levels of the periaxonally located myelin proteins, MAG and CNP, were found to increase more rapidly than MBP during myelination in the rat optic nerve and the medulla, as determined by a sensitive radioimmunoassay (51).

4. <u>Shared epitope between neural cell adhesion molecules and MAG</u>. MAG from several species, including human MAG, shares a highly immunogenic carbohydrate epitope with the known neural cell adhesion molecules N-CAM, L1 (or Ng-CAM), and J1 (63,55,85). The hypothesis that the presence of this epitope is a distinctive feature of a family of cell adhesion molecules was proposed by Schachner and co-workers and tested using anti-carbohydrate epitope antibodies to isolate novel proteins.

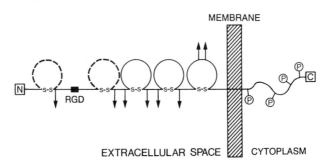

Figure 4   Potential three-dimensional structure of MAG. The transmembrane protein MAG is shown with its amino terminus (N) projecting into the extracellular space and the carboxyl end (C)located within the cytoplasm. The immunoglobulin-like homology units are represented as disulphide-linked domains. The homology units with the least homology (domains I and II) are shown as dotted circles. The positions of the fibronectin-like tripeptide sequence RGD, of the asparagine-linked carbohydrate chains (arrows), and of potential phosphorylation sites (P) are indicated.

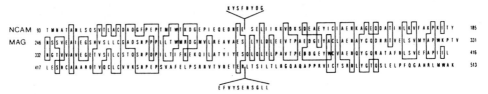

Figure 5   Homology between the three major repeated domains in MAG (domains III, IV and V) and the neural cell adhesion molecule (N-CAM). Line 1 shows the third of the five internal repeats of N-CAM (70) and the residue numbers shown are according to Hemperly et al (40). Lines 2, 3, and 4 contain the contiguous amino acid sequence of MAG between amino acids 246 and 513, aligned to show maximum internal homology (domains III, IV, and V). Amino acid positions are indicated in the left and right margins. Amino acid identities are boxed.

Their hypothesis was supported by the fact that one of the proteins isolated, J1, was subsequently found to mediate astrocyte/neuron interactions (55). The finding of a cell adhesive property for the J1 glycoprotein suggests the possibility that MAG may be involved in cell-cell interactions in the nervous system.

5. Role of MAG in disease. The human demyelinating diseases in the CNS include multiple sclerosis, progressive multifocal leukoencephalopathy, and subacute sclerosing panencephalitis. The peripheral neuropathies are a heterogenous group of disorders which may be characterized by primary or secondary demyelination in the PNS, resulting in motor and sensory dysfunction, (reviewed in 91). Despite the fact that experimental demyelination can be induced by immunization with whole myelin or certain myelin antigens such as myelin basic protein (MBP), proteolipid protein (PLP), cerebrosides, and the P_2 protein of the PNS (reviewed in 9), no autoantigen specific for demyelinating diseases such as multiple sclerosis has yet been found (50).

MAG is the only known myelin protein which was identified as an autoantigen in demyelinating peripheral neuropathies (8). Twenty percent of peripheral neuropathies are characterized by a plasma cell dyscrasia resulting in the production of high levels of a single monoclonal antibody directed to the previously described immunogenic carbohydrate epitope on MAG (58,8,31) in 60-70% of patients (37,68). Other autoantigens recognized by monoclonal antibodies associated with peripheral neuropathies have been reported and include unidentified gangliosides of CNS and PNS myelin (68). Since the carbohydrate epitope on MAG is also present on two small glycoproteins and two sphingolipids of the PNS (47), the possibility cannot be excluded that these antigens, rather than MAG, are originally responsible for the development of the disease.

The MAG-reactive IgM paraprotein isolated from patients is clearly capable of inducing peripheral demyelination upon intraneural injection into cats (38). A contrasting report by Bosch et al (7), who observed no significant demyelination upon injection of serum from three polyneuropathy patients into sciatic nerves of rats, can be explained by the known lack of cross-reaction of the human carbohydrate-specific monoclonal antibodies with rat MAG (59). However, it is not certain if the disease is actually caused by the MAG-reactive IgM or if the presence of the monoclonal antibody is a secondary effect of previous injury to the myelin sheath. In some patients, the neuropathy appears before the development of gammopathy (21), suggesting that the accumulation of monoclonal IgM may be a secondary event. Biopsy material from several patients with peripheral neuropathy was shown to contain IgM deposits on the outer mesaxon of the myelin sheaths and antibody damage was indicated by the widening of outside myelin lamellae (64). However, since no strict correlation was found between the level of IgM deposits and the severity of the disease, the IgM antibody may not be directly involved in the disease process (65). It has been suggested by R. Quarles (78) that the MAG-reactive IgM antibodies, although not the actual cause of the neuropathy, may influence the clinical course of the disease by preventing remyelination.

Several independent studies of multiple sclerosis (MS) patients (48,34,52) have shown that MAG is significantly decreased relative to other myelin proteins in the region surrounding demyelinated plaques. Thus, in brains of six MS patients the quantitative reduction of MAG, detected by radioimmunoassay, was greater than that of MBP or CNP in the normal appearing white matter, as well as the periplaque and plaque regions (52). However, since similar observations were made in patients with peripheral multifocal leukoencephalopathy (49), the early loss of MAG may merely be an indicator of the death of oligodendrocytes induced by the disease. A humoral immune reaction to MAG seems unlikely in MS since most studies have failed to detect anti-MAG antibodies in the cerebrospinal fluid (CSF) of MS patients (29,67). On the other hand, Johnson et al (53) have reported recently that 30% of MS patients studied (9 out of 30) exhibited a cell-mediated immune response to MAG, compared to 9.8% (8 out of 81) responding to MBP in vitro. Although the level of the cell-mediated response to MAG exhibited by lymphocytes from individual patients was lower than to MBP, the number of patients reacting to MAG was significantly higher than controls. The significance, if any, of this finding with respect to the disease process in MS is not yet clear. The sum of the evidence accumulated up to date does not support a direct role for MAG in the development of MS.

Cloning and Sequencing of cDNAs Coding for MAG

The unique properties of MAG suggested that it might play an important role during the process of myelination. No data were available on the sequence or structure of MAG, due to the low abundance of the protein in myelin (<1% in CNS and <0.1% in PNS) and because of blockage of its amino terminus (77). To provide a basis for future studies of the function of MAG, we set out to obtain reliable sequence information by cloning the cDNA coding for MAG. We used two MAG peptide-specific monoclonal antibodies, GenS1 and GenS3, to screen a library of adult rat brain cDNA cloned in the expression vector λgt-11 (102). This vector permits the expression of any cloned insert in the form of a B-galactosidase fusion protein and the detection of the correct hybrid protein with the help of specific antibodies. Seventeen different clones were isolated from a screen of 10^6 recombinants with the pooled monoclonal antibodies.

Several criteria were applied to distinguish between the genuine MAG clones and those detected by spurious antibody reactions. The genuine monoclonal antibody-reactive MAG clones should contain part of the same DNA sequence, and are therefore expected to cross-hybridize in Southern blots. Two main cross-hybridizing groups and several clones that failed to cross-hybridize with any other recombinant were observed (Table 1). The seven cross-hybridizing clones showed the pattern of expression predicted for MAG-specific clones, since they were expressed in brain but not liver and were present at high abundance during the time of peak myelination in brain. Two of the seven brain-specific clones (1.2 and 6.2) were further shown to be myelin-specific by demonstrating that the level of hybridizing message was greatly diminished in the brain of the hypomyelinating mouse mutant jimpy when compared to its normal littermates. Although the primary genetic defect in this mutant has been mapped to the gene coding for proteolipid protein (19), the expression of the MAG gene is known to be greatly reduced in jimpy brain, as demonstrated by in vitro translation and immunoprecipitation of translation products with MAG-specific antibodies (32). Furthermore, clones 1.2 and 6.2 were the only recombinants out of the 17 clones originally isolated with the monoclonal antibodies that gave a positive immune reaction in a second-stage screen with two different MAG-specific polyclonal antibodies.

The critical piece of evidence for the identification of the cDNAs was obtained by the determination of the amino-terminal amino acid sequences of two MAG peptides, C2 and C4, derived from gel-purified MAG by cyanogen bromide cleavage. Confirmation that the peptides originated from MAG was provided by demonstrating that they reacted with anti-MAG polyclonal antisera and that one of these peptides (C2) also reacted with

70

Table 1   CHARACTERIZATION OF ANTIBODY-REACTIVE GT-11 CLONES

Clones[a]	2	5	6	4.3	5.1	5.2	1.3	3.1	3.2	3.3	6.1	1.2	6.2
Antibody reaction													
moAb GenS1	+	+	+	+	+	+	+	+	+	+	+	+	+
moAb GenS3	+	+	+	-	+	-	+	+	+	+	+	+	+
antiserum R3B10	-	-	-	-	-	-	-	-	-	-	-	+	+
antiserum RL	-	-	-	-	-	-	-	-	-	-	-	+	+
Cross-hybridization													
group 1	-	-	-	-	-	-	+	+	+	+	+	-	-
group 2	-	-	-	-	-	-	-	-	-	-	-	+	+
mRNA expression													
in brain	-	-	-	-	+	-	+	+	+	+	+	+	+
in liver	-	-	-	-	+	-	-	-	-	-	-	-	-
myelin-specific[b]							-	-	-	-	-	+	+
mRNA size (kb)					0.5		3.5	3.5	3.5	3.5	3.5	2.5 3.0	2.5 3.0

[a]The inserts of thirteen out of seventeen cDNA clones reactive with monoclonal antibodies GenS1 and GenS3 were purified for Southern and Northern analysis.

[b]As determined from the abundance of the specific transcript in the brain of the hypomyelinating mutant jimpy.

an anti-MAG monoclonal antibody. Phenylthiohydantoin derivatives of amino acids were identified at 13 of the first 15 cycles of Edman degradation of peptide C2 and 9 of 9 cycles for peptide C4 (Fig. 3). These sequences exactly matched the amino acid sequence deduced from the DNA sequence of clones 1.2 and 6.2 at all assigned positions (Fig. 2). Taken together, these data unambiguously demonstrate the correct identity of the cDNA clones MAG 1.2 and MAG 6.2.

Features of the Deduced MAG Amino Acid Sequence

A search of the DNA sequence databank Genbank revealed that the MAG clones 1.2 and 6.2 overlapped with part of an unidentified brain-specific cDNA clone, designated 1B236 (89). Since the amino acid sequence of MAG was previously not known, the identity of clone 1B236 remained undetermined until we reported a sequence overlap between clone 1B236 and the proven MAG cDNA clones 1.2 and 6.2 (2).

The DNA sequence of MAG 6.2, which represents 91% of the total coding sequence of MAG, combined with part of the sequence of the random clone 1B236, resulted in the complete open reading frame for MAG (Fig. 2). The single open reading frame codes for 626 amino acids with a calculated molecular weight of 69 274, which matches size estimates obtained from the electrophoretic mobility of in vitro translated, non-glycosylated MAG polypeptides in polyacrylamide gels (30). The MAG coding sequence shown in Fig. 2 was precisely confirmed by Salzer et al (84) and Lai et al (56) several months after our original publication.

A hydrophobicity profile of the translated open reading frame reveals two areas of strong hydrophobicity, a stretch of 17 residues at the extreme amino terminus and a hydrophobic region between residues 517 and 536 (Fig. 2). The amino terminal hydrophobic region probably represents the leader sequence, which is required during translation of the mRNA coding for many integral membrane proteins and secreted proteins (83) to ensure the correct insertion of the nascent peptide into the endoplasmic reticulum. The second hydrophobic stretch encoded by the cDNA most likely represents the transmembrane domain and is characterized by the presence of mostly small uncharged residues capable of forming an alpha helix. A group of basic amino acids (Arg-Arg-Lys-Lys) lies at the junction of the transmembrane segment and the putative cytoplasmic domain between positions 537 and 540, as is commonly observed in many integral membrane proteins. These residues are thought to interact with the charged phospholipid head groups and represent a "halt transfer" signal, operative during the translational insertion of nascent polypeptides into the membrane of the endoplasmic reticulum (83). The membrane spanning region contains a single cysteine residue which may serve as a site for fatty acid attachment as described for the heavy (a) chains of some human MHC antigens (54). At present, there are, however, no experimental data available to support this possibility.

The MAG coding region is divided into a putative 499 amino acids long extracellular domain and a 90 amino acids long cytoplasmic domain (Fig. 4). Consensus sequences for eight N-linked glycosylation sites (Asn-X-Ser/Thr) are present in the extracytoplasmic domain, in confirmation of the approximate number of oligosaccharide chains estimated to be present in the MAG protein (30).

A fibronectin-like tripeptide sequence Arg-Gly-Glu (RGD) is found approximately 100 amino acids from the amino terminal end of MAG, at residues 118-120 (Fig. 2). This tripeptide has been shown to play a crucial role in the interaction of the extracellular matrix proteins fibronectin (71,100), vitronectin (76), and osteopontin (69) with their respective cell surface receptors. The amino terminal disposition of the RGD sequence in MAG would allow for a potential interaction of the extracellular domain of the glycoprotein with a cell surface receptor in the closely apposed axonal membrane. However, the presence of the RGD sequence in a protein may not always be indicative of a function. Therefore, the possibility that the fibronectin-like sequence of MAG may play a role in the interaction between the glial cell and the axon will require experimental confirmation.

MAG is a Member of the Immunoglobulin Supergene Family

The most striking feature of the extracellular domain of MAG is the presence of up to five tandemly repeated units of sequence homology (Figure 4). Domains III and IV, between residues 246 and 331 and 332 and 416, respectively, share the highest internal homology, consisting of 45% identical amino acids or 74% combined identical and conserved residues (Fig. 5). In contrast, the internal homology in domains I, II, and V is mostly restricted to the amino terminal half of each unit.

The homology units are approximately 90 amino acids in length, contain two invariant cysteines (Fig. 5), and share significant homology (20-25%) with the variable and constant domains of immunoglobulins (Table 2). Immunoglobulin domains were shown by x-ray crystallography to consist of two B-pleated sheets, held together by the disulphide-linked conserved cysteines to form a structure called the antibody fold (1). The two B-pleated sheets are formed by seven antiparallel B strands in the constant domain and by nine B strands in the variable domain (42,74).

Analysis of the secondary structure characteristics of the MAG homology units using Pepplot (35), a computer program based on the method of Chou and Fasman (14), indicates that the amino acids surrounding the two cysteines have the potential to form B-pleated sheets. The presence of intrachain disulphide bonds in MAG has been suggested by Gulcher et al (36), based on the preliminary observation that the mobility of unreduced MAG in polyacrylamide gels is slightly greater than that of reduced MAG. It is, therefore, likely that the MAG repeat units form immunoglobulin-like domains that are held together by disulphide bonds (Fig. 4).

The predicted number of B pleated strands indicates that the MAG homology units are more closely related to constant immunoglobulin domains, as judged by secondary structure characteristics. On the other hand, the pattern of conserved amino acids, exhibited especially by domains III and IV, shows characteristics of the sequence of variable domains. The most striking variable-like sequence conservation is represented by the amino acid stretch E-D-(X)-G-X-Y-X-(C), found on the amino terminal side of the second cysteine of domains III and IV, at positions 299-305 and 385-392 (Fig. 5).

Most potential sites for N-glycosylation are situated on the carboxy terminal side of the second cysteine of each MAG homology unit, placing the oligosaccharide chains between adjacent domains (Fig. 4). Exceptions are two N-glycosylation sites which lie in the center of the fifth homology domain. The sequence stretch separating units I and II is larger than the average distance between the other homology domains and contains the fibronectin-like sequence RGD in its center.

The internal homology domains in MAG are related to five tandem homology units found in the extracellular domain of the neural cell adhesion molecule (N-CAM (40,70) (Fig. 6). There is a maximum degree of homology of 30% between the third domain of N-CAM and the third domain of MAG (Fig. 5). The relatedness of MAG and a known cell adhesion molecule in the nervous system provides further evidence for a cell-adhesive role of MAG during myelination. As in MAG, the homology units of N-CAM display sequence characteristics similar to variable domains and secondary structure characteristics typical of constant domains of immunoglobulins. Furthermore, similar to MAG, the first of the five tandem homology units lies close to the amino terminus of N-CAM (70,5). Most importantly, the first two, and possibly the third, amino terminal domains of N-CAM have been shown to mediate the homophilic binding between two cell adhesion molecules (16). By analogy with N-CAM, the possibility is strong that the homology units of MAG may mediate the recognition and specific binding of a putative receptor on the axonal membrane, resulting in myelin/axon adhesion.

Many proteins with a role in recognition in the immune system have been found to contain at least one immunoglobulin-like homology unit and have, therefore, been classified as members of the immunoglobulin supergene family (45). After N-CAM, MAG is the second protein of the nervous system shown to belong to this family of recognition molecules (Table 2). This supergene family includes eight multigene families and several single-gene representatives. The multigene group consists of the families of light (k and λ) and heavy chain immunoglobulin genes, of the a, B, and Y chain genes of the T cell receptor and of the class I and class II genes of the major histocompatibility complex (MHC). Single gene members include accessory molecules to the T cell receptor involved in class I (CD8) and class II (CD4) MHC restriction, as well as in ion channel formation (T3g, T3Ë), the poly-Ig receptor, and B_2 microglobulin (associated with class I MHC molecules) and the surface markers Thy-1 and Ox-2, present on both lymphocytes and neurons (42,45).

A common feature of the genes specifying all members of the
immunoglobulin supergene family, except N-CAM, is the fact that the
coding information for each homology unit is contained within a single
exon (42,70). Sutcliffe and collaborators have just published the exon-
intron structure of the 1B236 gene (56), which we had shown previously to
be identical to MAG (2). The entire extracellular domain including the
transmembrane region of the glycoprotein is encoded by a total of five
exons (exons 5-9). The exon boundaries of the MAG gene lie within ten
residues or less of the approximate borders of the homology domains at
positions 28, 147, 246, 332, and 417. It is, therefore, clear that, in
contrast to the N-CAM gene where each unit is represented by two exons
(70), the homology domains of MAG are encoded in their entirety by single
exons. This gene structure lends support to the view that the presence
of the immunoglobulin homology domain in many diverse proteins can be
attributed to the repeated duplication during evolution of a common
ancestral gene unit which may have encoded a primordial cell surface
receptor (42).

Phosphorylation of MAG

The cytoplasmic domain of MAG contains several potential
phosphorylation sites. The sequence at the carboxyl terminus of MAG
surrounding tyrosine-620 is strongly homologous with the major site of
autophosphorylation (tyrosine-1137) at the carboxyl terminus of the
epidermal growth factor (EGF) receptor (22) (Fig. 7). In this region, 7

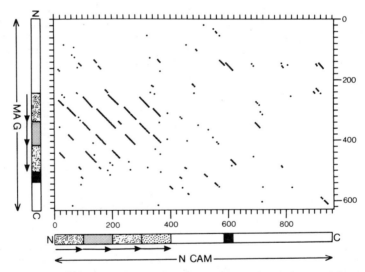

Figure 6   Homology between MAG and N-CAM. The amino acid sequence of
MAG (vertical axis) was compared to that of N-CAM (horizontal axis) (40)
using the Sequence Analysis Package designed by the University of
Wisconsin Genetics Computer Group (UWGCG). Homology between the MAG and
the N-CAM amino acid sequence (across a window of 30 amino acids at a
stringency of 32.0 according to the Staden protein comparison matrix) is
indicated as dots. The three major domains of MAG (domains III, IV, and
V) that are homologous to the N-CAM tandem repeat units (of which four
are shown) are indicated as patterned boxes. The membrane-spanning
domains are represented as black boxes and the amino (N) and carboxyl (C)
termini of the proteins are shown. The map position of amino acid
residues is indicated.

of 12 residues of the MAG sequence (between residues of 614-625) are identical to the EGF receptor sequence, if a two-residue gap is included. There is evidence that autophosphorylation of tyrosine-1137 regulates the function of the EGF receptor by stimulating its intrinsic kinase activity. It is unclear at present what relevance, if any, this sequence homology has to the function of MAG, since its cytoplasmic portion lacks a tyrosine kinase-related domain (46). In addition, there are several potential serine and threonine phophorylation sites in the cytoplasmic

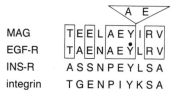

MAG	T E E L A E Y I R V
EGF-R	T A E N A E Y L R V
INS-R	A S S N P E Y L S A
integrin	T G E N P I Y K S A

Figure 7   Homology between the cytoplasmic domains of MAG and of the receptors for EGF (EGF-R) insulin (INS-R) and fibronectin (integrin band 3). The carboxyl terminal MAG residues 614-625, including a two-residue gap after tyr-620 are shown in the standard letter code. The major autophosphorylation site in the EGF receptor (tyr-1137) is marked with a filled-in circle. Homologies between MAG and the EGF receptor are boxed in. Cytoplasmic segments of INS-R and integrin band 3, with less extensive homology to the EGF-R than MAG, are shown for comparison.

domain of MAG (Fig. 2), as indicated by consensus sequences for the calcium calmodulin-dependent or cAMP-dependent protein kinases at residues 537-543 (R-R-K-K-N-V-T) and for protein kinase C at residues 575-582 (K-R-L-G-S-E-R-R) and 604-608 (K-R-P-T-K).

We have shown in preliminary experiments that purified MAG is an excellent substrate for in vitro phosphorylation both by the src and the fps tyrosine kinases. It is not clear at present if the phosphorylated site is identical with the tyrosine-620 within the EGF receptor-like sequence. The possibility that tyr-620 may be the major site of in vitro phosphorylation in MAG was, however, suggested indirectly by the finding that the phosphorylation of MAG was almost completely inhibited by the addition of a 16-residue synthetic MAG peptide, containing residues 609 to 624, to the tyrosine kinase reaction. However, since not all proteins that are phosphorylated in vitro by tyrosine kinases are in vivo substrates (46), the relevance of this observation to the in vivo situation needs to be studied further.

Phosphorylation in the cytoplasmic domain of MAG might have various effects. It might induce a conformational change which, upon transmission to the extracellular domain, may influence the putative binding interactions of MAG. Such an effect has been observed for growth factor receptors, as well as the B-adrenergic receptor, where phosphorylation of serines or threonines in the cytoplasmic domain reduces the binding affinity for ligands (86). However, contrary to MAG, these receptors either contain more than one subunit or, as has been shown for the EGF receptor, dimerize upon ligand binding.

Alternatively, phosphorylation might affect the interaction of MAG with the cytoskeleton. Preliminary immunocytochemical evidence indeed suggests that MAG may interact with the cytoskeletal protein spectrin, as indicated by the co-localization of MAG and spectrin in the non-compacted areas of myelin (J. Griffin, unpublished observation). Further evidence that MAG may interact with the cytoskeleton comes from the observation of homology between a region in the center of the cytoplasmic domain of MAG (residues 550-573) and the carboxyl terminal sequence of the band 3 component of the fibronectin receptor integrin (90,84). Since the small intracellular domain of the band 3 protein is known to interact with actin, probably through the intermediary of spectrin, the homologous sequence in MAG may represent the site of interaction of the glycoprotein with spectrin.

Possible Functional Implications of the Structure of MAG

The contact between the axon and the myelin membrane may be mediated by the large external domain of MAG, which is thought to project from the innermost myelin leaflet into the periaxonal space (78) and may contact a specific receptor on the axonal membrane. The immunoglobulin-like tandem repeat units of MAG may be important in the binding between the glycoprotein and its putative receptor. This possibility is strongly suggested by analogy with the neural cell adhesion molecule N-CAM, where several similar repeat units form part of the homophilic binding domain of the molecule (16).

MAG may be involved in the initial recognition of axons by the myelinating cell, as well as during the process of myelination. It is plausible that during active myelination the short-lived association of individual MAG molecules with specific MAG receptors on the axon supports the continued circumnavigation of the innermost myelin leaflet around the axon. The continued movement of the inner lip requires that the interaction between MAG and its putative receptor, as well as the contact between adjacent myelin leaflets, be of low affinity and, therefore, reversible. This would permit the sliding of compact myelin lamellae past each other while new membrane material is added.

After completion of myelin synthesis, MAG continues to be expressed in the adult nervous system. This underlines the importance of its function, not only during myelin synthesis but also in the maintenance of the myelin sheath.

Mapping of the MAG Locus

In a collaborative effort, Uta Francke and coworkers used the clone MAG 1.2 in Southern blot analyses of rodent x human and hamster x mouse somatic cell hybrids to assign the human locus for MAG to chromosome 19 (HSA19) and the mouse locus to chromosome 7 (MMU7) (6). The location of the MAG gene was mapped more precisely on the human chromosome by showing that a hybrid that expresses human glucose phosphate isomerase (mapped to human 19cen--<q13) but contains no microscopically undetectable chromosome 19, was positive for the MAG gene. The region containing the locus for glucose phosphate isomerase (GPI) is conserved in evolution between the human chromosome 19 and the mouse chromosome 7 (57) and includes other homologous loci, such as peptidase D (PEPD and Pep-4), luteinizing hormone B subunit (LHB), and transforming growth factor B. One neurological mutant, quivering (qv), is part of this linkage group (Fig. 8) and is flanked by Gpi and Pep-4 at a distance of about 1 cM (20). Quivering are known to suffer from auditory impairment (43,44) and exhibit varying levels of paralysis which may be due to a defect in myelination. The chromosomal mapping data of Uta Francke and co-workers suggested that the defect in quivering could involve the MAG gene.

To test this hypothesis, the quivering mutant was analyzed for a possible defect in the structure or the expression of the MAG gene. Southern analysis of the MAG gene in the quivering mutant indicated the absence of a gross deletion or DNA rearrangement. Furthermore, the abundance and the size of both the MAG transcript and of the MAG protein were normal in the quivering mutant (6). These results suggest that if the MAG gene is affected in the quivering mutant, the defect is due to an undetected small deletion or point mutation. Such a mutation may impair the function of one or both forms of the MAG gene. It is conceivable that no large structural disruption in the MAG gene was observed because a major alteration would result in a lethal mutation, and would, therefore, not be present amongst the panel of viable neurological mutants. Alternatively, the MAG gene may not be identical with the qv locus, despite the close map location of the two loci. Resolution of this issue will require a more detailed analysis of the MAG gene isolated from the quivering mutant.

Table 2   PROTEIN HOMOLOGIES OF MAG

	Identity %	MAG Region (residue)	Gaps
N-CAM (93 --> 182)[a]	30.5	246 --> 327	1
HLA class II, DRa (119 --> 190)[b]	28.7	246 --> 326	3
HLA class II, SBa (104 --> 142)[c]	45.9	254 --> 290	1
IgA heavy chain C region (140 --> 199)[d]	22	254 --> 312	2
T-cell receptor a, V region (28 --> 73)[e]	30	247 --> 297	1
BOLF1 protein of EBV (25 --> 106)[f]	47.5	246 --> 327	1

A search of the National Biomedical Research Foundation (NBRF) protein data base with the amino acid sequence of MAG indicated homology between the third domain of MAG and several members of the immunoglobulin gene family. These include (a) chicken N-CAM (40); (b) HLA class II histocompatibility antigen, DRa chain precursor (18); (c) HLA class II histocompatibility antigen, SBa chain precursor (3); (d) Iga-1 chain, constant region (75); and (e) T-cell receptor a chain precursor, V region (101). The hypothetical BOLF1 protein of unknown function from Epstein Barr virus (4) is included (f) due to its high homology across the entire third MAG homology domain and the presence of two invariant cysteine residues, at a distance of 45 amino acids. The numbers in parentheses represent the region of highest MAG homology within each protein.

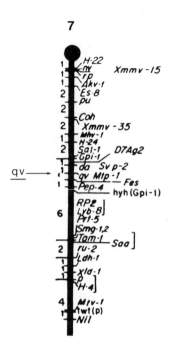

Figure 8   Chromosomal location of the murine quivering (qv) locus. The region of mouse chromosome 7 (35% of the total chromosome) containing the loci for quivering (qv), glucose phosphate isomerase (gpi) and peptidase (pep-4) is shown. The mag and qv loci are closely linked.

SUMMARY

The isolation and sequencing of two specific cDNAs from a rat brain cDNA library has allowed the deduction of the complete amino acid sequence of myelin-associated glycoprotein. The primary structure of MAG has revealed a striking structural relationship with the known neural cell adhesion molecule N-CAM and other members of the immunoglobulin supergene family, lending support to the hypothesis of a role of MAG in cell/cell interactions. The sequence characteristics exhibited by the immunoglobulin-like tandem repeat units found in the extracellular domain of MAG suggest that the units may be folded into domains consisting of two B-pleated sheets, held together by a disulphide bond.

Two developmentally regulated species of MAG are found in the nervous system. An early form (p72MAG) is produced at the onset and during myelination, whereas a shortened form of MAG (p67MAG), differing in the cytoplasmic domain, predominates in the adult animal. The two MAG species arise from the same gene by alternate splicing of the mRNA. The difference in the cytoplasmic domains of the two MAG polypeptides suggests that the function of MAG in mature myelin may be modulated through an altered interaction with the cytoskeleton of other cytoplasmic components, such as protein kinases. The amino acid sequence, as well as preliminary in vitro phosphorylation experiments, indicate that MAG may be a substrate for one or more protein kinases. However, the importance of phosphorylation to the function of MAG remains to be established.

The gene coding for MAG is a single-copy gene and was mapped to chromosome 19 in the human and to chromosome 7 in the mouse, in the vicinity of the locus for the neurological mutation quivering. Recent experiments, however, suggest that the MAG gene may not be identical with the qv locus.

The structural information obtained from the cloned MAG-specific cDNAs has provided a new direction to the study of the function of the glycoprotein in myelination. It is hoped that with the availability of the amino acid sequence and the cDNA for MAG, our understanding of myelination can be advanced significantly in the near future.

ACKNOWLEDGEMENTS

We thank N. Latov (Columbia University), R. Lazzarini (NIH), Don Frail and Peter Braun (McGill University) for their generous gifts of antibodies, Tony Pawson for his assistance with studies on phosphorylation, V. Auld and J. Marshall (Toronto) for the cDNA libraries and all of the above for helpful discussions. The help of Martha Garrett with the MAG preparations and nucleotide sequencing is gratefully acknowledged. This work was supported by the Multiple Sclerosis Society and the Medical Research Council of Canada program grant on "The Molecular Biology of Myelination".

REFERENCES

1. Amzel LM, Poljak RJ: Three-dimensional structure of immunoglobulins. Ann Rev Biochem 48:961, 1979
2. Arquint M, Roder J, Chia L-S, Down J, Wilkinson D, Bayley H, Braun P, Dunn R: The molecular cloning and primary structure of myelin associated glycoprotein. Proc Natl Acad Sci USA 84:600, 1987
3. Auffrey C, Lillie JW, Arnot D, Grossberger D, Kappes D, Strominger JL: Isotypic and allotypic variation of human class II histocompatibility antigen alpha-chain genes. Nature 308:327, 1984
4. Baer R, Bankier AT, Biggin MD, Deininger PL, Farrell PJ, Gibson TJ, Hatfull G, Hudson GS, Satchwell SC, Sequin C, Tuffnell PS, Barrell BG: DNA sequence and expression of the B95-8 Epstein-Barr virus genome. Nature 310:207, 1984
5. Barthels D, Santoni M-J, Willie W, Ruppert C, Chaix J-C, Hirsch M-R, Fontecilla-Camps JC, Goridis C: Isolation and nucleotide sequence of mouse N-CAM cDNA that codes for a MW 79 000 polypeptide without membrane-spanning region. EMBO J 6:907, 1987
6. Barton DE, Arquint M, Roder J, Dunn R, Francke U: The myelin-associated glycoprotein gene: mapping to human chromosome 19 and mouse chromosome 7 and expression in quivering mice. Genomics, in press
7. Bosch EP, Ansbacher LE, Goeken JA, Cancilla PA: Peripheral neuropathy associated with monoclonal gammopathy. Studies of intraneural injections of monoclonal immunoglobulin sera. J Neuropathol and Exp Neurol 41:446, 1982
8. Braun PE, Frail DE, Latov N: Myelin-associated glycoprotein is the antigen for a monoclonal IgM in polyneuropathy. J Neurochem 39:1261, 1982
9. Brostoff SW: Immunological responses to myelin and myelin components. In Morell P (ed): Myelin. New York, Plenum Press, 1984, pg 405
10. Bunge MB, Bunge RP, Pappas GD: Electron microscopic demonstrations of connections between glia and myelin sheaths in the developing mammalian central nervous system. J Cell Biol 12:448, 1962

11. Bunge RP: Glial cells and the central myelin sheath. Physiol Rev 48:197, 1968

12. Bung RP, Bunge MB: Tissue culture observations relating to peripheral nerve development, regeneration, and disease. In Dyck PJ, Thomas PK, Lambert EH, Bunge R (eds): Peripheral Neuropathy, Vol I (2nd ed). Philadelphia, WB Saunders Co, 1984, pg 378

13. Bunge RP, Bunge MB, Eldridge CF: Linkage between axonal ensheathment and basal lamina production by Schwann cells. Ann Rev Neurosci 9:305, 1986

14. Chou PY, Fasman GD: Prediction of the secondary structure of proteins from their amino acid sequence. In Meister A (ed): Advances in Enzymol. New York, Interscience Publications, John Wiley and Sons, 1978, 47:45

15. Cunningham BA, Hoffman S, Rutishauser U, Hemperly JJ, Edelman GM: Molecular topography of the neural cell adhesion molecule N-CAM: Surface orientation and location of sialic acid-rich and binding regions. Proc Natl Acad Sci USA 80:3116, 1983

16. Cunningham BA, Hemperly JJ, Murray BA, Prediger EA, Brackenbury R, Edelman GM: Neural cell adhesion molecule: structure, immunoglobulin-like domains, cell surface modulation, and alternative RNA splicing. Science 236:799, 1987

17. D'Eustachio P, Owens GC, Edelman GM, Cunningham BA: Chromosomal location of the gene encoding the neural cell adhesion molecule (N-CAM) in the mouse. Proc Natl Acad Sci USA 82:7631, 1985

18. Das HK, Lawrance SK, Weissman SM: Structure and nucleotide sequence of the heavy chain gene of HLA-DR. Proc Natl Acad Sci USA 80:3543, 1983

19. Dautigny A, Mattei M-G, Morello D, Alliel PM, Pham-Dinh D, Amar L, Arnaud D, Simon D, Mattei J-F, Guenet J-L, Jolles P, Avner P: The structural gene coding for myelin-associated proteolipid protein is mutated in jimpy mice. Nature 321:867, 1986

20. Davisson MT, Roderick TH: Linkage map of the mouse. Mouse News Letter 75:11, 1986

21. Dellagi K, Dupouey P, Brouet JC, Billecocq A, Gomez D, Clauvel JP, Seligmann M: Waldenstroem's macroglobulinemia and peripheral neuropathy: a clinical and immunologic study of 25 patients. Blood 62:280, 1983

22. Downward J, Parker P, Waterfield MD: Autophosphorylation sites on the epidermal growth factor receptor. Nature 311:483, 1984

23. Dubois-Dalcq M, Behar T, Hudson L, Lazzarini RA: Emergence of three myelin proteins in oligodendrocytes cultured without neurons. J Cell Biol 102:384, 1986

24. Edelman GM: Specific cell adhesion in histogenesis and morphogenesis. In Edelman GM, Thiery J-P (eds): The cell in contact. Adhesions and junctions as morphogenetic determinants. New York, John Wiley & Sons, 1985a, pg 139

25. Edelman GM: Cell adhesion and the molecular processes of morphogenesis. Ann Rev Biochem 54:135, 1985b

26. Edelman GM: Cell adhesion molecules in the regulation of animal form and tissue pattern. Ann Rev Cell Biol 2:81, 1986

27. Eldridge CF, Bunge MB, Bunge RP, Wood PM: Differentiation of axon-related Schwann cells in vitro: I. Ascorbic acid regulates basal lamina assembly and myelin formation. J Cell Biol (in press)

28. Favilla JT, Frail DE, Palkovitz CG, Stoner GL, Braun PE, Webster H deF: Myelin-associated glycoprotein (MAG) distribution in human central nervous tissue studied immunocytochemically with monoclonal antibody. J Neuroimmunol 6:19, 1984

29. Frail D: The myelin-associated glycoprotein in development and disease. PhD Thesis, Department of Biochemistry, McGill University, Montreal, 1984

30. Frail DE, Braun PE: Two developmentally regulated messsenger RNAs differing in their coding region may exist for the myelin-associated glycoprotein. J Biol Chem 259:1485, 1984

31. Frail DE, Edwards AM, Braun PE: Molecular characteristics of the epitope in myelin-associated glycoprotein that is recognized by a monoclonal IgM in human neuropathy patients. Mol Immunol 21:721, 1984

32. Frail DE, Braun PE: Abnormal expression of the myelin-associated glycoprotein in the central nervous system of dysmyelinating mutant mice. J Neurochem 45:1071, 1985

33. Frail DE, Webster H deF, Braun PE: Development expression of the myelin-associated glycoprotein in the peripheral nervous system is different from that in the central nervous system. J Neurochem 45:1308, 1985

34. Gendelman HE, Pezeshkpour GH, Pressman NJ, Wolinsky JS, Quarles RH, Dobersen MJ, Trapp BD, Kitt CA, Aksamit A, Johnson RT: A quantitation of myelin-associated glycoprotein and myelin basic protein loss in different demyelinating diseases. Ann Neurol 18:324, 1985

35. Gribskov M, Burgess RR, Devereux J: PEPPLOT, a protein secondary structure analysis program for the UWGCG sequence analysis software package. Nucleic Acids Res 14(1):327, 1986

36. Gulcher JR, Marton LS, Stefansson K: Two large glycosylated polypeptides found in myelinating oligodendrocytes but not in myelin. Proc Natl Acad Sci USA 83:2118, 1986

37. Hafler DA, Johnson D, Kelly JJ, Panitch H, Kyle R, Weiner HL: Monoclonal gammopathy and neuropathy: myelin-associated glycoprotein reactivity and clinical characteristics. Neruol 36:75, 1986

38. Hays AP, Latov N, Takatsu M, Sherman WH: Experimental demyelination of nerve induced by serum of patients with neuropathy and an anti-MAG IgM M-protein. Neurology 37:242, 1987

39. He H-T, Barbet J, Chaix H-C, Goridis C: Phosphatidylinositol is involved in the membrane attachment of NCAM-120, the smallest component of the neural cell adhesion molecule. EMBO J 5:2489, 1986

40. Hemperly JJ, Murray BA, Edelman GM, Cunningham BA: Sequence of a cDNA clone encoding the polysialic acid-rich and cytoplasmic domains of the neural cell adhesion molecule N-CAM. Proc Natl Acad Sci USA 83:3037, 1986

41. Hirano A, Dembitzer HM: A structural analysis of the myelin sheath in the central nervous system. J Cell Biol 34:555, 1967

42. Hood L, Kronenberg M, Hunkapiller T: T cell antigen receptors and the immunoglobulin supergene family. Cell 40:225, 1985

43. Horner KC, Bock GR: Single unit responses in the cochlear nucleus of the deaf quivering mouse. Hear Res 13:63, 1984

44. Horner KC, Bock GR: Combined electrophysiological and autoradiographic delimitation of retrocochlear dysfunction in a mouse mutant. Brain Res 331:217, 1985

45. Hunkapiller T, Hood L: The growing immunoglobulin gene superfamily. Nature 323:15, 1986

46. Hunter T, Cooper JA: Protein-tyrosine kinases. Ann Rev Biochem 54:897, 1985

47. Ilyas AA, Quarles RH, Brady RO: The monoclonal antibody HNK-1 reacts with a peripheral nerve ganglioside. Biochem Biophys Res Commun. 122:1206, 1984

48. Itoyama Y, Sternberger NH, Webster H deF, Quarles RH, Cohen SR, Richardson EP Jr: Immunocytochemical observations on the distribution of myelin-associated glycoprotein and myelin basic protein in multiple sclerosis lesions. Ann Neurol 7:167, 1980

49. Itoyama Y, Webster HD, Sternberger NH, Richardson EP Jr, Walker DL, Quarles RH, Padgett BL: Distribution of papovavirus, myelin-associated glycoprotein, and myelin basic protein in progressive multifocal leukoencephalopathy lesions. Ann Neurol 11:396, 1982

50. Jen Chou C-H, Chou FC-H, Tourtelotte WW, Kibler RF: Search for a multiple sclerosis-specific brain antigen. Neurol 33:1300, 1983

51. Johnson D, Quarles RH: Deposition of the myelin-associated glycoprotein in specific regions of the developing rat central nervous system. Dev Brain Res 28:263, 1986

52. Johnson D, Sato S, Quarles RH, Inuzuka T, Brady RO, Tourtelotte WW: Quantitation of the myelin-associated glycoprotein in human nervous tissue from controls and multiple sclerosis patients. J Neurochem 46:1086, 1986a

53. Johnson D, Hafler DA, Fallis RJ, Lees MB, Brady RO, Quarles RH, Weiner HL: Cell-mediated immunity to myelin-associated glycoprotein, proteolipid protein, and myelin basic protein in multiple sclerosis. J Neuroimmunol 13:99, 1986b

54. Kaufman JF, Krangel MS, Strominger JL: Cysteines in the transmembrane region of major histocompatibility complex antigens are fatty acylated via thioester bonds. J Biol Chem 259:7230, 1984

55. Kruse J, Keilhauer G, Faissner A, Timpl R, Schachner M: The J1 glycoprotein - a novel nervous system cell adhesion molecule of the L2/HNK-1 family. Nature 316:146, 1985

56. Lai C, Brow MA, Nave K-A, Noronha AB, Quarles RH, Bloom FE, Milner RJ, Sutcliffe JG: Two forms of 1B236/myelin-associated glycoprotein, a cell adhesion molecule for postnatal neural development, are produced by alternative splicing. Proc Natl Acad Sci USA 84:4337, 1987

57. Lalley PA, McKusick VA: Report of the Committee on Comparative Mapping. Cytogenet Cell Genet 40:536, 1985

58. Latov N, Braun PE, Gross RB, Sherman WH, Penn AS, Chess L: Plasma cell dyscrasia and peripheral neuropathy: Identification of the myelin antigens that react with human paraproteins. Proc Natl Acad Sci USA 78:7139, 1981a

59. Latov N, Gross RB, Kastelman J, Flanagan T, Lamme S, Alkaitis D, Olarte MR, Sherman WH, Chess L, Penn AS: Complement-fixing antiperipheral nerve myelin antibodies in patients with inflammatory polyneuritis and with polyneuropathy and paraproteinemias. Neurol 31:1530, 1981b

60. Martini R, Schachner M: Immunoelectron microscopic localization of neural cell adhesion molecules (L1, N-CAM, and MAG) and their shared carbohydrate epitope and myelin basic protein in developing sciatic nerve. J Cell Biol 103:2439, 1986

61. Matthieu J-M, Quarles RH, Poduslo J, Brady RO: [^{35}S] sulfate incorporation into myelin glycoproteins. I. Central nervous system. Biochem Biophys Acta 392:159, 1975

62. Matthieu JM, Waehneldt TV, Eschmann, N: Myelin-associated glycoprotein and myelin basic protein are present in central and peripheral nerve myelin throughout phylogeny. Neurochem Int 8:521, 1986

63. McGarry RC, Riopelle RJ, Frail DE, Edwards AM, Braun PE, Roder JC: The characterization and cellular distribution of a family of antigens related to myelin associated glycoprotein in the developing nervous system. J Neuroimmunol 10:101, 1985

64. Melmed C, Frail D, Duncan I, Braun P, Danoff D, Finlayson M, Stewart J: Peripheral neuropathy with IgM kappa monoclonal immunoglobulin directed against myelin-associated glycoprotein. Neurol 33:1397, 1983

65. Mendell JR, Sahenk Z, Whitaker JN, Trapp BD, Yates AJ, Griggs RC, Quarles RH: Polyneuropathy and IgM monoclonal gammopathy: studies on the pathogenetic role of anti-MAG antibody. Ann Neurol 17:243, 1985

66. Nguyen C, Mattei MG, Goridis C, Mattei JF, Jordan BR: Localization of the human N-CAM gene to chromosome 11 by in situ hybridization with a murine N-CAM cDNA probe. Cytogenet Cell Genet 40:713, 1985

67. Nobile-Orazio E, Spagnol G, Scarlato G: Failure to detect anti-MAG antibodies by RIA in CSF of patients with multiple sclerosis. J Neuroimmunol 11:165, 1986

68. O'Shannessy DJ, Ilyas AA, Dalakas MC, Mendell JR, Quarles RH: Specificity of human IgM monoclonal antibodies from patients with peripheral neuropathy. J Neuroimmunol 11:131, 1986

69. Oldberg A, Franzen A, Heinegard D: Cloning and sequence analysis of rat bone sialoprotein (osteopontin) cDNA reveals an Arg-Gly-Asp- cell-binding sequence. Proc Natl Acad Sci USA 83:8819, 1986

70. Owens GC, Edelman GM, Cunningham BA: Organization of the neural cell adhesion molecule (N-CAM) gene: alternative exon usage as the basis for different membrane-associated domains. Proc Natl Acad Sci USA 84:294, 1987

71. Pierschbacher MD, Ruoslahti E: The cell attachment activity of fibronectin can be duplicated by small synthetic fragments of the molecule Nature 309:30, 1984

72. Pollerberg GE, Sadoul R, Goridis C, Schachner M: Selective expression of the 180 kD component of the neural cell adhesion molecule N-CAM during development. J Cell Biol 101:1921, 1985

73. Porter S, Clark MB, Glaser L, Bunge RP: Schwann cells stimulated to proliferate in the absence of neurons retain full functional capability. J Neurosci 6:3070, 1986

74. Pumphrey R: Computer models of the human immunoglobulins. Shape and segmental flexibility. Immunol Today 7:174, 1986

75. Putnam FW, Liu Y-SV, Low TLK: Primary structure of a human IgA1 immunoglobulin. IV. Streptococcal IgA1 protease digestion, Fab and Fc fragments, and the complete amino acid sequence of the alpha 1 heavy chain. J Biol Chem 254:2865, 1979

76. Pytela R, Pierschbacher MD, Ruoslahti: A 125/115 kDa cell surface receptor specific for vitronectin interacts with the arginine-glycine-aspartic acid adhesion sequence derived from fibronectin. Proc Natl Acad Sci USA 82:5766, 1986

77. Quarles RH, Barbarash GR, Figlewicz DA, McIntyre LJ: Purification and partial characterization of the myelin-associated glycoprotein from adult rat brain. Biochem Biophys Acta 757:140, 1983

78. Quarles RH: Myelin-associated glycoprotein in development and disease. Dev Neurosci 6:285, 1984

79. Raine CS: Morphology of myelin and myelination. In Morell (ed): Myelin, New York, Plenum Press, 1984, pp 1-41

80. Raine CS: Oligodendrocytes and central nervous system myelin. In Davis L, Robertson DM (eds): Textbook of Neuropathology 1985, p. 92

81. Reiger F, Daniloff JK, Pincon-Raymond M, Crossin KL, Grumet M, Edelman GM: Neuronal cell adhesion molecules and cytoactin are colocalized at the node of Ranvier. J Cell Biol 103:379, 1986

82. Rutishauser U: Molecular and biological properties of a neural cell adhesion molecule. Cold Spring Harbor Symp Quant Biol 48:501, 1983

83. Sabatini DD, Kreibich G, Morimoto T, Adesnick M: Mechanisms for the incorporation of proteins in membranes and organelles. J Cell Biol 92:1, 1982

84. Salzer JL, Holmes WP, Colman DR: The amino acid sequences of the myelin-associated glycoproteins: homology to the immunoglobulin gene superfamily. J Cell Biol 104:957, 1987

85. Schachner M, Faissner A, Fischer G, Keilhauer G, Kruse J, Kunemund V, Lindner J, Wernecke H: Functional and structural aspects of the cell surface in mammalian nervous system development. In Edelman GM, Thiery J-P (eds): The cell in contact. Adhesions and junctions as morphogenetic determinants. New York, John Wiley & Sons, 1985, pg 257

86. Sibley DR, Benovic JL, Caron MG, Lefkowitz RJ: Regulation of transmembrane signaling by receptor phosphorylation. Cell 48:913, 1987

87. Sorkin BC, Hoffman S. Edelman GM, Cunningham BA: Sulfation and

phosphorylation of the neural cell adhesion molecule, N-CAM. Science 225:1476, 1984

88. Sternberger NH, Quarles RH, Itoyama Y, Webster H deF: Myelin-associated glycoprotein demonstrated immunocytochemically in myelin and myelin forming cells of developing rat. Proc Natl Acad Sci USA 76:1510, 1979

89. Sutcliffe JG, Milner RJ, Shinnick TM, Bloom FE: Identifying the protein products of brain-specific genes with antibodies to chemically synthesized peptides. Cell 33:671, 1983

90. Tamkun JW, DeSimone DW, Fonda D, Patel RS, Buck C, Horwitz AF, Hynes RO: Structure of integrin, a glycoprotein involved in the transmembrane linkage between fibronectin and actin. Cell 46:271, 1986

91. Thomas PK: Clinical features and differential diagnosis. In Dyck PJ, Thomas PK, Lambert EH, Bunge RP (eds): Peripheral Neuropathy, Vol II (2nd ed). Philadelphia, WB Saunders Co, 1984, pg 1169

92. Thor G, Pollerberg GE, Schachner M: Molecular association of two neural cell adhesion molecules within the surface membrane of cultured mouse neuroblastoma cells. Neurosci Lett 66:121, 1986

93. Trapp BD, Quarles RH: Presence of myelin-associated glycoprotein correlates with alterations in the periodicity of peripheral myelin. J Cell Biol 92:877, 1982

94. Trapp BD, Quarles RH: Immunocytochemical localization of the myelin-associated glycoprotein. Fact or artifact? J Neuroimmunol 6:231, 1984

95. Trapp BD, Quarles RH, Suzuki K: Immunocytochemical studies of Quaking mice support a role for the myelin-associated glycoprotein in forming and maintaining the periaxonal space and periaxonal cytoplasmic collar in myelinating Schwann cells. J Cell Biol 99:595, 1984

96. Webster H deF, Palkovits CG, Stoner GL, Favilla JT, Frail DE, Braun PE: Myelin-associated glycoprotein: electron microscopic immunocytochemical localization in compact developing and adult central nervous system myelin. J Neurochem 41:1469, 1983

97. Wood PM: Separation of functional Schwann cells and neurons from normal peripheral nerve tissue. Brain Res 115:361, 1976

98. Wood PM, Williams AK: Oligodendrocyte proliferation and CNS myelination in cultures containing dissociated embryonic neuroglia and dorsal root ganglion neurons. Brain Res 314:225, 1984

99. Wood PM, Bunge RP: Myelination of cultured dorsal root ganglion neurons by oligodendrocytes obtained from adult rats. J Neurol Sci 74:153, 1986

100. Yamada KM, Humphries MJ, Hasegawa T, Hasegawa E, Olden K, Chen W-T, Akiyama SK: Fibronectin: molecular approaches to analyzing cell interactions with the extracellular matrix. In Edelman GM, Thiery J-P (eds): The cell in contact. Adhesions and junctions as morphogenetic determinants. New York, John Wiley & Sons, 1985, pg 303

101. Yanagi Y, Chan A, Chin B, Minden M, Mak TW: Analysis of cDNA clones specific for human T cells and the alpha and beta chains of the T-cell receptor heterodimer from a human T-cell line. Proc Natl Acad Sci USA 82:3430, 1985

102. Young RA, Davis RW: Yeast RNA polymerase II genes: isolation with antibody probes. Science 222:778, 1983

6

MOLECULAR GENETICS OF MYELIN PROTEOLIPID PROTEINS AND THEIR

EXPRESSION IN NORMAL AND JIMPY MICE

K. Mikoshiba[1,2], C. Shiota[1], and K. Ikenaka[1]

[1]Division of Regulation of Macromolecular Function, Osaka
University, Osaka, Japan
[2]Division of Behavior and Neurobiology, National Institute for
Basic Biology, Aichi, Japan

INTRODUCTION

Myelin proteolipid protein (PLP) is one of the most abundant proteins
in myelin of the central nervous system (CNS), and together with myelin
basic protein (MBP), constitutes 80% of the myelin membrane proteins of
the CNS. It is called "proteolipid" because of its unusual property to
dissolve in organic solvents (5). There is another less abundant
proteolipid, DM20, which shows immunological cross-reactivity with the
major proteolipid protein (13). Most investigators have obtained
molecular weights of 24,000-26,000 for PLP by SDS-PAGE, and approximately
5,000 less than PLP for DM20 (13). Lipids are covalently attached to PLP
via an O-ester linkage (28). Because amino acid and nucleotide sequences
of PLP are remarkably well conserved among mouse, rat, cow, and human
(4,6,8,9,14,16,20,21,25) and because severe dysmyelination is observed in
the animals whose PLP-synthesis are affected (7), PLP is thought to play
a crucial role in myelination in the CNS; probably by promoting the
apposition of extracellular surfaces of the myelin lamellae.

Recently molecular genetics of PLP has made a great advance and many
interesting aspects on synthesis and function of PLP have been reported.
In this chapter, we focus on this recent advance in PLP-molecular
genetics and the characterization of the X-linked dysmyelinating jimpy
mutant mouse, whose PLP synthesis is affected.

GENE ORGANIZATION OF PLP

Structure of PLP-mRNA

Rat (16), bovine (20), mouse (8), and human (25) PLP cDNA have been
cloned and the nucleotide sequences were determined. The coding and the
5'-untranslated region were highly conserved among species at the amino
acid and nucleotide level.

Northern blot analysis of rat brain RNA using labelled PLP cDNA as a
probe indicated that the rat PLP-mRNAs occur as two abundant families of
approximately 3.2 and 1.6 kilobase, and a less abundant family of

approximately 2.4 kilobase in length (16). It has been shown that 3.2 and 1.6 kilobase, and probably 2.4 kilobase PLP-mRNAs were produced by alternate usage of the polyadenylation sites (16) (see Fig. 3). Relative abundance of the three PLP-mRNAs seems to differ among species; i.e., mouse has much lower level of 1.6 kb and higher level of 2.4 kb RNA (3,6,8,16,21), whereas human apparently contains only the longest 2.8 kb RNA (25). The nucleotide sequences around the polyadenylation sites and possible polyadenylation signals of 1.6 kb-PLP-mRNA from the three species aligned in Fig. 1 show high degree of homology among them, although there are some mismatching as shown in the figure. These changes might account for the decreased efficiency of the polyadenylation, however, the reason for this still remains to be elucidated. No homologous sequence was found which corresponds to the polyadenylation site for 2.4 kb RNA in human (25).

```
                                        • • • • •
rat   : AAATATATTCTCTTTGATGCACAAAA
mouse: AAATAcATTCTCTTTGATGCACAAAA
human: AAATATATTCTCTTTGgTGtACAAAA
```

Figure 1   Alignment of the nucleotide sequences surrounding rat, mouse, and human PLP-mRNA (1600 bases) polyadenylation sites. Putative polyadenylation signals are underlined. Poly(A) addition occurred at either one of the dotted bases. Small letters show the bases mismatching with the rat PLP-mRNA sequence.

Structure of DM-20-mRNA

DM20 is also a proteolipid found in the normal CNS and it has been shown to have a high degree of structural similarity with PLP (13). Morello et al (17) and Hudson et al (8) suggested that DM20-mRNA was produced by alternative splicing of the PLP-mRNA precursor and that a part of the PLP-mRNA was deleted to form DM20-mRNA. Nave et al proved this hypothesis to be true by directly cloning and sequencing the mouse DM20 cDNA (22). They showed that 105 nucleotides were deleted from PLP-mRNA to form the DM20-mRNA, which correspond to the deletion of 35-amino acid residues (116-150) at the amino acid level. An oligonucleotide complementary to the junction of the deleted region (5'-TGCCCACAAACGTTGCGCTC-3') was synthesized, 5'end-labelled by ^{32}P and used as a probe for Northern transfer analysis of total brain RNA from normal mouse, fractionated on 1.5% formaldehyde agarose gel. After autoradiography, bands showing the same mobility as the PLP-mRNA were detected (9). Therefore, mouse DM20-mRNA is also present as three families of RNA, probably formed by the alternate usage of the same polyadenylation sites as the PLP-mRNA.

The PLP Gene Structure

The genes for the mouse (9,14) and human (4) PLP have been isolated and both of them were shown to consist of seven exons (Fig. 2). The exon-intron junctions of PLP are completely conserved between mouse and human. Exon 1 contains the 5' untranslated region of PLP-mRNA, translation initiation codon "ATG", and the 5'-G of the triplet for glycine, the N-terminus of mature PLP (Fig. 3). Homology between mouse and human PLP untranslated region (94%) and their upstream of the major cap site for at least 100 bases (89%) is remarkably high. This suggests some very strong functional constraints on these regions that limit the variation allowed in these sequences (14). An interesting aspect of the

Figure 2   Restriction of the mouse PLP gene. Open squares represent
exons (I to VII). PLP genes were cloned in two clones in our laboratory
(9), however, they did not overlap with each other. The uncloned region
is shown by a dotted line. The distance between the BamHi site within
the first exon and the EcoRI site just down stream of the dotted line
were estimated from the result of Macklin et al (14). BamHI(B),
EcoRI(E), HindIII(H), and PstI(P) sites are indicated.

exon usage is that four of the five hydrophobic domains are encoded in
individual exons (4). Southern transfer analysis revealed that a single
gene encode for PLP and DM20 (3,16). The 3' end of the deleted portion
in DM20-mRNA corresponds to the junction point of exons 3 and 4 of PLP-
mRNA, while the 5' end of the deletion starts in the middle of exon 3
(9,22). A sequence "ACG/GTAAC" was found in exon 3 (underlined in Fig.
3) where the deletion apparently begins. It is homologous to the
consensus 5' splice junction sequence "(C or A)AG/GTRAG", where R
represents purine and "/" shows the splicing site. Therefore, alternate
usage of the splice donor sites is involved in formation of PLP and DM20-
mRNA. All the polyadenylation sites are present in exon 7 (Fig. 3, Exon
7).

Transcription Initiation Sites

     Transcription of the rat (14,16), mouse (9,14,16), human (14), and
baboon (14) PLP gene was shown to start from multiple sites, by several
methods involving primer extension analysis (9,16), S1 nuclease
protection (14), T4 polymerase primer extension (9), and RNase mapping
(9). The exact position of the initiation sites are indicated in Fig. 3,
although the position of the most frequently used sites of mouse PLP gene
detected by the following two methods were inconsistent with each other;
primer extension analysis predicted a "C" residue at position +7, while
T4 DNA polymerase primer extension predicted a "G" residue at +16 (9).
There were no common sequences found upstream of each initiation site. A
possible "TATA" and "CAAT" box were found at position -26 and -10,
respectively, although we cannot find the "TATA" box corresponding to the
"CAAT" box. The unique feature of the sequence of the 5'-flanking region
of the PLP gene is that it has four tandemly repeated eleven-base-pair
sequences (indicated as a box in Fig. 3). A homology search against
GenBank revealed that the repeated sequence shared high degree of
homology (9/11) with a herpes simplex virus repeated sequence (DR2),
which is believed to be involved in the cleavage and packaging reactions
of the virus (2,30) (Fig. 4). At present, however, the function of this
sequence is not known.

     The upstream sequence of the transcription initiating site, including
the promoter sequence, is well known to play an important role in the
regulation of gene expression. Recently, the nucleotide sequence of the
5'-flanking region of the mouse MBP transcription have been mapped (Miura
et al, manuscript in preparation). PLP and MBP are expressed
specifically in the oligodendrocyte in the central nervous system and the
developmental changes in their expression are also quite similar. When
the nucleotide sequences of the 5'-flanking regions of both genes were
compared, four homologous regions were found, as shown in Fig. 5 (9).
The distance between each box and the transcription initiation site of
the PLP gene were approximately the same as the distance between the

87

Exon 1 and 5'-Flanking Region

```
cagctggttc tatctgcatt cttctaaaac caaagctttt gaagaaatta tttttaaatga ctttttttct tccccattgt gtttccagtg ccaggaagag  -473
aaagaatgct tttttttgctt acagaggaaa ggaaaggttc catggtcaag ggcaacgagc agtgagagtt gggtgcggtg tgtttggtag tatagtaagt  -373
aggcttttga ttcagaccc cttctcatca gggctactat ttcacatgac tttacatgct cagacccagg tatgacacat ttaaatggac ccaaggatca  -273
tttgggagga ttcaagaacc cctccattca atttacaccc ctaattcaca cttcctgatt tatttaaagc aaaatgaaat tctagagaag ctttagggg  -173
gaaaagagag agaaagaaaa aaaacaattg ggagtgaaaa ggcataaaga gaagatggag ccccttaaaga gggagtgatc ccaaaggagt ggggacaag  -73
```

```
                                                                          +1              O
ggggagagaa agggaggagg agaggaggag gggaacgagc ctgtctcttt aagggggttg gctgtcaatc agAAAGCCCT TTTCATTGCA GGAGAAGAGG  +28
ACAAAGATCT TCAGAGAGAA AAAGTAAAGG ACAGAAGAAG GAGACTGGAG AGACCAGGAT CCTTCCAGCT GAGCAAAGTC AGCCGCAAAA CAGACTAGCC  +128
```

```
                             MetG
AACAGGCTAC AATTGGAGTC AGAGTGCCAA AGACATGGt gagtttcaaa aactccagca tcaaagatgc aggcacagga gttcaacttt ggggcttttgg  +166
```

Exon 2

```
cacactctgt gcttggtaac atgggctgct tggcccagca gtctagtgtg agtggatgag ttacctcgta tgcgctacct gactttctct ttcttcttcc
```

```
           lyLeuL euGluCysCy sAlaArgCys LeuValGlyA laProPheAl aSerLeuVal AlaThrGlyL euCysPhePh eGlyValAla LeuPheCysG
ccagGCTTGT TAGAGTGTTG TGCTAGATGT CTGGTAGGGG CCCCCTTTGC TTCCCTGGTG GCCACTGGAT TGTGTTTCTT TGGAGTGGCA CTGTTCTGTG
```

```
lyCysGlyHi sGluAlaLeu ThrGlyThrG luLysLeuIl eGluThrTyr PheSerLysA snTyrGlnAs pTyrGluTyr LeuIleAsnV a
GATGTGGACA TGAAGCTCTC ACTGGTACAG AAAAGCTAAT TGAGACCTAT TTCTCCAAAA ACTACCAGGA CTATGAGTAT TCATTAATG Tgtaagtacg
tctcaccact cagcta
```

Exon 3

```
                                          lIleH isAlaPheGl nTyrValIle TyrGlyThrA laSerPhePh ePheLeuTyr
tgctctctga ataaggttat ctggatttc tgtctgtcca tgcagGATTC ATGCTTTCCA GTATGTCATC TATGGAACTG CCTCTTTCTT CTTCCTTTAT
```

```
GlyAlaLeuL euLeuAlaGl uGlyPheTyr ThrThrGlyA laValArgGl nIlePheGly AspTyrLysT hrThrIleCy sGlyLysGly LeuSerAlaT
GGGGCCCTCC TGCTGGCTGA GGGCTTCTAC ACCACCGGCG CTGTCAGGCA GATCTTTGGC GACTACAAGA CCACCATCTG CGGCAAGGGC CTGAGCGCCA
```

```
hrValThrGl yGlyGlnLys GlyArgGlyS erArgGlyGl nHisGlnAla HisSerLeuG luArgValCy sHisCysSerLeu GlyLysTrpL euGlyHisPr
CGGTAACAGG GGGCCAGAAG GGGAGGGGTT CCAGAGGCCA ACATCAAGCT CATTCTTTGG AGCGGGTGTG TCATTGTTTG GGAAAATGGC TAGGACATCC
```

```
         /DM20
oAspLys
CGACAAGgtg atcatcctca ggattttgtg gcaataataa ggggtggggg acaattggga gtgagtctgt agcctgatcc ccacccaagg ttgggtcctc
```

Exon 4

```
                                          PheValGlyI leThrTyrAl aLeuThrVal ValTrpLeuL euValPheAl aCysSerAla
ttgatgctga tttttaacca ctccatgtca attgttttag TTTGTGGGCA TCACCTATGC CCTGACTGTT GTATGGCTCC TGGTGTTTGC CTGTTCTGCC
```

```
ValProValT yrIleTyrPh eAsnThrTrp ThrThrCysG lnSerIleAl aPheProSer LysThrSerA laSerIleGl ySerLeuCys AlaAspAlaA
GTACCTGTGT ACATTTACTT CAATACCTGG ACCACCTGTC AGTCTATTGC CTTCCCTAGC AAGACCTCTG CCAGTATAGG CAGTCTCTGC GCTGATGCCA
```

```
rgMerTyrG
GAATGTATGg tgagttgaat gtgggga
```

Exon 5

```
                                                      lyValLeuPr oTrpAsnAla PheProGlyL
          ggcctcta gccttatgaa gtttactctg gctgcttta tgtatcttaG GTGTTCTCCC ATGGAATGCT TTCCCTGGCA
```

```
ysValCysGl ySerAsnLeu LeuSerIleC ysLysThrAl aGlu
AGGTTTGTGG CTCCAACCTT CTGTCCATCT GCAAAACAGC TGAGgtaagt gaatgagaag agtgctttt aaaaaataga ttggctagac atggagg
```

Exon 6

```
                                                      PheGl nMerThrPhe HisLeuPheI leAlaAlaPh eValGlyAla
ttgtgcttgc ttttctgttc taagaaataa ttctctctca tacatcttct tgcagTTCA AATGACCTTC CACCTGTTTA TTGCTGCGTT TGTGGGTGCT
```

```
AlaAlaThrL euValSerLe u
GCGGCCACAC TAGTTTCCCT Ggtaagttat tttaagataa tattagaaaa gaagtggtcc agggatagca ttaggccgaa agactagcag agagactcct
```

Exon 7

```
                                          Le uThrPheMet IleAlaAlaT hrTyrAsnPh eAlaValLeu LysLeuMetG lyArgGlyTh rLysPhe***
cttaccttct tttctctgtt ccctacagCT CACCTTCATG ATTGCTGCCA CTTACAACTT CGCCGTCCTT AAACTCATGG GCCGAGGCAC CAAGTTCTGA
GCTCCCATAG AAACTCCCCT TTGTCTAATA GCAAGGCTCT AACCACACAG CCTACAGTGT TGTGTTTTAA CTCTGCCTTT GCCACTGATT GGCCCTCTTC
TTACTTGATG AGTATAACAA GAAAGGAGAG TCTTGCAGTG ATTAATCTCT CTCTGTGGAC TCTCCCTCTT AGTACCTCTT TTAGTCATTT TGCTCCACAG
CAGGCTCCTG CTAGAAATGG GGGATGCCTG AGAAGGTGAC TCCCCAGCTG CAAGTCGCAG AGGAGTGAAA GCTCTAATTG ATTTTGCAAG CATCTCCTGA
AGACCAGGAT GGGTTCCTTT CTCAAAGGGC ACTTCCAACT GAGGAGAGCA GAACGGAAAG GTTCTCAGGT AGAGAGCAGA AATGTCCCTG GTCTTCTTGC
CATCGTAGG AGTCAAATAC ATTCTCTTTG ATGCACAAAA CCAAGAACTC ACTCTTACCT TCCTGTTTCC ACTGAAGACA GAAGAAAATA AAAAGAATGC
TAGCAGAACA ATATAGCATT TGCCCAAATC TGCCTCCTGC AGCTGGGAGA AGGGTGTCAA AGCAAGGATC TTTCGCCCTT AGAAAGAGAG CTCTGAGCCC
AGTGGCAATG GACTATTTAA GCCCTAACTC AGCCAACCTT CCTACGGCAA TTAGGGAGCA CAGTGCCTGT ATAGACAAAG CGGGGCGGAG GGGGGGGGCA
TCATCTGTCC TTATAGCTCA TTAGGAAGAG AAACAGTGTT GTCAGGATCA TCTCACTCCC TTCTCCTTGA TAACAGCTAC CATGACAACC TTGTGGTTTC
CAAGGAGCTG AGAATAGAAA GGAACTAGCT TATTTGAAAT AAGACTGTGA CCTAAGGAGC ATCAGTTGGT GGATGCTAAA GGTGTTAATT GAAATGGCCT
TCGGGTAAAT CCAAGATACT TAACTCTTTG GATAGCATGT GTTCTTCCCC CACCCCTATC CGCTAGTTCT GGCCCCTGGC CTCTGGCATA ATATCTTCAC
AATGGTGCTT TTTTTCCTGG GGTTTTATCC ATTCACTCAT AGCAGTGAT TAGACGATCT TGATTAGTTT CATATTTCCC AATTGTTTAT GAAATGGCCT
GAGTTGTATC AGAAAGACCT GGAGGATGAT TCTTTGAGCA TAGTTCTTTT TGAAAACAAG AAAGAGAAAC TGGGCAGAAA GCATCACAAA AATATTTGAA
ATTGTACGGT CCCATGAAAT TATTGGGAAT TCCCCCAAGT AGTCTACCAT TTGTAGAACT AGGCTTGATA AATTTGAACC TAATTGGTCT
GGTATTTTCT TTTCTAATAA ATGCACAGATG ATTTTACTTG CTAATATTAT CTCAGCATTT TGATAATTTA GGCTTACCAT AGAAGTTACT GTCTCTTGGT
ATATATAGGT CACATAATAG ATTCTGCCAG CTGTTAGCTG TTCAGTTCAT AAGCTTCCAT AGAGCTCTGG AGCCGGCAAG AGACCAGGCA GAATTTGAAA
CCTAAAGAAC TCCCAGATTT CAGGCTTATC CTGTATTTGT TAACTTTGGG TGAAAGAAAG AAAGAAAGAA AGAAAGAAAG AAAGAAATGAA AGAAAGAAAG
AAAAAAGCCC CTGATCGAAT TTCCTGGAGG AAAAGTTATT GTAGCTGTTT CATTGTAGAT TTGTGCTGTC ATTCCCCAAA GTGCTTTCTG CTGTGTTGAA
AGAGATTATA GAATTTACAA GAAGACACTT GAGACTTGTT CTTGGGCCAA TATATAAGGT AAACAAGCAG CATGCACAAG AGTGAGGACA GCTAAAAGAA
CATGTAAGAA ACCAATCAAG ATCAAGGAAG GTGAAATAAT CTATATCTTT TATTTTGTTT TGGTTTAATA TAACAGATAA CCAACCATTC CCTTAAAAAA
TCTCACATGC ACACACACAC ACACACACAC ACACACGTAC AAAGAAGAGTT AATCAACTGC AGTGTTTCCC TTCATTTCTG ATACAGAATT TTGATTTTAA
CAACATAAGG GATACTTTTA GAAACTTTTA GAAACTCATC TTACAAAATG TATTTTATAA AATTAAAGAA AATAAATTA AGAATGTTCT Caatcaaaca
tcgtgtcctt tgagtgaatt gttctatttg acctcaataa caggtactta attatagtta attatggtcg
```

Figure 3   Nucleotide sequence of the mouse PLP gene. The nucleotide sequence of the exons and their flanking regions are shown. Bold capital letters represent the nucleotide sequences in exons and small letters show the sequences of introns and 5'-flanking region. The four tandemly-repeated sequences are surrounded by boxes. The transcription initiation sites determined by primer extension analysis are shown by closed circles. The most frequently used initiation site mapped by the T4 DNA polymerase primer extension method (shown by an open circle) did not coincide with the result obtained by regular primer extension (+7). Numbers indicate the relative position from the most upstream transcription initiation site.

A putative slice site of DM20-mRNA is underlined in exon 3, and the "G" residue mutated in jimpy is enclosed in a circle.

Putative three polyadenylation signals of PLP-mRNA are indicated by double lines.

corresponding homologous box and the initiation site of the MBP gene. One of them was homologous to the repeated sequence found upstream of the PLP gene. However, the overall homology was quite small in the region we compared the DNA sequence (715 bp and 572 bp from upstream the MBP and PLP transcription initiation sites, respectively), although it is possible that the element determining the tissue specificity does not lie in this region.

SYNTHESIS OF PLP

Location of PLP and MBP-mRNA Within an Oligodendrocyte in Culture by In Situ RNA Hybridization

As mentioned above, PLP is synthesized exclusively in oligodendrocytes and its synthesis concomitantly increase when myelin is actively formed. Since recent development of culture techniques has enabled us to analyze the oligodendrocyte that expresses phenotypic characteristics of myelin lamella formation and myelin-specific proteins (11,19), expression of the PLP and MBP-mRNA within an oligodendrocyte was analyzed. When the oligodendrocyte is immature, both mRNAs are located

1: **GTGGGGACAAG**
2: **GGGAGGAGAAG**
3: **GGGAGGAGGAG**
4: **AGGAGGAGGGA**

Consensus: **(GGGAGGAG^A_GAG)**

DR2: **GGGAGGAGCGG**

Figure 4   Homology between the 11-base-pair repeate in the upstream region of the mouse PLP gene and the DR2 sequence of herpes simplex virus genome.

Nucleotide sequence of each repeated box (1 to 4) and the consensus sequence of the four are given together with the nucleotide sequence of the DR2 box.

throughout the oligodendrocyte cell bodies. But when the cells differentiate, PLP mRNA is located only in the cell body of the cultured oligodendrocyte, whereas MBP mRNA is present over the length of the slender processes as well as in the cell body (Fig. 6). Though PLP is present in all parts of the cell in primary culture as is the case with MBP, PLP-mRNA is located exclusively in the cell body (Shiota C, Miura M, and Mikoshiba K: submitted).

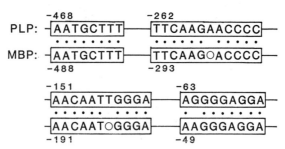

PLP: ┤AATGCTTT├────┤TTCAAGAACCCC├──
      • • • • • • • •   • • • • • •
MBP: ┤AATGCTTT├────┤TTCAAG○ACCCC├──

┤AACAATTGGGA├────┤AGGGGAGGA├──
 • • • • • •   • • • •   • • • • • • •
┤AACAAT○GGGA├────┤AAGGGAGGA├──

Figure 5   Nucleotide sequences homologous between the 5'-flanking regions of the PLP and MBP genes.

    The nucleotide sequences of the 5'-flanking regions of the PLP and MBP genes were aligned, and the homologous sequences were searched. An open circle indicates a deletion in the sequence. The numbers indicate the relative position from the transcription initiation site. In the case of PLP, the most upstream starting point was chosen as +1.

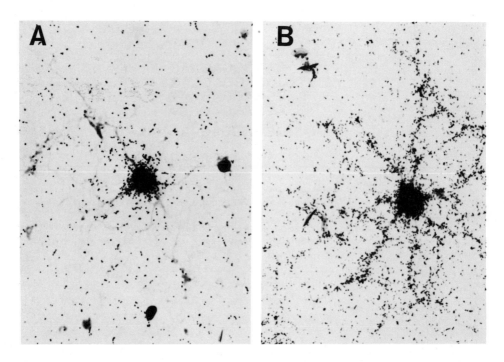

Figure 6   Localization of PLP and MBP-mRNAs in differentiated oligodendrocytes in culture. Autoradiograms after _in situ_ RNA hybridization with [^{34}S]-labelled PLP cDNA probe (A) or with MBP cDNA probe.

Electron microscopic immunohistochemical study demonstrated that the location of PLP is restricted to the rough endoplasmic reticulum, the Golgi apparatus and apparent Golgi vesicles (25). It is clear that PLP is synthesized in the rough endoplasmic reticulum in the perikaryon and transported to the myelin sheath. On the other hand, MBP-mRNA itself seems to be transported to the myelin sheath, where MBP is synthesized in free ribosomes (29, Shiota et al, submitted).

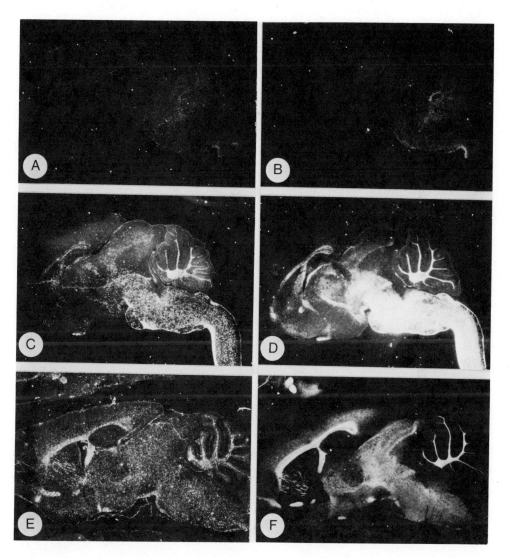

Figure 7   Developmental profile of PLP-mRNA (A-C) and MBP-mRNA (D-F) on the parasagittal section of the mouse brain detected by <u>in situ</u> hybridization technic. A and D: the 6th postnatal day. B and: the 12th postnatal day. C and F: the 30th postnatal day.

Several reports on quantitation of PLP or MBP-mRNA extracted from whole mouse brain using PLP and MBP cDNA probes have been published (3,6,8,16,24,27). However, blot analysis gives us the average value of the parts of the tissue examined. In situ hybridization has made it possible to visualize the gene expression even at the cellular level at any region of the brain. We observed great differences in the intensity of expression of the mRNAs from region to region and also in the timing of the appearance of the mRNAs (Fig. 7). Both PLP and MBP-mRNAs were detected first in the pons and the medulla oblongata, then in the cerebellum and later in the cerebrum. It is now clear that the myelin-specific gene expression proceeds generally in the caudo-cranial direction at the mRNA level (12,16,29).

After examination of the sections by in situ hybridization, we found that the pattern of the PLP gene expression reflects the number of labelled cells in the region examined. The number of actively myelinating cells which are positive for the mRNA deposition peaks at 12 to 18 days after birth in the cerebellum and the posterior region of the brain stem, and probably after 30 days in the anterior part of the cerebrum. On the 30th postnatal day, the number of PLP mRNA-positive cells is decreased in the pons, the medulla oblongata and the brain stem.

At the early stages of myelination of each oligodendrocyte, MBP-mRNA was detected in the cell body as was the case for PLP-mRNA. In these stages, therefore, it is possible to count the number of cells expressing MBP gene. Additionally, rather small number of labelled cells at the early developmental stages enabled us to compare the number of MBP-mRNA-positive cells on the 3rd and 6th postnatal days (compare Fig. 6 A and D). In agreement with the results of the in vitro study, these results indicate that expression of MBP gene precedes that of PLP gene in vivo.

EXPRESSION OF THE PLP GENE IN JIMPY MUTANT MOUSE

PLP Production is Severely Affected in the Jimpy Mutant Mouse

The X-linked recessive mutant mouse, jimpy, is characterized by abnormal myelin formation in the central, but not in the peripheral, nervous system and is considered to be an animal model for human Pelizaeus-Merzbacher disease (7). Many histological and biochemical studies have revealed a drastic decrease in all of the myelin components in the mutant brain (7). This is apparently caused by the severe degeneration of the oligodendrocyte which forms myelin in the CNS. The abnormality was shown in studies of chimeric mice (1) to be intrinsic to the oligodendrocyte itself. Recently, it has been shown by chromosomal mapping that the structural gene for PLP lies near the jimpy locus (3,31). This prompted us and other investigators to study the PLP by immunizing rabbits with a synthetic peptide (corresponding to residues 110-127) conjugated with bovine serum albumin, and PLP as well as MBP production in mice brains was studied immunohistochemically. As shown in Fig. 8 MBP production in jimpy brain was much less than that in wild type control, showing the extent of oligodendrocyte degeneration in jimpy. However, synthesis of PLP in jimpy seemed to be more severely affected and PLP was almost undetectable in the jimpy brain.

PLP-mRNA in Jimpy Brain

Cloning of the PLP cDNA enabled us to study PLP expression in jimpy

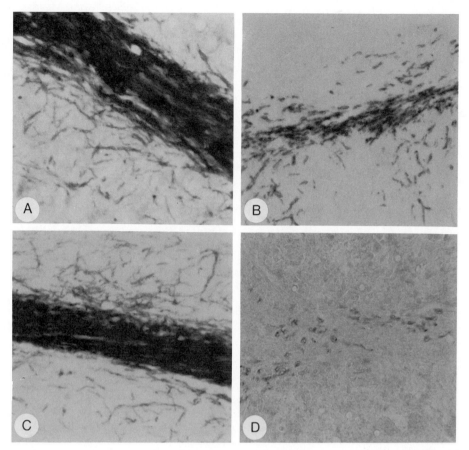

Figure 8   Immunohistochemical staining of normal and jimpy mice brains. Sagittal sections of cerebella from normal (A and C) and jimpy (B and D) mice were prepared and were stained immunohistochemically using antibodies against MBP (A and B) or PLP (C and D).

brain at the nucleotide level. PLP-mRNA was detectable in jimpy although its level was greatly reduced (3,6,8,17,18,21) and this contrasted with the observation at the protein level. No mutation was detected by Southern blot analysis, however, evidence indicating abnormality in the jimpy PLP-mRNA had been accumulated (6,17), and finally, 74 bp deletion in the jimpy PLP was detected by Nave et al (21). This deleted sequence was identical to that of the fifth exon of the PLP-mRNA, and thus the fifth exon of the PLP gene was not utilized in jimpy (18).

Nature of the Jimpy Mutation

     In order to understand the mechanism underlying the aberrant splicing in jimpy PLP-mRNA, cloning of the jimpy PLP gene was performed. An A to G conversion at the conserved "AG" residues of the 3'-splice site was found (15,23), which apparently resulted in the fifth exon deletion of the PLP-mRNA (Fig. 3). Protein encoded by this mRNA would have an altered C-terminal structure owing to the reading frame shift (21). The jimpy type PLP has not been detected yet, however, its mutated C-terminal is expected to be extraordinary rich in cysteine residues (8/36). When we synthesized an oligopeptide according to its sequence, we found that the peptide in aqueous solution soon formed aggregates and became insoluble, even in the presence of 1 mM dithiothreitol (unpublished observation). Since a characteristic feature of the _jimpy_ brain is the degeneration of oligodendrocytes, it must be determined whether this unusual PLP exerts toxic effect on the cells or the absence of the normal PLP is fatal.

PROSPECTS

PLP is specifically synthesized in the oligodendrocytes in the CNS and its regulation was shown to occur at the transcriptional level. At present elements determining the tissue restricted expression of PLP have not been characterized. Most cells in the nervous system do not divide after its final differentiation and this limits use of regular _in vitro_ transfection promoter assay. However, recent development in the techniques for producing transgenic mice should be helpful in determining the cis-acting elements for tissue-restricted expression of PLP. Transgenic mice synthesizing PLP in a tissue-restricted manner from the introduced gene, are also useful in analyzing the cause of the early death of the oligodendrocytes in _jimpy_. If the _jimpy_ phenotype was cured by the trans-gene, it can be assumed that PLP is indispensable for mature oligodendrocytes and its absence is fatal to the cells. In this case, we can have hope for developing gene therapy for human Pelizaeus-Merzbacher disease.

REFERENCES

1.   Beraducci A, Peterson AC, Aguayo AJ: Oligodendrocyte mosaicism in the CNS of _jimpy_ normal mouse chimaerass. Neurology 31:118, 1981
2.   Chou J, Roizman B: Isomerization of Herpes Simplex Virus 1 genome: Identification of the _cis_-acting and recombination sites within the domain of the _a_ sequence. Cell 41:803, 1985
3.   Dautigny A, Maattei M-G, Morello D, Alliel PM, Pham-Dinh D, Amar L, Arnoud D, Simon D, Mattei J-F, Guenet J-L, Jolles P, Avner P: The structural gene coding for myelin-associated proteolipid protein is mutated in _jimpy_ mice. Nature 321:867, 1986
4.   Diehl H-J, Schaich M, Budzinski R-M, Stoffel W: Individual exons encode the integral membrane domains of human myelin proteolipid protein. Proc Natl Acad Sci 83:9807, 1986
5.   Folch J, Lees M: Proteolipids: a new type of tissue lipoproteins. J Biol Chem 191:807, 1951
6.   Gardinier MV, Macklin WB, Diniak AJ, Deininger PL: Characterization of myelin proteolipid mRNAs in normal and jimpy mice. Mol Cell Biol 6:3755, 1986
7.   Hogan EL, Greenfield S: Animal models of genetic disorders of myelin. In Myelin (Morell P, ed.) Plenum Press, New York, pp 489-534, 1984
8.   Hudson LD, Berndt JA, Puckett C, Kozak CA, Lazzarini RA: Aberrant splicing of proteolipid protein mRNA in the dysmyelinating jimpy mutant mouse. Proc Natl Acad Sci 84:1454, 1987
9.   Ikenaka K, Furuichi T, Iwasaki Y, Moriguchi A, Okano H, Mikoshiba K: Myelin proteolipid protein gene structure and its regulation of expression in normal and _jimpy_ mutant mice. J Mol Biol 199:587, 1988
10.   Knapp PE, Skoff RP, Redstone DW: Oligodendroglial cell death in jimpy mice: An explanation for the myelin deficit. J Neurosci 6:2813, 1986

11. Knapp PE, Bartlett WP, Skoff RP: Cultured oligodendrocytes mimic _in vivo_ phenotypic characteristics: Cell shape, expression of myelin-specific antigens, and membrane production. Develop Biol 120:356, 1987

12. Kristensson K, Zeller NK, Dubois-Dalcq ME, Lazzarini RA: Expression of myelin basic protein gene in the developing rat brain as revealed by in situ hybridization. J Histochem Cytochem 34:467, 1986

13. Lees MB, Brostoff SW: Proteins of myelin. In Myelin (Morell P, ed.) Plenum Press, New York, pp 197-217, 1984

14. Macklin WB, Campagnoni CW, Deininger PL, Gardinier MV: Structure and expression of the mouse myelin proteolipid protein gene. J Neurosci Res 18:383, 1987

15. Macklin WB, Gardinier MV, King KD, Kampf K: An AG-GG transition at a splice site in the myelin proteolipid protein gene in jimpy mice results in the removal of an exon. FEBS Letters 223:417, 1987

16. Milner RJ, Lai C, Nave K-A, Lenoir D, Ogata J, Sutcliffe JG: Nucleotide sequences of two mRNAs for rat brain myelin proteolipid protein. Cell 42:931, 1985

17. Morello D, Pham-Dinh D, Jolles P: Myelin proteolipid protein (PLP and DM-20) transcripts are deleted in _jimpy_ mutant mice. EMBO J 5:3489, 1986

18. Moriguchi A, Ikenaka K, Furuichi T, Okano H, Iwasaki Y, Mikoshiba K: The fifth exon of the myelin proteolipid protein-coding gene is not utilized in the brain of _jimpy_ mutant mice. Gene 55:333, 1987

19. Nagaike K, Mikoshiba K, Tsukada Y: Dysmyelination of Shiverer mutant mice _in vivo_ and _in vitro_. J Neurochem 39:1235, 1982

20. Naismith AL, Hoffman-Chudzik E, Tsui L-C, Riordan JR: Study of the expression of myelin proteolipid protein (lipophilin) using a cloned complementary DNA. Nucleic Acids Res 13:7413, 1985

21. Nave K-A, Lai C, Bloom FE, Milner RJ: Jimpy mutant mouse: A 74-base deletion in the mRNA for myelin proteolipid protein and evidence for a primary defect in RNA splicing. Proc Natl Acad Sci 83:9264, 1986

22. Nave K-L, Lai C, Bloom FE, Milner RJ: Splice site selection in the proteolipid protein (PLP) gene transcript and primary structure of the DM-20 protein of central nervous system myelin. Proc Natl Acad Sci 84:5665, 1987

23. Nave K-L, Bloom FE, Milner RJ: A single nucleotide difference in the gene for myelin proteolipid protein defines the _jimpy_ mutation in mouse. J Neurochem 49:1873, 1987

24. Okano H, Miura M, Moriguchi A, Ikenaka K, Tsukada Y, Mikoshiba K: Inefficient transcription of the myelin basic protein gene possibly causes hypomyelination in _myelin deficient_ mutant mice. J Neurochem 48:470, 1987

25. Puckett C, Hudson L, Ono K, Friedrich V, Benecke J, Dubois-Dalcq M, Lazzarini RA: Myelin-specific proteolipid protein is expressed in myelinating Schwann cells but is not incorporated into myelin sheaths. J Neurosci Res 18:511, 1987

26. Schwob VS, Clark HB, Agrawal D, Agrawal HC: Electron microscopic immunocytochemical localization of myelin proteolipid protein and myelin basic protein to oligodendrocytes in rat brain during myelination. J Neurochem 45:559, 1985

27. Sorg BA, Smith MM, Campagnoni AT: Developmental expression of the myelin proteolipid protein and basic protein mRNAs in normal and dysmyelinating mutant mice. J Neurochem 49:1146, 1987

28. Stoffyn P, Folch J: On the type of linkage binding fatty acids present in brain white matter proteolipid apoprotein. Biochem Biophys Res Commun 44:157, 1971

29.   Trapp BD, Moench T, Pulley M, Barbosa E, Tennekoon, Griffin J: Spatial segregation of mRNA encoding myelin-specific proteins.   Proc Natl Acad Sci 84:7773, 1987

30.   Varmuza SL, Smiley JR: Signals for site-specific cleavage of HSV DNA: Maturation involves two separate cleavage events at sites distal to the recognition sequences.   Cell 41:793, 802, 1985

31.   Willard HF, Riordan JR: Assignment of the gene for myelin proteolipid protein to the X chromosome: implications for X-linked myelin disorders.   Science 230:940, 1985

7

THE IONIC CHANNELS OF GLIAL CELLS

C. Krieger and S.U. Kim

Division of Neurology, Department of Medicine
University of British Columbia
Vancouver, Canada

INTRODUCTION

Considerable recent experimentation has demonstrated the presence of numerous types of ionic channels in the cell membranes of all classes of glial cells. Some of these ionic channels are largely voltage-insensitive, others show prominent voltage-sensitivity and still other channels are activated by pharmacological agents. As a whole these channels appear to play prominent roles in the activities of glial cells, such as potassium ion regulation or possibly in the intercellular transfer of substances to axons. What has become increasingly clear is that the distribution of some of these channels in both axons and in glia is interdependent. Abnormal neuronal electrophysiological function occurs as a result of demyelination, and neurons themselves appear to influence the electrophysiological properties of glial cells. Furthermore, in demyelination and remyelination the electrical properties of neurons and glia are altered in ways that contribute extensively to the pathophysiology and clinical symptomatology of these processes. This brief review will attempt to summarize some of the information available concerning the ionic channels of glia.

General Description of Ionic Channels

With the advent of newer types of electrophysiological recording particularly patch clamp recording (51) it has been possible to dissect out the individual elements contributing to the action potential and other active properties of various cell types. It is now believed that this electrical behavior is due to the opening and closing of independent ion channels within the cell membrane (31). The channel itself is thought to be a transmembrane protein which is glycosylated on its external surface. Within the protein is an aqueous pore through which ions pass when they are permitted to do so by a voltage-sensitive gate. Some restriction on the nature of the ions entering the pore is made by a selectivity filter which permits entry to the pore on the basis of the charge and size of the ion. This filter would, for example, exclude large divalent cations like calcium from entering a sodium channel. Whether or not an ion will move through a channel and thereby produce a small current depends both on the type of ion and the driving force. The driving force is related both to the relative concentrations of the ions inside and outside the cell as well as the transmembrane potential. The large number of different types of ion channel permit a wide range of control on the excitability of cells (31).

Voltage-dependent Channels of Astrocytes

Until relatively recently it was believed that astrocytes were an inexcitable cell type and had only very modest numbers of channels ("conductances") (30,62). Surprisingly, both type 1 and type 2 astrocytes have a large number of voltage-sensitive channels including sodium and chloride channels as well as at least two types of calcium channel (in the case of type 2 astrocytes) and three types of channels for potassium ions (4,5,6,7,13,48,56,60,70). Although the currents generated by sodium channels differ in some ways from those in neurons (5) in most respects they are quite similar. For example, sodium current can be abolished by tetrodotoxin although neuronal sodium channels are more sensitive to this agent, and the shape of the voltage-dependence of activation of sodium currents are similar although not identical (5). As in neurons, type 2 astrocytes can be induced to fire action potentials by depolarizing stimuli (4). These stimuli, however, are about ten times larger than those required to produce action potentials in neurons. This difference in stimulus is probably accounted for by the much lower density of sodium channels in type 2 astrocytes and therefore a proportionately greater amount of glial membrane and correspondingly greater number of sodium channels must be activated in order to achieve a sufficient depolarization for regenerative spike activity.

Calcium channels have been demonstrated in type 2 astrocytes and like neurons several types of calcium channels have been observed (4). In particular, T (transient) and L (long-lasting) channels have ben isolated in different individual astrocytes. These two currents differ in the duration that the channel will stay open in response to a stimulus although both permit the entry of calcium. The behavior of these different calcium channels is however, very similar to their behavior in neurons (4). When potassium channels are blocked these calcium channels are capable of supporting calcium-dependent action potentials both spontaneously generated and following depolarizing current pulses (48).

At least three distinct types of potassium channel are present on the membranes of type 1 and type 2 astrocytes (4,5,56,60,70). These three correspond to the delayed rectifier current (I_K), as well as the transient outward current (I_A) and a calcium-dependent potassium current (I_{KCa}). Voltage-dependent chloride channels have also been observed by a number of workers (4,5,56).

One may ask why astrocytes would possess this wide array of ion channels and what could be the physiological requirement for spiking astrocytes. Shrager and colleagues (69) have suggested that astrocytes as well as Schwann cells may produce ionic channels which are eventually incorporated into the neuronal membrane at the nodes of Ranvier. This theory has some support in that the kinetics and pharmacological sensitivity of astrocytic and axonal sodium and potassium channels are similar, however at present, there is no direct evidence for this proposal (82).

The observation of regenerative activity in astrocytes could be physiologically irrelevant in itself and may reflect only the presence of calcium and sodium channels in a cell with a relatively high resistance. In order to observe calcium-dependent action potentials MacVicar (48) applied barium and tetraethylammonium ions in order to block potassium channels; a comparable block of outward potassium current may be unlikely to arise in the intact animal. Although not necessarily producing regenerative activity the entry of calcium ions may permit the outgrowth of astrocytic processes. The morphological differentiation of astrocytes by cyclic adenosine monophosphate (c-AMP) can be blocked by the use of cadmium and cobalt, both of which block calcium channels (49). MacVicar (49) has proposed that calcium entry may produce changes in the shape of the protein, actin, a calcium-sensitive contractile element in astrocytes. Actin contractility could then alter cell shape and produce the resulting morphological differentiation. Calcium ions have also been implicated in growth cone formation in neurons (1).

The observation of action potential production in astrocytes where potassium channels were not blocked by pharmacological agents is more puzzling (4). It must be remembered however that the stimulating currents used to produce action potentials were very high, much higher than those necessary to produce spiking in neurons so it may be questionable whether astrocyte action potentials are ever present in an intact animal. Although the possibility exists that astrocytes have some form of neurotransmitter release, synaptic vesicles have not been identified within these cells.

The most accepted hypothesis for the function of potassium channels in astrocytes as well as other glial cells is that these channels are involved in the transfer of extracellular potassium away from actively firing neurons to more distant regions (47,57,58,73,74). To elaborate on this function further it should be recalled that in both neurons and glial cells the resting membrane potential is determined by the unequal concentration of potassium ions between the inside and outside of the cell. In neurons, resting membrane potentials are not very sensitive to small changes in potassium ion concentration. In contrast, glial cells are very sensitive to changes in potassium concentration and will alter their membrane potential in a manner which closely follows the ratio of the transmembrane potassium concentration as given by the Nernst relation (57). This occurs because glial cells in general have a high permeability to potassium and a low permeability to other ions. In the Muller cell of the retina, a specialized astrocyte-like glial cell, the permeability for potassium ions is about 500 times greater than for sodium ions (54).

The astrocyte interior, as in most cells, has a high potassium concentration (on the order of 150 mM) compared to the cell exterior (the order of 5 mM), elevated extracellular potassium will decrease the transmembrane concentration difference and decrease the transmembrane voltage thus depolarizing the glial cell. Depolarization opens ionic channels, including potassium channels and consequently potassium ions will flow out of glial cells down their electrochemical gradient into the extracellular space. If extracellular potassium is raised, less potassium will flow out of the cell as the driving force for potassium movement out of the cell is less. Under the hypothetical condition where an astrocyte could be held at a membrane potential of 0 millivolts by applying positive charge into the cell, a rise in extracellular potassium to around 150 mM would not be accompanied by any movement of potassium ions even if the potassium channels were open. This is because there would not be any driving force, this force being generated under normal conditions both by the concentration difference in potassium ions on either side of the cell membrane and by the fact that the cell interior has a slight negative charge with respect to the cell exterior.

The question of how high a concentration of external potassium is achieved during extensive repetitive nerve activity is not precisely known however estimates have placed this value at between 10 and 20 mM

(57,74,75) in the dorsal horn of frog spinal cord (but see 80). This release of potassium from active neurons and its accumulation in the periaxonal regions serves as a trigger for glial cells to increase their metabolism in a number of ways (57). Some of these metabolic effects serve to buffer extracellular potassium concentration. General mechanisms of potassium buffering might include: 1) a diffusion of extracellular potassium through the extracellular space, 2) a passive re-uptake of potassium by neurons, 3) an active re-uptake of potassium by neurons and glia, 4) a passive uptake of potassium by glia accompanied by water and a counter ion to maintain electroneutrality, and 5) a 'spatial buffering' of potassium by glial cells (54). Although the relative importance of each of these mechanisms is unknown, considerable recent attention has been directed to the importance of spatial buffering as originally conceived by Orkand and colleagues (57).

In no part of the nervous system has spatial buffering been more well studied than in the retina (15,52,53,54,57,58). The cells of Muller are the major type of glial cells in the vertebrate retina and are similar to other types of astrocyte. The Muller cell lies across the full width of retina with its cell body positioned over the level of the inner nuclear layer and two thick radial processes extending toward the limiting membrane on the inner and outer sides of the retina. Thus, one of these processes, a so-called distal process is in intimate association with the inner segments of photoreceptors. As even small changes in the external potassium concentration can produce significant effects on the resting membrane potentials of photoreceptors and adjacent cells and consequently on their function it is important that potassium ions be buffered. The Muller cell appears to be central to this process. The elevated external potassium ion concentrations present around photoreceptors will also depolarize nearby Muller cells. As mentioned previously, this depolarization will open potassium channels and potassium ions will flow out of the cell down their concentration gradient. If these potassium channels were located in the photoreceptor region they would compound the problem of the elevated potassium concentration as they too would dump even more potassium into the oversupplied area. In fact, the majority of potassium channels are located some distance away at the end feet of the Muller cell at the opposite end of the cell (52,53). In response to depolarization by elevated potassium the Muller cell will release potassium to the extracellular space in a region quite distant from the source of elevated potassium. Effectively this will shunt or siphon potassium away from an area of high concentration, around the photoreceptor region in this case, to an area where the potassium ions will not alter important processes (54).

In an exciting extension of this line of reasoning Newman (53) has evidence that a similar regulatory mechanism may apply to pericapillary astrocytes in vertebrate brain. In the salamander brain the endfeet of some astrocytes are closely applied to the blood vessel endothelium. The end feet of these astrocytes appear to have many more potassium channels than elsewhere in these cells. As in the retina, astrocyte depolarization produced by elevated potassium concentrations will lead to an efflux of potassium from the astrocyte. As many of these channels are located in the perivascular endfeet, most of the potassium flux will be siphoned off into the bloodstream.

Voltage-dependent Channels of Schwann Cells

Like astrocytes, Schwann cells have a fairly wide range of voltage-dependent ion channels including sodium channels (17), potassium channels (17) and several non-selective channels (27). These latter channels include a calcium-dependent cation channel which is not selective between monovalent cations like sodium and potassium (27) and an anion channel which may permit the movement of cell metabolites through its pore (27).

It must be noted that there is considerable species variability in the density of voltage-dependent ion channels in Schwann cells. Rat Schwann cells have only a few sodium channels as inferred from a low number of saxitoxin binding sites (17,27,82). Rabbit Schwann cells have a much higher sodium channel density and have been extensively used for patch-clamp experiments (17,18,27). These Schwann cell sodium channels have substantially similar properties as do neuronal sodium channels (17,18,69). This gave rise to the speculation mentioned earlier for astrocytes that glial cells produce neuronal ion channels for assimilation into the axolemmal membrane (69). Chiu (17) has extended these studies on Schwann cell sodium channels by examining their ionic properties following a mild enzymatic digestion designed not to disrupt Schwann cell-neuron contact. Patch-clamp recordings made of such axon-associated Schwann cells showed that sodium current was only detected in those Schwann cells which were apposed to unmyelinated nerve fibers. No sodium current was observed in Schwann cells associated with myelinated axons (17). These results were interpreted as reflecting either an effect of axons on the membrane properties of the Schwann cell or a consequence of the extensive myelin layering which Schwann cells produce upon myelinated axons. Under the latter conditions the myelin spirals would electrically insulate the Schwann cell body and 'hide' the ionic channels present there (17). The alternate interpretation, however, also derives support from immunocytochemical studies which suggest that unmyelinated and myelinated Schwann cells differ in some of their surface and internal proteins (38).

More recently, Chiu (18) has studied the changes in the ion channel distribution in Schwann cells following nerve transection. In brief, the sodium and potassium currents of Schwann cells associated with myelinated nerve increase following transection; these currents are unaltered in Schwann cells associated with transected unmyelinated nerve. These results are consistent with both the above mentioned hypotheses; that is following transection myelinating Schwann cells could lose their myelin insulating layer and 'expose' their ionic currents. Alternatively, non-myelinating and myelinating Schwann cells could be sufficiently different in their properties so as to respond differently to nerve transection.

Potassium channels in Schwann cells appear to comprise a delayed rectifier potassium current (I_K) which is similar to its counterpart in axons both in its kinetics and in its sensitivity to 4-aminopyridine (17). As in neurons the main function of this channel may be to repolarize the cell following sodium-dependent depolarizations. Although calcium channels do not appear to have been observed in Schwann cells Chiu (17) has described a slowly activating inward current which appears with large depolarizations and shows little inactivation over 200-500 milliseconds. Although such currents could be due to calcium other ions could produce a similar response.

Voltage-dependent Channels of Oligodendrocytes

In contrast to astrocytes and Schwann cells, oligodendrocytes appear to possess only a modest number of ionic channels and show no evidence of having sodium or calcium channels. The predominant ion channel and the only potassium channel type observed in several voltage-clamp studies was an inward rectifier conductance (I_{IR}) (see 51) (4,39,44,). In a study of voltage-gated potassium currents in cultured ovine oligodendrocytes conductances are observed. An inward rectifier current, a transient outward current and a non-inactivating outward current (71). The inward rectifier current may result from two different types of inward rectifier channels differing only in the impedance of the channel (4). These inward rectifying potassium channels are open at the resting membrane potential of the oligodendrocyte and consequently will permit the internal buffering of potassium ions present in the extracellular fluid (3,24,26). As described in more detail elsewhere in this volume (51) this channel allows movement of potassium into the oligodendrocyte but restricts outward movement of this ion. This potassium buffering may function in conjunction with the spatial buffering mechanism described above for astrocytes. Like astrocytes, oligodendrocytes also have chloride channels (4).

In addition to passive internal buffering and spatial buffering, glial cells are believed to actively accumulate potassium, usually through an energy-dependent transport process such as the sodium-potassium pump or potassium-chloride co-transport (3,26). An additional mechanism has been hypothesized for potassium accumulation by oligodendrocytes and possibly other glial cells involving the passive movement of potassium and chloride ions through voltage-dependent channels (39). This passive uptake mechanism has its basis in the demonstration that the oligodendrocyte has both potassium as well as chloride channels (4,39). With oligodendrocyte depolarization both of these channel types will open. A net influx of chloride will occur due to the higher concentration of this ion in the extracellular space. Kettenmann (39) maintains that this chloride movement will be accompanied by net potassium movement into the cell to maintain the electroneutrality of ionic influx. As mentioned above the oligodendrocyte has numerous inward rectifier channels open at around rest potential to produce this potassium ion influx.

Evidence obtained by Kettenmann and colleagues (39,44) supports the above view. In the presence of raised extracellular potassium oligodendrocytes will accumulate potassium and chloride as demonstrated by ion sensitive electrodes and this effect is not blocked by agents which inhibit active re-uptake mechanisms (39,80). A comparable pathway is believed to exist in astrocytes and this pathway can be interrupted by potassium channel blockers (3,26). The oligodendrocytes are not as extensively coupled as are astrocytes (42) and do not possess the same anatomical relationships to blood vessels and other obvious sites for potassium ion disposal. It may be then that the spatial buffering ability of astrocytes is greater than oligodendrocytes (39).

The potassium channels of oligodendrocytes probably serve other important functions beside potassium buffering. Soliven and colleagues (71,72) have studied single channel currents in cultured oligodendrocytes using patch clamp recording methods. They have observed that both forskolin, a stimulator of adenylate cyclase and protein kinase C activators alter the amplitude and kinetics of several types of potassium channel. More specifically, forskolin appeared to decrease both the probability that two different types of potassium channel would open as well as reduce the time the channels remained open. These observations are of interest in that these same pharmacological agents are involved in the initiation of myelin metabolism and therefore the signal to myelinate may involve phosphorylation of ion channels (72).

Summary of Voltage-dependent Channels of Glial Cells

It should be evident from this brief review that all types of glial cells have a wide range of voltage-dependent ion channels within the cell membrane. In both astrocytes, oligodendrocytes and perhaps Schwann cells potassium and chloride channels are present to control the concentration

of extracellular potassium. This occurs both by spatial buffering of
potassium from regions of high extracellular potassium to regions of low
concentration and inward rectifier potassium channels which permit the
entry of potassium into the cell but restrict its exit (51). In
addition, active, as well as passive intracellular accumulation of
potassium occurs. The passive accumulation being aided by concurrent
chloride flux through voltage-dependent potassium and chloride channels.

Glial cells have other channel types and these may be responsible for
process outgrowth, such as the calcium channels of astrocytes. Ionic
current may be modified by myelination, such as with the disappearance of
Schwann cell sodium current when Schwann cells are associated with
myelinated axons. Lastly, currents may be modified by intracellular
signals, such as for myelination in oligodendrocytes and presumably other
important signalling functions are mediated by ion channels.

Receptor Mediated Channels of Glial Cells

As distinct from electrically excitable channels many ionic channels
are largely electrically inexcitable but can respond to neurotransmitter
substances. Their basic structure may be similar to voltage-dependent
channels but instead of a voltage-sensitive sensor they may have a
transmitter receptor coupled directly to a gate for ion movement. In
some cases it appears that a second messenger is involved to control the
ionic channel which may be remote from the receptor site (31).

These transmitter mediated ion channels have been well studied in
neurons (31,55) however it was believed until recently that glial cells
were insensitive to the action of neurotransmitters. Occasionally, when
glial cells were seen to depolarize following the application of these
substances it was thought that this depolarization was secondary to the
rise in extracellular potassium as a result of neuronal discharges
triggered by the neurotransmitter (32,73). This view has now been shown
to be incorrect and furthermore all types of glial cells have been
demonstrated to possess neurotransmitter activated channels. In addition
to having such channels, glia are well recognized to have re-uptake
mechanisms for many neurotransmitters released by neurons or during
metabolic activity (29). Numerous types of neurotransmitter receptors
have been identified on the surface of astrocytes and other glial cells
including receptors for adrenergic agents (29,36) histamine (35),
dopamine (33), serotonin (33,34), substance P (78), angiotensin II (61),
benzodiazepines (29), neuropeptides (29, as well as other agents
(12,13,25,29,40,41,43,76). In many cases these neurotransmitter
receptors are not linked to ion channels but presumably mediate their
effect through intracellular second messengers such as c-AMP, cyclic
guanosine monophosphate (c-GMP), or inositol phosphates.
Neurotransmitters such as acetylcholine, bradykinin and others have also
been shown to depolarize glioma cell lines (28,65).

Receptor Mediated Channels of Astrocytes

Astrocytes respond to a variety of neurotransmitter substances with
cell depolarization. These neurotransmitters include aspartate (40),
glutamate, gamma-amino-butyric acid (GABA) (40) and other agents
(33,34,35,36,43). In contrast, these cells are insensitive to taurine
and glycine (40). The GABA mediated depolarization of astrocytes is
produced by opening chloride channels, an effect virtually identical to
the neuronal $GABA_A$ receptor response. This chloride channel opening
permits the efflux of chloride ions from the astrocyte, thus depolarizing
the membrane. These astrocyte GABA receptors share other features with
neuronal GABA receptors such as their temperature sensitivity, their

ability to be blocked by the GABA receptor antagonists bicuculline and picrotoxin and their stimulation by muscimol (43). This suggests that these GABA receptors must be a similar if not identical molecule. The biological significance of these receptor mediated channels is obscure however. Since it is believed that glial cells do not produce neurotransmitters themselves these receptor mediated effects may exert some fine control over the spatial buffering properties or other functions of astrocytes.

Astrocytes also possess binding sites for serotonin on their surface as demonstrated radioautographically (34) and application of this agent produces a hyperpolarization in cultured astrocytes which can be abolished by the serotonin antagonist ketanserin (33). Dopamine will also hyperpolarize astrocytes, an effect which is abolished by the antagonists cis-flupenthixol and domperidone (33). Electrophysiological responses to adrenergic stimulation of astrocytes has also been observed (36).

Perhaps the most exciting recent development in the pharmacology of astrocytes in the observation that phorbol esters will induce oscillations of the resistance of these cells (50). Astrocytes in kainate treated hippocampal slices are activated by phorbol esters to produce a rhythmic oscillations of membrane potential of about 5-10 mV in height and 1 Hz in frequency. Inactive phorbol esters do not evoke such depolarizations and rarely untreated astrocytes also demonstrate such behavior. These changes in membrane potential may be produced by changes in the potassium permeability although the exact mechanism is unknown.

Receptor Mediated Channels of Schwann Cells

An extensive literature exists on the neurotransmitter sensitivity of Schwann cells, much contributed by Villegas and coworkers (79). These authors have studied the neuronal glial interactions between the nerve fiber and surrounding Schwann cells of the giant squid axon. In this species, Schwann cell membrane potential can be modified by changes in cyclic adenosine monophosphate (c-AMP) levels produced by nicotinic cholinergic receptors and octopamine receptors (79). Neuropeptide substances can also produce membrane potential changes but these appear to be independent of c-AMP. Some of the neuropeptides producing these de- and hyperpolarizing responses on Schwann cells include Vasoactive intestinal polypeptide (VIP), substance P and somatostatin (6). Little information is available however about the neurotransmitter sensitivity of mammalian Schwann cells. This probably derives from technical difficulties arising from the recording of such cells with micropipettes.

Receptor Mediated Channels of Oligodendrocytes

Cultured mouse oligodendrocytes have been demonstrated to respond to glutamate and GABA with depolarization (7). This does not occur with every oligodendrocyte and only about one-third will respond. Depolarization is also seen with the GABA agonist muscimol and GABA antagonists such as bicuculline will block this response. However, as for astrocytes the physiological significance of these depolarizations are unclear (7). Other transmitter substances such as aspartate, glycine, adrenaline, nor-adrenaline, acetylcholine, bradykinin or 5-hydroxytryptamine do not affect these cells (7).

Summary of Receptor Mediated Channels of Glial Cells

Glial cells have neurotransmitter receptors on their surface and can respond to the application of some of these agents by a change in potential. Both mammalian astrocytes and oligodendroglia can respond to GABA and glutamate and in addition astrocytes can also respond to aspartate and nor-adrenaline as well as other agents. Little information is available about the transmitter sensitivity of mammalian Schwann cells however squid Schwann cells respond to acetylcholine, octopamine as well as many neuropeptides. When detailed information is available about these ion channels, as is the case for the astrocyte GABA receptor channels, they appear to resemble neuronal GABA receptors. The function of these channels is enigmatic but it has been suggested that there could be regional variability in the transmitter sensitivity of glial cells depending on the area of the neuraxis from which they come (25).

Voltage-insensitive Ion Channels of Glial Cells

Mention should be made of a variety of voltage-insensitive ion channels which are responsible for solute movement by passive diffusion. These channels presumably lack the voltage or receptor-activated gate which controls the entrance or egress of ions across the transmembrane pore. Thus the channels permit flow of ions of the appropriate size or charge but without the specific control of the gating system of the channel. Although it might be imagined that glial cells would possess many such channels, in fact, it is not clear whether glial cells have appreciable numbers of these channels. Although their membrane impedances are quite low when studied in slice preparations or in the intact animal (50), when studied with patch clamp recording their resistance is significantly higher (71).

Ion flux in glial cells is maintained to a great degree by carrier mediated ion transport systems such as the sodium-potassium pump, potassium and chloride co-transport, sodium-hydrogen exchange add other mechanisms (39,80). These metabolic pumps are not mediated by ionic channels per se and the net flow of ions occurs through regions of membrane different from the previously mentioned channels. It should be emphasized however that these metabolic ion pumps are very important in many glial cell functions particularly those concerned with potassium concentration and cell volume regulation (57,58,74,80).

Gap Junctions

A notable feature of glial cells is the presence of cell coupling through gap junctions. These intercellular connections allow small water soluble molecules to pass directly between cells as well as electrical activity to spread to coupled cells. Thus sugars, amino acids, vitamins and other small molecules can traverse the junction and supply adjacent cells. Coupling has been observed in cultured oligodendrocytes and astrocytes from rats and mice (42) as well as from other species (42,57). Coupling can also occur between cell types, such as between oligodendrocytes, astrocytes and fibroblasts or between Schwann cells and dorsal root ganglion neurons (42). In many cases cell coupling has been observed between cells which were not immediately adjacent and sometimes could be as far as 300 microns from one another (42). These gap junctions probably serve to improve the spatial buffering of potassium concentration by glial cells. Additionally, they may serve as a form of cell-cell communication.

Influence of Glial Cells on the Ionic Properties of Neurons

Of critical importance in demyelination and remyelination is the interdependence of nerve and glial cell. Glial cells have prominent effects on the ionic properties of neurons. Conversely, neurons have been shown to exert some effects on the ion channels of glial cells.

The most striking effect of glial cells on neuronal ion channels is the channel redistribution which occurs following demyelination of nerve fiber axons. Early work on saltatory conduction in myelinated nerve fibers demonstrated that under normal conditions excitation is confined to the nodes of Ranvier (37,77). This conclusion has been supported by more recent evidence showing potassium channels to be located largely in the internodal region while sodium channels are present largely within the nodal membrane (2,9,10,19,20,21,45,46,64,66,67,81,82). This has been based on labelled saxitoxin binding studies (68), osmotic shock studies of the nodes of Ranvier under voltage clamp, anatomical and immunocytochemical evidence (16). Following demyelination secondary to diphtheria toxin or crush Bostock, Sears and colleagues have shown that the internodal membrane can become excitable and conduct impulses (8,9,10,11). Most ventral roots exposed to diphtheria toxin will show slowed saltatory conduction and in sites of inward current flow corresponding to the nodes of Ranvier the time required for current flow will be increased (8). This accounts for the slowed conduction velocities of these demyelinated fibers. In a few fibers, however, the normal step-wise pattern of impulse generation was lost and the axonal sodium current appeared to progressively increase in latency along the course off the fiber corresponding to a 'continuous' type of conduction like that seen in non-myelinated fiber (8,9,10,23).

Probably in most axons the density of sodium channels in the internodal axon is too small to permit the development of sufficient sodium current to cause continuous conduction to occur immediately following the loss of the myelin sheath (82). After several days have elapsed following a demyelinating lesion such as a crush injury significant numbers of sodium channels can appear in areas corresponding to the internodal regions and provide the basis for the observed continuous conduction. It is not clear, however, whether these internodal channels are produced de novo or are channels which have migrated into the internodal membrane (23). The regulation of the distribution of these sodium channels may be under the control of Schwann or other glial cells although at present there is no direct proof for this. In the mouse strain dy/dy an abnormal nerve fiber ion channel distribution is found concurrently with abnormal myelination, however this does not prove that glial cells directly regulate axonal ion channels (63).

This glial-neuronal interaction could behave in a manner which has been compared to the interaction between nerve and muscle at the neuromuscular junction (82). At the neuromuscular junction the presynaptic nerve fiber is associated with post-synaptic acetylcholine receptors within the junction itself. Following the loss of the presynaptic fiber, acetylcholine receptors are found outside the junction just as internodal sodium channels are found outside the node of Ranvier following demyelination.

It should be noted that the electrophysiological features of demyelination have been well studied including some of the clinical consequences. The interested reader is referred to a number of recent articles on this subject (11,14,45,46,59,60,63,67,81,82).

Influence of Neurons on the Ionic Properties of Glial Cells

There is only very limited information about the influence of neurons on the ionic channels of glia and this data is derived from studies of the effect of myelinated or unmyelinated axons on the ion currents in Schwann cells (17,18). As mentioned above it has been reported that sodium currents in Schwann cells are only detectable if they are associated with nonmyelinated nerve fibers. Additionally, when myelinated fibers are transected Schwann cells which were associated with those fibers will demonstrate potassium and sodium current. The time course of appearance of sodium and potassium current seemed to follow the extent of myelin loss. Schwann cells which were not associated with myelinated nerve showed some decline in the amplitude of their sodium and potassium dependent currents. These results were interpreted as implying that the regulation of sodium and potassium current is dependent on some interaction with degenerating myelin and axons. Axonal degeneration was believed to lead to a loss of Schwann cell sodium current whereas products of myelin degradation were suggested to produce the appearance of this current (18).

CONCLUSION

From this brief review it should be evident that glial cells are not the inexcitable, unresponsive cells which they were formerly believed to be. As a group they possess a wide variety of voltage-dependent and receptor activated ion channels. Undoubtedly the list of channels they contain will continue to grow as further research is done in this area. At this point in time it is perhaps more useful to think of the similarities between neurons and glia rather than their differences, coming as they do from common neuroepithelial precursor cells. Sadly, our knowledge of the functions of many of these channels is limited however as has been reviewed here it seems likely that at least some are related to the modulation of axon-glial relations which occur in demyelination and remyelination.

ACKNOWLEDGEMENTS

Supported by the Medical Research Council of Canada.

REFERENCES

1.   Anglister L, Farber IC, Shahar A, Grinvald A: Localization of voltage-dependent calcium channels along developing neurites: Their possible role in regulating neurite elongation. Dev Biol 94:351-365, 1982
2.   Baker M, Bostock H, Grafe P, Martius P: Function and distribution of three types of rectifying channel in rat spinal root myelinated axons. J Physiol 383:45-67, 1987
3.   Ballanyi K, Grafe P, ten Bruggencate G: Ion activities and potassium uptake mechanisms of glial cells in guinea-pig olfactory cortex slices. J Physiol 382:159-174, 1986
4.   Barres BA, Chun LLY, Corey DP: Ion channel expression by white matter glia: 1. Type 2 astrocytes and oligodendrocytes. Glia 1:10-30, 1988
5.   Bevan S, Chiu SY, Gray PTA, Ritchie JM: Voltage-gated ion channels in rat cultured astrocytes. In: Ion channels in neural membranes. Eds. JM Ritchie, RD Keynes L Bolis, A Liss, New York, pp 159-174, 1986
6.   Bevan S, Chiu SY, Gray PTA, Ritchie JM: The presence of voltage-gated sodium, potassium and chloride channels in rat cultured astrocytes. Proc R Soc London B 225:299-313, 1985
7.   Bevan S, Raff M: Voltage-dependent potassium currents in cultured astrocytes. Nature 315:229-232, 1985
8.   Bostock H, Sears TA: Continuous conduction in demyelinated

mammalian nerve fibers. Nature 263:786-787, 1976

9. Bostock H, Sears TA: The internodal axon membrane: Electrical excitability and continuous conduction in segmental demyelination. J Physiol 280:273-301, 1978

10. Bostock H, Sears TA, Sherrat RM: The spatial distribution of excitability and membrane current in normal and demyelinated mammalian nerve fibers. J Physiol 341:41-58,1983

11. Bostock H, Sherrat RM, Sears TA: Overcoming conduction failure in demyelinated nerve fibers by prolonging action potentials. Nature 274:385-387, 1978

12. Bowman CL, Kimelberg HK: Excitatory amino acids directly depolarize rat brain astrocytes in primary culture. Nature 311:656-659, 1984

13. Bowman CL, Kimelberg HK, Frangakis MV, Berwald-Netter Y, Edwards C: Astrocytes in primary culture have chemically activated sodium channels. J Neurosci 4:1527-1534, 1984

14. Bray GM, Rasminsky M, Aguayo AJ: Interactions between axons and their sheath cells. Ann Rev Neurosci 4:127-162, 1981

15. Brew H, Gray PTA, Mobbs P, Attwell D: Endfeet of retinal glial cells have higher densities of ion channels that mediate K^+ buffering. Nature 324:466-468, 1986

16. Catterall WA: The molecular basis of neuronal excitability. Science 223:653-661, 1984

17. Chiu SY: Sodium currents in axon-associated Schwann cells from adult rabbits. J Physiol 386:181-203, 1987

18. Chiu SY: Changes in excitable membrane properties in Schwann cells of adult rabbit sciatic nerves following nerve transection. J Physiol 396:173-188, 1988

19. Chiu SY, Ritchie JM: Potassium channels in nodal and internodal axon membrane of mammalian myelinated fibers. Nature 284:170-171, 1980

20. Chiu SY, Ritchie JM: Evidence for the presence of potassium channels in the paranodal region of acutely demyelinated mammalian single nerve fibers. J Physiol 313:415-437, 1981

21. Chiu SY, Ritchie JM: On the physiological role of internodal potassium channels and the security of conduction in myelinated nerve fibers. Proc R Soc London B 220:415-422, 1984

22. Evans PD, Reale V, Villegas J: Peptidergic modulation of the membrane potential of the Schwann cell of the squid giant nerve fiber. J Physiol 379:61-82, 1986

23. Feasby TE, Bostock H, Sears TA: Conduction in regenerating dorsal root fibers. J Neurol Sci 49:439-454, 1981

24. Gardiner-Medwin AR: A study of the mechanisms by which potassium moves through brain tissue in the rat. J Physiol 335:353-374, 1983

25. Gilbert P, Kettenmann H, Schachner M: Gamma-aminobutyric acid directly depolarizes cultured oligodendrocytes. J Neurosci 4: 561-569, 1984

26. Grafe P, Ballanyi K: Cellular mechanisms of potassium homeostasis in the mammalian nervous system. Can J Physiol Pharmacol 65:1038-1042, 1987

27. Gray PTA, Bevan S, Chiu SY, Shrager P, Ritchie JM: Ionic conductances in mammalian Schwann cells. In: Ion channels in neural membranes. Eds. JM Ritchie, RD Keynes, L Bolis, A Liss, New York, pp 145-157, 1986

28. Hamprecht B, Kemper W, Amano T: Electrical response of glioma cells to acetylcholine. Brain Res 101:129-135, 1976

29. Hertz L, Schousboe I, Hertz L, Schousboe A: Receptor expression in primary cultures of neurons or astrocytes. Prog Neuro-psychopharmacol Biol Psychiat 8:521-527, 1984

30. Hild W, Tasaki I: Morphological and physiological properties of neurons and glial cells in tissue culture. J Neurophysiol 25:277-304, 1962

31. Hille B: Ionic channels of excitable membrane. Sinauer, Sunderland, MA, 1984

32. Hosli L, Hosli E, Andres PF, Landolt H: Evidence that the depolarization of glial cells by inhibitory amino acids is caused by an efflux of K+ from neurons. Exp Brain Res 42:443-448, 1981

33. Hosli L, Hosli E, Baggi M, Bassetti C, Uhr M: Action of dopamine and serotonin on the membrane potential of cultured astrocytes. Exp Brain Res 65:482-485, 1987

34. Hosli E, Hosli L: Autoradiographic localization of binding sites for ^{3}H-serotonin and ^{3}H-ketaserin on neurons and astrocytes of cultured rat brain stem and spinal cord. Exp Brain Res 65:486-490, 1987

35. Hosli L, Hosli E, Schneider U, Wiget W: Evidence for the existence of histamine H_1- and H_2-receptors on astrocytes of cultured rat central nervous system. Neurosci Lett 48:287-291, 1984

36. Hosli L, Hosli E, Zehntner C, Lehmann R, Lutz TW: Evidence for the existence of alpha and beta-adrenoreceptors on cultured glial cells-an electrophysiological study. Neuroscience 7:2867-2872, 1982

37. Huxley AF, Stampfli: Evidence for saltatory conduction in peripheral myelinated nerve fibers. J Physiol 108:315-339, 1949

38. Jessen KR, Mirsky R: Nonmyelin-forming Schwann cells coexpress surface proteins and intermediate filaments not found in myelin-forming cells: a study of Ran-2, A5E3 antigen and glial fibrillary acidic protein. J Neurocytol 13:923-934, 1984

39. Kettenmann H: K+ and Cl- uptake by cultured oligodendrocytes. Can J Physiol Pharmacol 65:1033-1037, 1987

40. Kettenmann H, Backus KH, Schachner M: Aspartate, glutamate and aminobutyric acid depolarize cultured astrocytes. Neurosci Lett 52:25-29, 1984

41. Kettenmann H, Backus KH, Schachner M: Gamma-aminobutyric acid opens Cl- channels in cultured astrocytes. Brain Res 404:1-9, 1987

42. Kettenmann H, Orkand RK, Schachner M: Coupling among identified cells in mammalian nervous system cultures. J Neurosci 3:506-516, 1983

43. Kettenmann H, Schachner M: Pharmacological properties of GABA, glutamate and aspartate induced depolarizations in cultured astrocytes. J Neurosci 5:3295-3301, 1985

44. Kettenmann H, Sonnhof U, Schachner M: Exclusive potassium dependence of the membrane potential in cultured mouse oligodendrocytes. J Neurosci 3:500-505, 1983

45. Kocsis JD: Functional characteristics of potassium channels of normal and pathological mammalian axons. In: Ion channels in Neural Membranes. Eds. JM Ritchie, RD Keynes, L Bolis, A Liss, New York, pp 123-144, 1986

46. Kocsis JD, Waxman SG: Long-term regenerated fibers retain sensitivity to potassium channel blocking agents. Nature 304:640-642, 1983

47. Kuffler SW: Neuroglial cells: physiological properties and a potassium mediated effect of neuronal activity on the glial membrane potential. Proc R Soc London B 168:1-21, 1967

48. MacVicar BA: Voltage-dependent calcium channels in glial cells. Science 226:1345-1347, 1984

49. MacVicar BA: Morphological differentiation of cultured astrocytes is blocked by cadmium or cobalt. Brain Res 420:175-177, 1987

50. MacVicar BA, Crichton SA, Burnard DM, Tse FWY: Membrane conductance oscillations in astrocytes induced by phorbol ester. Nature 329:242-243, 1987

51. McLarnon JG, Kim SU: Physiological function of inward potassium currents in bovine oligodendrocytes in culture. This volume.

52. Newman EA: Regional specialization of retinal glial cell membrane. Nature 309:155-157, 1984

53. Newman EA: High potassium conductance in astrocyte endfeet. Science 233:453-454, 1986

54. Newman EA: Regulation of potassium levels by Muller cells in the vertebrate retina. Can J Physiol Pharmacol 65:1028-1034, 1987

55. Nicoll RA: The coupling of neurotransmitter receptors to ion channels in the brain. Science 241:545-551, 1988

56. Nowak L, Ascher P, Berwald-Netter Y: Ionic channels in mouse astrocytes in culture. J Neurosci 7:101-109, 1987

57. Orkand RK: Signalling between neuronal and glial cells. IN: Neuronal-glial cell inter-relationships, ED. TA Sears, Dahlem Konferenzen, Springer-Verlag, Berlin, pp 147-158, 1982

58. Orkand RK, Dietzel I, Coles JA: Light-induced changes in extracellular volume in the retina of the drone, Apis mellifera. Neurosci Lett 45:273-278, 1984

59. Pender MP, Sears TA: Conduction block in the peripheral nervous system in experimental allergic encephalomyelitis. Nature, 296:860-862, 1982

60. Quandt FN, MacVicar BA: Calcium activated potassium channels in cultured astrocytes. Neuroscience 19:29-41, 1986

61. Raizada MK, Phillips MI, Crews FT, Sumners C: Distinct angiotensin II receptors in primary cultures of glial cells from rat brain. Proc Natl Acad Sci USA 84:4655-4659, 1987

62. Ransom B, Neale R, Henkart M, Bullock P, Nelson PG: Mouse spinal cord in cell culture. I. Morphology and intrinsic neuronal electrophysiological properties. J Neurophysiol 40:1132-1150, 1977

63. Rasminsky M: Ectopic impulse generation in pathological nerve fibers. Trend Neurosci 6:388-390, 1983

64. Rasminsky M, Sears TA: Internodal conduction in undissected demyelinated nerve fibers. J Physiol 227:323-350, 1972

65. Reiser G, Hamprecht B: Bradykinin induces hyperpolarizations in rat glioma cells and in neuroblastoma X glioma hybrid cells. Brain Res 239:191-199, 1982

66. Ritchie JM: Sodium and potassium channels in regenerating and developing mammalian myelinated nerves. Proc R Soc London B 215:273-287, 1982

67. Ritchie JM: The distribution of sodium and potassium channels in mammalian myelinated nerve. In: Ion channels in Neural Membranes. Eds. JM Ritchie, RD Keynes, L Bolis, A Liss, New York, pp 105-122, 1986

68. Ritchie JM, Rogart RB: The density of sodium channels in mammalian myelinated nerve fibers and the nature of the axonal membrane under the myelin sheath. Proc Natl Acad Sci USA 74:211-215, 1977

69. Schrager P, Chiu SY, Ritchie JM: Voltage-dependent sodium and potassium channels in mammalian cultured Schwann cells. Proc Natl Acad Sci USA 82:948-952, 1985

70. Seager MJ, Deprez P, Martin-Moutot N, Couraud F: Detection and photoaffinity labelling of the Ca^{2+}-activated K+ channel-associated apamin receptor in cultured astrocytes from brain. Brain Res 411:226-230, 1987

71. Soliven B, Szuchet S, Arnason BGW, Nelson DJ: Voltage-gated potassium currents in cultured ovine oligodendrocytes. J Neurosci 8:2131-2141, 1988

72. Soliven B, Szuchet S, Nelson DJ: Protein Phosphorylation modulated single K channel kinetics in cultured oligodendrocytes. Neurology 38:(suppl 1) 232, 1988

73. Somjen GG: Extracellular potassium in the mammalian nervous system. Annu Rev Physiol 41:159-177, 1979

74. Sykova E: Extracellular K+ accumulation in the central nervous system. Prog Biophys Molec Biol 42:135-189, 1983

75. Sykova E: Modulation of spinal cord transmission by changes in extracellular K+ activity and extracellular volume. Can J Physiol Pharmacol 65:1058-1066, 1987

76. Tang CC-M, Orkand RK: Glutamate directly depolarizes glial cells. Neuroscience 7:208, 1982

77. Tasaki I: Physiology and electrochemistry of nerve fibers. Academic, New York, 1982

78. Torrens Y, Beaujouan JC, Saffroy M, Daguet de Montety C, Bergstrom L, Glowinski J: Substance P receptors in primary cultures of cortical astrocytes from the mouse. Proc Natl Acad Sci USA 83:9216-9220, 1986

79. Villegas J: Axon-Schwann cell relationship. In: Current Topics in Membranes and Transport, Vol. 22, The Squid axon. Ed P Baker, Academic Orlando Fl, pp 547-567, 1984

80. Walz W: Swelling and potassium uptake in cultured astrocytes. Can J Physiol Pharmacol 65:1051-1057, 1987

81. Waxman SG, Ritchie JM (eds): Demyelinating diseases, Basic and Clinical Electrophysiology. Raven Press, New York, 1981

82. Waxman SG, Ritchie JM: Organization of ion channels in the myelinated nerve fiber. Science 228:1502-1507, 1985

PHYSIOLOGICAL FUNCTION OF INWARD POTASSIUM CURRENTS IN BOVINE

OLIGODENDROCYTES IN CULTURE

James G. McLarnon[1] and Seung U. Kim[2]

Department of Pharmacology and Therapeutics[1]
and Division of Neurology[2]
University of British Columbia
Vancouver, Canada

INTRODUCTION

The elaboration of the electrophysiological properties of glial cells has recently proliferated through the use of the patch clamp technique. Measurements of macroscopic currents using the whole-cell mode and of single channel currents have shown that astrocytes and oligodendrocytes possess a variety of ion channels which selectively regulate ionic flow through the cellular membrane. At present however, physiological roles for these channels largely remain to be determined. Most of the studies undertaken to date have employed cultured cells from embryonic rat or mouse brain. The work to be described in this chapter considers oligodendrocytes obtained from adult bovine brains and studied for periods up to several months after the initial culture preparation.

The bulk isolation and culture preparation of bovine oligodendrocytes were performed following the methods previously described (1,2) and cover slips with essentially pure oligodendrocyte (95-98% purity) could be utilized for electrophysiological experiments. Culture medium was composed of 5% horse serum, 5 mg/ml glucose and 20 ug/ml gentamicin in Eagle's minimum essential medium. Immunofluorescence microscopic identification of oligodendrocytes was performed using monoclonal antibody directed against galactocerebroside, a cell type-specific marker for oligodendrocyte, and fluorescein-conjugated goat anti-mouse immunoglobulin (2,3). Galactocerebroside-positive oligodendrocytes constituted better than 95% of total cell population during the whole cell culture period for up to 6 months. The bovine oligodendrocytes are characteristically round in appearance with diameters close to 10 um and appear individually or in clusters; in most cells multipolar processes are clearly defined (Fig. 1).

In these experiments the cell-attached patch clamp configuration has been used to isolate small patches of cellular membrane and determine the kinetic properties of the single channels resident in the patch (4). Fortuitously, many of the patches appear to contain only a single active channel which is convenient for the analysis of the open and closed times for the channel. Furthermore, the amplitudes of the currents are reproducible from experiment to experiment which suggests that only one species of channel is present in the patches. An important consideration in the use of the cell-attached configuration is that intracellular factors which may regulate channel function are supposedly intact in the experiments; this point is relevant to the channel properties studied presently. One significant disadvantage to the use of the cell-attached method, however, is the general inability to use pharmacological maneuvers which could be useful in the characterization of ion channels.

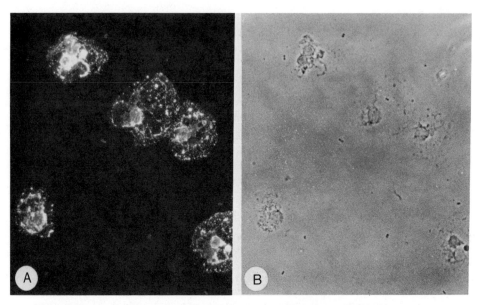

Figure 1   Galactocerebroside immunostaining of bovine oligodendrocytes in culture grown for 2 months.   A: Fluorescence microscopy.   B: Phase contrast microscopy.

In these experiments the bathing solution consisted of the following (in mM):   NaCl 140, KCl 5, CaCl$_2$ 0.2, and HEPES 10.   The pipettes were fire-polished and had resistances between 5-8 M   were filled with the following (in mM):   KCl 140, NaCl 5, CaCl$_2$ 0.2, and HEPES 5.   In all cases the solutions were filtered with 0.22 um Millipore filters. Following initial contact between the pipette and the cell, gentle suction was applied and resulted in the formation of giga-ohm seals in about one-third of the attempts; older cells in particular often resisted a tight-seal contact.   In the majority of experiments channel activation was observed after seal formation with the pipette potential held at zero (Fig. 2).   Increasing the driving force with the application of hyperpolarizing steps (i.e. increasing the magnitude of the pipette potential) increased the amplitude of the single channel currents as shown in Figure 2. Similar records to those shown in Fig. 2 have been obtained in other experiments.

A current-voltage plot for the channel in Fig. 2 is shown in Fig. 3; the slope of the graph is a measure of the conductance of the channel and was found to be 30 pS. The zero-current potential from the plot was -54 mV which is close to the expected resting potential in oligodendrocytes (5,6). This channel is primarily selective for K$^+$ since with similar concentrations of K$^+$ across the membrane the reversal potential would correspond to the resting level. Since the concentration of Cl is presumably low inside the cell a reversal potential for this ion would be hyperpolarizing to the resting level which is far from the

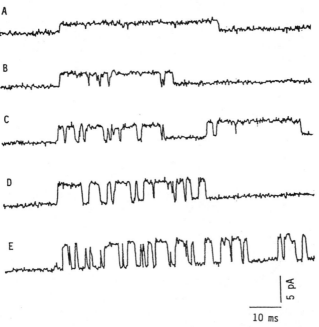

Figure 2   Records of inward current using the cell-attached configuration. Channel openings are indicated by upward deflections from the baseline. The pipette potential was held at the following values: (A) 0 mV; (B) 20 mV; (C) 40 mV; (D) 60 mV; (E) 80 mV. All data were filtered at 2 kHz.

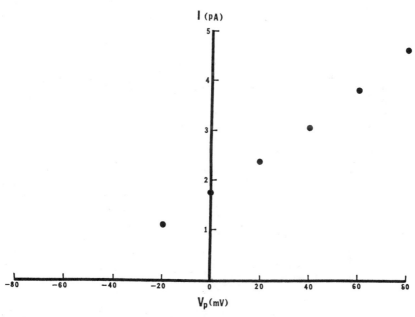

Figure 3   Current-voltage relation for the channel shown in Fig. 2. The channel conductance was 30 pS and the extrapolated reversal potential was -54 mV.

measured value. Several experiments were undertaken in which 70 mM KCl in the pipette was replaced with 70 mM NaCl so the pipette contained equal concentrations of KCl and NaCl; the measured reversal potentials were consistent with a K^+ selective pore. A contribution of inward Na^+ or Ca^{2+} to the observed currents is also unlikely from considerations of the expected and measured values for the zero-current potential; previous studies have found no evidence for such channels in oligodendrocytes (7,8).

An interesting aspect of the K^+ behavior is the lack of inactivation with time; sustained channel openings could be recorded for periods in excess of ten minutes. In contrast, outward current does not flow through this channel. When large negative pipette potentials were applied (i.e. Vp = -130 mV) no distinct channel activity was elicited. An inward rectifying K^+ channel has recently been described in cultured oligodendrocytes from optic nerves (9); the properties of this channel would seem to be qualitatively similar to the inward rectifier current in the bovine cells.

It is clear from the records shown in Fig. 2 that the open times of the K^+ channel are strongly modulated by the membrane potential. As the membrane is hyperpolarized to a greater extent the channel exhibits very rapid bursts of openings (flickers) from a mainly closed state. Most of the transitions were easily resolvable with low-pass filtering at 2 kHz. Histograms for the open and closed time distributions from the records of the cell shown previously are presented in Fig. 4 (pipette potential of +80 mV) and Fig. 5 (pipette potential of +40 mV). The open time histograms were well fit with a single exponential whereas the closed time distributions required two-component exponential fits The contributions to the two-components in the closed time distributions are clear from the records shown in Fig. 6 where a continuous record over 400 ms is shown with the pipette potential held at +80 mV. The longer component is associated with the prolonged periods between channel bursts and the shorter component is a consequence of the rapid closures evinced during the bursts. A number of channels exhibit two-component exponential fits for the closed time distributions which could reflect two closed states for the system with one of these states isomerizing to an open level.

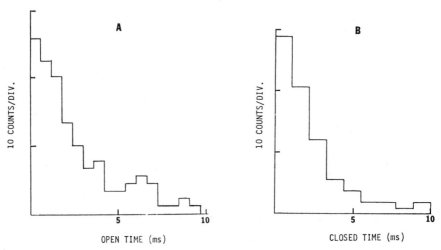

Figure 4   Open (A) and closed (B) time distributions for the channel shown in Fig. 2. The time constant from the open time distribution was 2.5 ms and the pipette potential was +80 mV. Approximately 200 events were included in the distributions.

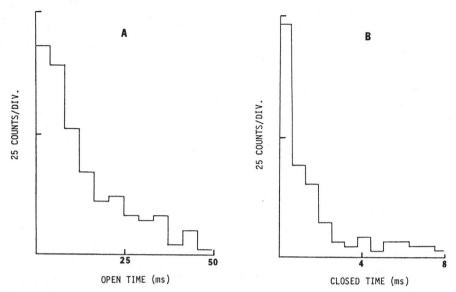

Figure 5   Open (A) and closed (B) time distributions with the pipette
potential held at +40 mV.   The time constant from the open time
distribution was 12.5 ms.

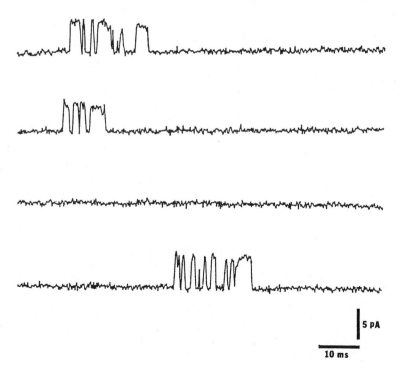

5 pA

10 ms

Figure 6   Contiguous traces of channel openings for a pipette voltage
of +80 mV showing the channel opening bursts and prolonged closed times.
The data were filtered at 2 kHz.

117

Withdrawal of the patch pipette from the cellular membrane results in an excised patch with the inner membrane surface in apposition to the bathing solution (inside-out patch clamp configuration). In these experiments this procedure would be expected to yield inward currents (with the pipette potential at zero) since a high concentration of K^+ is contained in the pipette. If the channel was selective for K^+ then a pipette potential of -85 mV should be close to the zero-current potential whereas in the cell-attached mode the comparable zero-current potential was approximately equal to the cell resting potential. In fact after the formation of a seal in the inside-out mode channel activity was absent in about one-half of the patches. Large values of positive and negative pipette potentials failed to elicit channel openings in these cases.

Several factors could account for the absence of channel openings observed after excision in some experiments. Intra-cellular components may modulate channel kinetics and the loss of such entities could cause channel inactivation. Many precedents for loss of channel activity after excision of the patch from the cellular membrane have been documented. For example Trube and Hescheler found that an inward rectifying K^+ channel in heart cell membrane was inactivated when the patch was withdrawn from the cell (10). In this case the intra-cellular factor was suggested to be ATP; restoration of channel openings was found after application of 4 mM ATP to the bathing medium in the inside-out patch clamp mode.

Other factors which could be involved are the concentrations of calcium and potassium. The calcium concentration inside the cell is buffered to a low value; after the inside-out mode has been formed however the inside of the cellular membrane is bathed in 0.2 mM calcium. A substantial change in K^+ concentrations is also evident for the two modes of patch clamp recording. With the cell-attached mode the internal K^+ is very high whereas after excision the inside of the cellular membrane is bathed in 5 mM K^+. Pharmacological methods can be used to determine if changes in the concentrations of these ions are involved in the altered channel activity. The calcium in the bath can be buffered, using EGTA, to values which are comparable to intra-cellular levels and the K^+ concentration in the bath can be increased to 140 mM. Preliminary experiments have not been successful in restoring channel activity using either of these pharmacological maneuvers. Another possibility is the formation of a membrane vesicle at the pipette tip when the pipette is withdrawn from the cell to form an inside-out configuration (11); previous studies on oligodendrocytes have noted that such vesicles were not uncommon (6). In some instances the vesicle can be removed by rapid movement of the patch through the solution-air interface. This procedure was not successful in restoring channel activity in these experiments even though the seal resistance was not materially affected by the procedure.

The properties of the K^+ channel which have been determined in this work are relevant to a physiological function of oligodendrocytes to take up K^+ which is released from active neurons (12,13). The channel passes inward K^+ current only, no outward current could be observed even for a large driving force for outward cationic current. The channel studied presently is available to pass inward current at the normal cell resting potential. The characteristics of the K^+ channel in the adult bovine oligodendrocytes are similar to an inward rectifier K^+ channel in neonatal rat oligodendrocytes as described by Barres et al (9). These authors also suggest that the large degree of inward rectification observed in their study is a consequence of magnesium binding to intra-cellular sites.

The dramatic modulation in the channel open times in the bovine oligodendrocytes has not been observed in other studies on rat oligodendrocytes. As the magnitude of the hyperpolarization was increased the channel open times decreased significantly to a point where only rapid flickers of openings were recorded. This property of the inward rectifier K^+ channel in the bovine cells may also be involved in the regulation of the concentration of K^+ across the cellular membrane. In adult bovine oligodendrocytes (and also in rat oligodendrocytes) K^+ channels can be detected which primarily pass outward current (outward K^+ current can be observed in these cells using the inside-out mode with high K^+ in the bath; large depolarizing steps are required for activation however). The requirement of large depolarizations to open channels which pass K^+ in an outward direction may also serve a regulatory role for oligodendrocytes. Further patch clamp studies will obviously be useful in the elucidation of the physiological function of the ion channels in oligodendrocytes.

ACKNOWLEDGEMENTS

This research was supported by grants from the Medical Research Council of Canada, Multiple Sclerosis Society of Canada, Jacob Cohen Fund for Research in Multiple Sclerosis, the Natural Sciences and Engineering Council of Canada, and a research scholar award to Dr. J. McLarnon from the British Columbia Health Care Research Foundation. The authors wish to thank Margaret Kim and Gordon Bird for technical assistance.

REFERENCES

1. Kim SU, Sato Y, Silberberg DH, Pleasure DE, Rorke LB: Long-term culture of human oligodendrocytes. J Neurol Sci 62:295-301, 1983
2. Kim SU: Antigen expression by glial cells grown in culture. J Neuroimmunol 8:225-282, 1985
3. Ranscht B, Clapshaw PA, Price J, Noble M, Siefert W: Development of oligodendrocytes and Schwann cells studied with a monoclonal antibody against galactocerebroside. Proc Natl Acad Sci 79:2709-2713, 1982
4. Hamill OP, Marty A, Neher E, Sakmann B, Sigworth F: Improved patch clamp techniques for high-resolution current recordings from cells and cell-free membrane patches. Pflugers Arch 391:85-100, 1981
5. Kettenman H, Sonnhof U, Schachner M: Exclusive potassium dependence of the membrane potential in cultured mouse oligodendrocytes. J Neurosci 3:500-505, 1983
6. Kettenman H, Orkand R, Lux H: Some properties of single potassium channels in oligodendrocytes. Pflugers Arch 400:215-221, 1984
7. Bevan S, Raff M: Voltage dependent potassium currents in cultured astrocytes. Nature 315:229-232, 1985
8. Kettenman H, Orkand R, Lux H, Schachner M: Single potassium currents in cultured mouse oligodendrocytes. Neurosci Lett 32:41-46, 1982
9. Barres B, Chun L, Corey D: Ion channel expression by white matter glia: type 2 astrocytes and oligodendrocytes. Glia 1:10-30, 1988
10. Trube G, Hescheler J: Inward-rectifying channels in isolated patches of the heart cell membrane: ATP-dependence and comparison with cell-attached patches. Pflugers Arch 401:178-184, 1984
11. Horn R, Patlak J: Single channel recordings from excised patches of muscle membrane. Proc Natl Acad Sci 77:6930-6934, 1980

12. Orkand R, Nicholls J, Kuffler S: Effect of nerve impulses on the membrane potential of glial cells in the central nervous system of amphibia. J Neurophys 29:788-806, 1966

13. Gardner-Medwin A: A study of the mechanisms by which potassium moves through brain tissue in rat. J Physiol 335:353-374, 1983

9

MOLECULAR DETERMINANTS IN AUTOIMMUNITY

Robert S. Fujinami

Department of Pathology
University of California
La Jolla, CA

INTRODUCTION

Many theories have been proposed to explain the pathogenesis of multiple sclerosis. From epidemiologic studies the occurrence of disease can be associated with genetic and environmental factors (reviewed by 1). Some of the environmental factors include infectious agents (37). Thus far, many agents including a whole host of viruses have been implicated, however, none have been shown definitively to be the causative agent (2-4, 12, 19, 21, 25, 29, 31, 41, 44). In addition, immunologic factors may play an important role in the pathogenesis (30). Along this premise some investigators have been using immunosuppressive regimes to modify the course of the disease (reviewed by 10). These studies have suggested modest improvement in some of the patients. Several models have been put forth to delineate the role of virus and immune responses in reproducing the observed pathologic features of multiple sclerosis.

One of these is experimental allergic encephalomyelitis (EAE) (32). This is an autoimmune disease of the central nervous system (CNS). Several CNS proteins have been determined to be encephalitogenic. It is when these proteins (derived usually from myelin) are injected into a suitable animal host with adjuvant an immune response ensues that cross-reacts with brain components. This disease is characterized by a perivascular mononuclear infiltrate that leads to demyelination. The disease is immunopathologically mediated. Many of the features of EAE can be transferred from animal to animal with the use of sensitized T lymphocytes. As yet the disease cannot be passively transferred with antibodies from EAE animals or animals immunized with myelin components. The relapsing and remitting form of chronic EAE has a lesion distribution and clinical course that can closely approximate multiple sclerosis (43).

Another model for multiple sclerosis is Theiler's murine encephalomyelitis virus (TMEV) infection of mice. Here, there are both genetic and virologic contributions to the demyelinating disease. This model was first championed by Lipton and colleagues in the 1970's as a system to study virus induced demyelination (22-24). They described a biphasic disease in Swiss outbred and later in SJL mice using the Daniels (DA) strain of TMEV. The early disease was an acute phase that resembled poliomyelitis where infection of neurons was prevalent and these cells went on to die. In the early phase mice became paralyzed and depending on the dose of the virus and strain of mouse would go on to die or would become persistently infected. In the SJL mouse strain the mice developed more of the chronic disease with less of the acute disease. This chronic late disease was characterized by a perivascular mononuclear cellular infiltrate and demyelination (8). Chronically infected mice have a spastic gait and difficulty righting themselves.

Various genetic components influence the production of clinical disease and demyelination. Both the susceptibility to disease and extent of demyelination have been correlated to H-2 and non H-2 regions in the mouse genome. In mice with the H-2^S alleles susceptibility correlated with the D region but not the K or I-A (5-7, 33). In addition, non H-2^S contributions of a region encoding for the constant portion of the ß-chain of the T cell receptor on mouse chromosome 6 can also contribute to disease production (27). Virus titers in the CNS from resistant and susceptible mouse strains correlating to the extent demyelination have been controversial. There is mounting evidence that delayed type hypersensitivity reactions which are class II restricted are involved in the demyelinating disease (Miller, 1978). Treatment of infected mice with anti-IA and L3T4 sera has been reported to influence the pattern of disease (13, 33, 42).

We have been combining the two aspects of viruses and autoimmune disease; i.e., exploring ways viruses could induce immune reactivities to "self". One hypothesis is that viruses in some way could trigger or initiate autoimmune attack on myelin components leading to demyelination. Along this line there is strong suggestive evidence that infections may precede relapses (37). Our initial studies were to define if viruses could have cross-reacting determinants with host cells. In raising monoclonal antibodies to measles virus it was noted that the monoclonal antibodies from the various fusions could be divided into three groups. First, and the largest group, contained monoclonal antibodies which were viral specific. These monoclonal antibodies reacted only with viral proteins by immunofluorescent staining, immunoprecipitation, or Western blot analysis. The second group of monoclonal antibodies was found to react with just self components and not viral proteins. These antibodies probably arose due to self reactive B cells present in the spleens of these mice. The last group of monoclonal antibodies which contained the smallest number of monoclonal antibodies, reacted with both viral proteins and host cell determinants. These monoclonal antibodies comprised approximately 1-3% of our total antibodies depending on the fusion.

One of these monoclonal antibodies was further investigated to characterize its specificity (16). It was first analyzed by immunofluorescent staining. HeLa cells infected with measles virus and mock infected cells were subjected to examination. The monoclonal antibody derived from mice immunized with measles virus was incubated with measles virus infected HeLa cells and immunofluorescent antibody procedure conducted. The pattern of staining was one of a cytoplasmic reaction. The monoclonal antibody bound to prominent cytoplasmic viral inclusions. This type of staining pattern was one typical of the distribution of measles virus phosphoprotein or nucleocapsid protein. When the monoclonal antibody was incubated with mock infected cells the staining pattern was one of a network-like appearance. In cells that were undergoing mitosis the pattern was markedly different, the uninfected cells had a speckled appearance. This type of staining was very characteristic of intermediate filament proteins, particularly vimentin or cytokeratin. Similar patterns of reactivity were found when this monoclonal antibody was tested in BHK and mouse L929 cells. Knowing this, a preparation enriched for intermediate filament proteins was prepared from uninfected HeLa cells and this preparation was used to adsorb with the monoclonal antibody preparation. The adsorption procedure removed the reactivity to infected cells. Thus, an intermediate filament preparation contained a common determinant with a viral protein.

To further define the viral and host proteins the monoclonal antibody reacted with biochemical methods were employed. Western blotting experiments were initially performed. First, an intermediate filament rich protein preparation from uninfected HeLa cells was prepared and the proteins separated by SDS-PAGE. The proteins were then transferred to nitrocellulose strips and strips incubated with the monoclonal antibody. The monoclonal antibody was found to react with a 52-54,000 molecular weight protein. With the use of other monoclonal antibodies to cytokeratin and vimentin, the cellular protein was identified as one of the cytokeratin proteins. Similarly, cytosols from infected HeLa cells were prepared in which the intermediate filaments proteins were depleted and these preparations were electrophoresed on SDS-polyacrylamide gels and proteins transferred to nitrocellulose paper. The monoclonal antibody which reacted with cytokeratin from uninfected cells was also found to bind to a 70,000 molecular weight viral protein, that co-migrated with measles virus phosphoprotein. Next, purified measles virus was electrophoreses and Western blot analysis performed. Again the monoclonal antibody reacted with the measles virus phosphoprotein that was incorporated into virions and not just present in infected cells. Therefore, a monoclonal antibody had the ability to define an epitope on a viral protein as well as a host cell component.

In like manner, additional monoclonal antibodies were studied: one against a herpes virus protein and another to vaccinia virus hemagglutinin (9, 16). The monoclonal antibody that reacted with a herpes virus protein was found to also bind with an intermediate filament protein. The vaccinia virus hemagglutinin monoclonal antibody bound to vimentin. Thus, additional viruses were shown to have common determinants with self proteins.

Further, Sairenji et al described a murine monoclonal antibody that recognized a filamentous structure in Epstein-Barr virus-producing lymphoblastoid cell lines (35). By immunofluorescent staining, the monoclonal antibody appeared to react with vimentin or a closely associated intermediate filament protein. The expression of this antigen was induced by superinfection with Epstein-Barr virus or treatment with tumor promoting agents, and its appearance may be similar to the induction of stress proteins (36). Along this line Sheshberadaran and Norrby described monoclonal antibodies against measles virus fusion protein that also reacted with cellular stress proteins. This was demonstrated by immunoprecipitation and immunofluorescent staining of infected and uninfected cells. This host stress protein was induced by infection of cells with paramyxoviruses, heat shock of uninfected HeLa cells, and treatment of various cell lines with 2-deoxyglucose, tunicamycin, L-canavanine. Other monoclonal antibodies have been identified to have similar reactivities in many viral systems. Recently, Srinivasappa et al reported that roughly 3-4% of all antiviral monoclonal antibodies reacted with host cells components (39).

As described, many of the monoclonal antibodies cross-react with intracellular determinants or filaments. This probably reflects the fact that viruses are intracellular parasites and assemble at very discrete sites within the infected cell (11). By having common sites or determinants with cellular proteins these viral proteins could be transported to similar areas as the intermediate filament proteins. Many viruses assemble and mature in association with intermediate filament proteins.

In producing monoclonal antibodies to another paramyxovirus, Goswami et al found that an antibody against the Simian virus 5HN glycoprotein bound to an antigen found in Purkinje cells of the adult rat brain (18). The monoclonal antibody had the ability to prevent infectivity. In addition, this monoclonal antibody could bind to white matter.

Tardieu et al have found a common determinant between reovirus types 1 and 3 and lymphocytes. The monoclonal antibody reacted with the Lyt 2,3 subset of murine lymphocytes (40). In addition, this monoclonal antibody had the ability to initiate complement dependent lysis of Lyt 2,3 positive lymphocytes (40).

From these data investigators have found cross-reacting determinants between viral determinants and host CNS and immune tissues. Using antibodies to define common determinants is very useful however it is very difficult to determine the actual epitope; i.e., the sequence involved in the cross-reaction. Thus, experiments defining common amino acids were instigated. These are described below.

A protein capable of inducing an autoimmune disease of the CNS was chosen since the encephalitogenic disease inducing determinants for a wide variety of species is known (20, 26, 38). This is myelin basic protein. It has been sequenced and has been widely studied. Using computer assisted analysis known viral protein sequences were compared to the encephalitogenic sites described for myelin basic protein (14). The original analysis revealed several sequence similarities between various animal viruses proteins and myelin basic protein. One of the best common sequence in tandem was between the myelin basic protein encephalogenic site for the rabbit and hepatitis virus B polymerase. This was:

        myelin basic protein T T H Y G S L P Q K
        hepatitis virus   I G C Y G S L P Q E

These peptides were synthesized. Studies looking for the production of autoantibody, cellular reactivity and disease production were undertaken.

Seven rabbits were immunized with the hepatitis virus peptide (HVP) and antibody monitored for the presence of antibody to HVP and myelin basic protein. Five of the seven animals made detectable antibody as measured by ELISA to whole myelin basic protein. Competitive inhibition experiments with increasing amounts of HVP blocked the binding of HBP antibodies to myelin basic protein in a dose dependent manner. As expected all seven rabbits made antibodies that reacted with HVP.

To test for cellular reactivity in rabbits, eight animals were immunized once with HVP and peripheral blood mononuclear cells were obtained. These mononuclear cells were then cultured in the presence of HVP or myelin basic protein. The peripheral blood mononuclear cells from all eight rabbits proliferated when cultured with HVP peptide. Peripheral blood mononuclear cells from half of the rabbits proliferated in the presence of myelin basic protein. Thus, 4/8 animals immunized with HVP reacted with myelin basic protein.

Histologic evaluation was performed in 11 rabbits immunized with HVP. Brain and cervical spinal cords of four animals had scattered lesions that consisted of perivascular mononuclear and meningeal infiltrates characteristic of experimental allergic encephalomyelitis. None of the rabbits immunized with the HVP developed clinical signs of experimental allergic encephalomyelitis. Similarly, one out of four animals injected with the encephalogenic peptide from myelin basic protein developed clinical signs and three of the four had histologic lesions of perivascular infiltrates in brain and spinal cord, thus, viral infection has the potential to trigger the production of autoantibodies and mononuclear cells that cross-react with self by a mechanism termed "molecular mimicry". The tissue injury from the viral initiated autoallergic event could take place in the absence of infectious virus.

Recently a monoclonal antibody against TMEV has been described to react with galactocerebroside (17). This virus has the ability to cause a chronic demyelinating disease in mice. The cross-reacting monoclonal antibody neutralizes the virus and binds to oligodendrocytes in newborn mouse CNS cultures. The presence of such an antibody could contribute to the observed demyelinating pattern of disease.

Similarly, a peptide (L G R P N E D S S S S S S S C) from the immediate-early region of human cytomegalovirus was analyzed by computer (17). It was found that the first five amino acids of this peptide had sequence similarity to the ß-chain of the human MHC HLA-DR protein. The common amino acids were located in a region that was conserved between the human and mouse histocompatibility antigens on the ß chain. The shared regions from the immediate-early region of human cytomegalovirus and HLA-DR had similar hydrophobicity and predicted ß-turn potential. This data suggested that the determinant would be on the surface of the protein. The IE-2 viral peptide induced antibodies that recognized the human DR ß-chain by western blot analysis. This suggests a mechanism to explain how human cytomegalovirus infection contributes to graft rejection and immunosuppression.

It is easy to speculate that in those instances where autoimmunity occurs, viral determinants reflect host cell determinants which have the capacity to induce disease, such as the encephalitogenic epitopes of myelin basic protein. Actual disease induction would not occur if the common site did not involve a disease inducing region. Immune responses involving non-disease inducing determinants may invoke autoantibody but actual autoimmune disease would not result. Another scenario pertains to those autoimmune afflictions that are of a chronic or relapsing and remitting nature. Here, viruses with the ability to persist may continually or cyclically express viral antigens. Even though the expression of the viral genome and therefore replication may be restricted, translation of the protein having the determinant in common with that of the host could continue. The resulting antigen, properly presented, may then evoke a smoldering immune response leading to chronic and progressive autoimmune disease.

ACKNOWLEDGEMENT

This research was supported by Public Health Service grant NS23162 from the National Institutes of Health and by National Multiple Sclerosis Society grant NMSS RG 1780A. The author would like to thank Diana Ferris for excellent manuscript preparation and Peggy Farness and Susan McClananhan for technical support.

REFERENCES

1.   Batchelar JR, Compston A, McDonald WI: The significance of the association between HLA and multiple sclerosis. Brit Med Bull 34:279, 1978

2. Burks JS, De Vald BL, Jankovsky LD, Geroes JC: Two coronaviruses isolated from central nervous system tissue of two multiple sclerosis patients. Science 209:933, 1980

3. Bychkova EN: Viruses isolated from patients with encephalomyelitis and multiple sclerosis. Communication I: Pathogenic and antigenic properties, Vopr. Virusol 9:173, 1964

4. Carp RI, Licursi PC, Merz PA: Multiple sclerosis associated agent. Lancet 2:814, 1977

5. Clatch RJ, Lipton HL, Miller SD: Characterization of Theiler's murine encephalomyelitis virus (TMEV)-specific delayed-type hypersensitivity response in TMEV-induced demyelinating disease: Correlation with clinical sign. J Immunol 136:920, 1986

6. Clatch RJ, Melvold RW, Dal Canto MC, Miller SD, Lipton HL: The Theiler's murine encephalomyelitis virus (TMEV) model for multiple sclerosis shows a strong influence of the murine Equivalents of HLA-A,B and D,J. Neuroimmunol 15:121, 1987

7. Clatch RJ, Melvold RW, Miller SD, Lipton HL: Theiler's murine encephalomyelitis virus (TMEV)-induced demyelinating disease in mice is influenced by the H-2D region: Correlating with TMEV-specific delayed-type hypersensitivity. J Immunol 135:1408, 1985

8. Dal Canto MC, Lipton HL: Primary demyelination in Theiler's virus infection. Lab Invest 33:626, 1975

9. Dales S, Fujinami RS, Oldstone MBA: Infection with vaccinia favors the selection of hybridomas synthesizing autoantibodies against intermediate filaments, one of them cross-reacting with the virus hemagglutinin. J Immunol 131:1546, 1983

10. Delmotte P, Hommes OR, Gonsette R, eds: Immunosuppressive treatment in multiple sclerosis. European Press Gen, 1977

11. Dorsett B, Cronin W, Chuma J, Ioachim HL: Anti-lymphocyte antibodies in patients with the acquired immune deficiency syndrome. Amer J Med 78:621, 1985

12. Field EJ, Cowshall S, Narang HK, Bell TM: Viruses in multiple sclerosis? Lancet 2:280, 1972

13. Friedmann A, Frankel G, Lorch Y, Steinman L: Monoclonal anti-I-A antibody reverses chronic paralysis and demyelination in Theiler's virus-infected mice: Critical importance of timing of treatment. J Virol 61:898, 1987

14. Fujinami RS, Oldstone MBA: Amino acid homology between the encephalitogenic site of myelin basic protein and virus: Mechanism for autoimmunity. Science 230:1043, 1985

15. Fujinami RS, Nelson JA, Walker L, Oldstone MBA: Sequence homology and immunologic cross-reactivity of human cytomegalovirus with HLA-DR chain: A means for graft rejection and immunosuppression. J Virol 62:100, 1988

16. Fujinami RS, Oldstone MBA, Wroblewska Z, Frankel ME, Koprowski H: Molecular mimicry in virus infection: Cross-reaction of measles virus phosphoprotein or of herpes simplex virus protein with human intermediate filaments. Proc Nat Acad Sci 80:2346, 1983

17. Fujinami RS, Zurbriggen A, Powell HC: Monoclonal antibody defines determinant between Theiler's virus and lipid-like structures. J Neuroimmunol. Submitted

18. Goswami KKA, Morris RJ, Rastogi SC, Lange LS, Russell WC: A neutralizing monoclonal antibody against a paramyxovirus reacts with a brain antigen. J Neuroimmunol 9:99, 1985

19. Gudnadottir M, Helgadottir H, Bjarnason O, Jonsdottir K: Virus isolated from the brain of a patient with multiple sclerosis. Exp Neurol 9:85, 1964

20. Hashim GA, Schilling FJ: Allergic encephalomyelitis: Characterization of the determinants for delayed type hypersensitivity. Biochem Biophys Res Comm 50:589, 1973

21. Koprowski H, DeFreitas EC, Harper ME, Sandberg-Wollheim M, Sheremata WA, Robert-Guzoff M, Saxinger CW, Feinberg MB, Wong-Staal F, Gallo RC: Multiple sclerosis and human T-cell lymphotropic retroviruses. Nature 318:155, 1985

22. Lipton HL, Dal Canto MC: Contrasting effects of immunosuppression on Theiler's virus infection in mice. Infect Immunity 15:903, 1977

23. Lipton HL, Dal Canto MC: Theiler's virus induced demyelination: prevention by immunosuppression. Science 192:62, 1976

24. Lipton HL: Theiler's virus infection in mice: an unusal biphasic disease proves leading to demyelination. Infect and immunity 11:1147, 1975

25. Margulis MS, Solojiew VD, Schublandze AK: Etiology and pathogenesis of acute sporadic disseminated encephalomyelitis and multiple sclerosis. J Neurol Neurosurg Psych 9:63, 1946

26. Martenson RE: The location of regions in guinea pig and bovine myelin basic proteins which induce experimental allergic encephalomyelitis in Lewis rats. J Immunol 114:592, 1975

27. Melvold ZW, Jokinen DM, Knobler RL, Lipton HL: Variations in genetic control of susceptibility to Theiler's murine encephalomyelitis virus (TMEV)-induced demyelinating disease I differences between susceptible SJL/J and resistant balbic strains map near the T cell ß-chain constant gene on chromosome 6. J Immunol 138:1429, 1987

28. Miller SD, Clatch RJ, Pewear DC, Troter JL, Lipton HL: Class V--restricted T cell responses in Theiler's murine encephalomyelitis virus (TMEV)-induced demyelinating disease I cross-specificity among TMEV substrains and related picornaviruses, but not myelin proteins. J Immunol 138:3776, 1987

29. Mitchell DN, Porterfield JS, Micheletti R, Lange LS, Goswami KKA, Tayler P, Jacobs J, Hockley DJ, Salsbury AJ: Isolation of an infectious agent from bone-marrows of patients with multiple sclerosis. Lancet 2:387, 1978

30. Oldstone MBA: Immunological aspects of diseases of the central nervous system. Lachmann PJ, Peters DK (eds). Clinical Aspects of Immunology. Blackwell, Oxford, 1982, pp. 1069-1103

31. Palsson PA, Pattison IH, Field EJ: Transmission experiments with multiple sclerosis. Gajdusek DC, Gibbs CJ Jr., Alpers M (eds). Slow, Latent and Temperate Virus Infections, U.S. Govern. Print Off, Wash. DC, 1965, pp. 49-59

32. Paterson PY: Neuroimmunologic diseases of animals and humans. Rev Inf Dis 1:468, 1979

33. Rodriguez M, LaFuse WP, Leibowitz J, David CS: Partial suppression of Theiler's virus-induced demyelination in vivo by administration of monoclonal antibodies to immune-response gene products (Io antigens). Neurology 36:964, 1986

34. Rodriguez M, Leibowitz J, David CS: Susceptibility to Theiler's virus-induced demyelination-mapping of the gene within the H-2D region. J Exp Med 163:620, 1986

35. Sairenji T, Nguyen QV, Woda B, Humphreys RE: Immune response to intermediate filament-associated Epstein-Barr virus-induced early antigen. J Immunol. In press

36. Sheshberadaran H, Norrby E: Three monoclonal antibodies against measles virus F protein cross-react with cellular stress proteins. J Virol 52:995, 1984

37. Sibley WA, Foley JM: Infection and immunization in multiple sclerosis. NY Acad Sci 122:457, 1965

38. Spitler LE: Experimental allergic encephalitis dissociation of cellular immunity to brain protein and disease production. J Exp Med 136:156, 1972

39. Srinivasappa J, Saegusa J, Prabhakar BS, Gentry MK, Buchmeier MJ, Wiktor TJ, Koprowski H, Oldstone MBA, Notkins AL: Molecular mimicry: Frequency of reactivity of monoclonal antiviral antibodies with normal tissues. J Virol 57:397, 1986

40. Tardieu M, Powers ML, Hafler DA, Hauser SL, Weiner HL: Autoimmunity following viral infection: Demonstration of monoclonal antibodies against normal tissue following infection of mice with reovirus and demonstration of shared antigenicity between virus and lymphocytes. Eur J Immunol 14:561, 1984

41. ter Meulen V, Koprowski H, Iwasake Y, Kackell YM, Muller D: Fusion of clustered multiple sclerosis brain cells with indicator cells: Presence of nucleolapsids and virions and isolation of parainfluenza type virus. Lancet 2:1, 1972

42. Welsh CJR, Tonks P, Nash AA, Blakemore WF: The effect of L3T4 T cell depletion on the pathogenesis of Theiler's murine encephalomyelitis virus infection in CBA mice. J Gen Virol 68:1659, 1987

43. Wisniewski HM, Keith AB: Chronic relapsing experimental allergic encephalomyelitis--an experimental model of multiple sclerosis. Ann Neurol 1:144, 1977

44. Wrobleska Z, Gilden D, Devlin M, Huang E-S, Rorke LB, Hamada T, Furukawa T, Cummins L, Kalter S, Koprowski H: Cytomegalovirus isolation from a chimpanzee with acute demyelinating disease after inoculation of multiple sclerosis brain cells. Infect and Immunity 25:1008, 1979

10

VIRUSES AND DEMYELINATION

Richard A. Shubin and Leslie P. Weiner

Department of Neurology
University of Southern California
Los Angeles, CA

INTRODUCTION

Multiple sclerosis (MS) has been recognized as a distinct clinical entity for over 100 years, but its etiology remains elusive. In all likelihood, a viral infection during childhood or adolescence triggers an autoimmune response to oligodendrocytes and/or myelin in susceptible individuals (1). Patients with MS are now being treated with cyclophosphamide (2,3), cyclosporine (4), azathioprine (5), whole body radiation (6,7), or plasmapheresis (8), on the assumption that MS is an autoimmune disease. These therapies have potentially serious hematologic, gastrointestinal, infectious, or neoplastic side effects (9). Immunosuppressive therapy, even if effective in stabilizing multiple sclerosis, is less than ideal because of the above mentioned side effects. Establishment of the etiology of MS may allow for more earlier, more specific, and less toxic treatment. In this paper we will review the epidemiologic evidence for a viral etiology of MS, the current state of candidate viruses, viral associated human demyelinating diseases other than MS, and the animal models of viral-induced demyelination.

EPIDEMIOLOGY OF MULTIPLE SCLEROSIS

The epidemiology of MS has been extensively studied in the hope of establishing its etiology. MS begins primarily, although not exclusively, in young adulthood. Risk rises steeply from early adolescence, reaches a peak about age 30, and declines to near zero by age 60 (10). This same unimodal age-specific onset curve is present in various ethnic groups (11) and in areas of high and low prevalence (12). Women are 1.4 times more likely to develop MS than men (13). Ethnic groups demonstrate a range of susceptibilities from the Anglo-Saxons, who are susceptible, to the Japanese, who are resistant. Multiple sclerosis is more common in temperate rather than tropical regions (30 to 80 per 100,000 versus 4 to 6 per 100,000) (12). This observation holds for both the Old and New Worlds as well as the northern and southern hemispheres (14). The risk of developing MS within one ethnic group varies by a factor of 10 to 20 based on latitude (12). There is an age-specific component to the effect of factors encountered at various latitudes. Migration from a high to low incidence region prior to age 15 reduces an individual's risk of contracting MS (15,16). Migration from a low to high incidence region prior to age 15 increases an individual's risk of contracting MS (12).

The risk of developing MS is in part due to the inheritance of specific histocompatibility (HLA) antigens. In Caucasians, the presence of the HLA A3 or B7 antigens increases an individual's risk of developing MS by two to three fold (17,18), while the presence of DR2 increases risk by four to five fold (19,20). An examination of the HLA of family members of patients with MS did not detect a single HLA haplotype which differed among affected and non-affected individuals (21). HLA provides a marker of susceptibility in Caucasians. There is, however, no consistent relationship between MS and HLA in other races (12).

A current hypothesis is that MS arises as a consequence of an abnormal immune response to a virus which occurs at a critical age in a susceptible individual. Rubella, measles and typhoid vaccinations occurred at a later age in MS patients compared to controls (21). Individuals who are at higher risk to develop MS because of their MHC haplotype were more likely to have measles or mumps at a later age than controls (24). MS may arise as a result of a susceptible individual contracting a common childhood infection at a point when there is a regulatory abnormality of immune system, which permits the development of autoimmunity against myelin and/or oligodendrocytes.

The increased incidence of MS in family members compared to the general population provides further evidence for the etiologic role of an environmental factor. MS is 6 to 8 times more frequent in siblings and 2 to 4 times more common in parents than unrelated controls (12). Monozygotic twins demonstrate a 50 percent concordance. The clinical signs of MS frequently develop in the same year, rather than at the same age, in siblings (23). This suggests MS may develop following a common triggering agent in susceptible individuals. Siblings discordant for MS have been shown to have fewer and less severe viral illnesses as children (24). Taken together, the increased rate of MS in family members suggest a common exposure to an environmental pathogen (25).

Further evidence for an environmental factor in the etiology of MS is provided by Kurtzke's studies of the epidemiology of Ms in the Faroe Islands. Prior to World War II, Faroe Islanders were not afflicted with MS. In contrast, MS is common in the British, who are of the same genetic stock and live at the same latitude (26). During World War II, the Faroe Islands were occupied by British soldiers. Subsequently, MS was diagnosed in native Faroe Islanders. A detailed examination of individuals who contracted MS revealed they had closer contact to the British forces than those who did not. Since World War II, MS has become endemic in native Faroe Islanders. MS appears to occur in a small percentage of individuals six to twelve years after the exposure to a presumably infectious agent (27).

VIRAL-INDUCED HUMAN DEMYELINATING DISEASES

A. Multiple Sclerosis

Intensive efforts have been made to isolate a virus from the brains of patients with MS. While a variety of viruses have been isolated, including rabies, herpes simplex, scrapie, parainfluenza I, measles, chimpanzee cytomegalovirus, simian viruses I and V, coronaviruses, MS-associated (Carp) agent, and the bone marrow (Mitchell) agent, all probably represent contaminants or adventitious, rather than causal agents (28). A variety of viruses have been identified in the brain of

MS patients by other techniques. Both measles (29) and herpes simplex type 1 (30) were found to _in situ_ hybridization. Coronavirus-like particles were detected by electron microscopy (31). It remains unclear if the presence of these viruses is causal or coincidental in the etiology of MS. Attempts to produce MS in primates by intracerebral injection of brain tissue from patients with MS have proven unsuccessful (32).

Antibodies to a variety of viruses have been found in the serum and cerebrospinal fluid (CSF) of patients with MS. Adams and Imagawa (32) found elevated levels of measles antibodies in the serum of MS patients compared to controls. Most subsequent studies have confirmed this observation (33). Increased levels of measles antibodies, however, are not found in every patient, and th absolute titer of measles antibodies is low (12). Antibodies against a variety of other viruses have been found in the cerebrospinal fluid (CSF) of patients with MS, but no virus has been detected universally (34,35). There is no consistent relationship between viral antibodies and the presence of oligoclonal bands in the CSF (36). The significance of viral antibodies in the serum or CSF of MS patients has recently bee reinterpreted. Elevated antibody titers to measles virus envelope, hemolysin, and hemagglutinin, antigens, Epstein-Barr virus capsid and nuclear antigen, and rubella hemagglutinin antigen were found in serum samples of patients with MS and rheumatoid arthritis compared to age and sex matched controls (37). The presence of elevated viral antibody titers may be a marker of abnormal immune regulation rather than being indicative of a specific etiologic agent.

The human lymphotropic virus type I (HTLV-I) was recently implicated as the etiologic agent of MS after antibodies to HTLV-I were identified in the CSF of MS patients from Sweden, and Key West, Florida (38). HTLV-I nucleotide sequences were also found in cells from the CSF by _in situ_ hybridization under low stringency conditions from these patients (38). A second group reported detecting antibodies to HTLV-I proteins in one quarter of Japanese patients with MS (39). In spite of these promising early reports, HTLV-I does not appear to play an etiologic role in MS. In subsequent studies, HTLV-I was not detected by enzyme-linked immunosorbent assay (ELISA) or _in situ_ hybridization techniques nor directly isolated from cultured lymphocytes, peripheral blood monocytes or brain tissues of patients with MS (40-43). Antibodies to HTLV-I, II, or III (human immuno-deficiency virus or HIV) do not occur more commonly in patients with MS compared to those with optic neuritis or other neurologic diseases (45). Finally, while HTLV-I is found in Japanese patients with MS, it was not statistically more common among patients with MS compared to those with other neurologic diseases and normal controls (44). At present, the weight of evidence is against HTLV-I being the "MS agent".

B. Tropical Spastic Paraparesis

HTLV-I was recently identified as the etiologic agent of tropical spastic paraparesis (TSP). HTLV-I associated TSP produces a slowly progressive, symmetrical, predominantly upper motor neuron disorder, characterized clinically by spastic paraparesis. It is associated with minimal sensory or autonomic dysfunction (49). Japanese (46) and Caribbean (50) patients with TSP have elevated serum antibodies to HTLV-I compared to controls. High levels of antibodies to HTLV-I are present in the CSF of patients with TSP (47) and these antibodies are synthesized intrathecally (48). Pathologic examination of the spinal cord reveals intense chronic meningomyelitis with demyelination; patchy collections of lymphocytes, plasma cells and macrophages are distributed in both grey and white matter. Demyelination is present predominantly in the pyramidal and dorsal medial columns. In chronic cases, spongiform changes develop in the white matter (51,52). TSP is diagnosed in the appropriate clinical setting by presence of antibodies against HTLV-I. Computerized tomography, magnetic resonance imaging, and/or myelogram, are normal. It has not been determined if demyelination is due to the direct effect of HTLV-I on oligodendrocytes and/or myelin, or if it is immune-mediated. Tentative evidence for the later mechanism is provided by the observation that some patients with TSP improved during immunosuppressive treatment with prednisone and subsequently deteriorated when prednisone was withdrawn (46). The identification of HTLV-I as the agent of TSP represents a major breakthrough and significantly enlarges the domain of human viral-induced demyelinating diseases.

C. Acquired Immunodeficiency Syndrome (AIDS)

While HIV (HTLV-III) has not been shown to be MS agent, it has been found to produce a variety of neurologic conditions, including vacuolar myelopathy, subacute encephalopathy, aseptic meningitis, sensory polyneuropathy and dysimmune motor polyneuropathy. CNS demyelination is a major feature of the first two syndromes. Vacuolar myelopathy is characterized clinically by paraparesis, ataxia and incontinence (53,54). Pathologic examination reveals demyelination, predominantly in the lateral and posterior columns of the thoracic spinal cord. Vacuolar myelopathy is found in up to 20 percent of patients with the acquired immunodeficiency syndrome (AIDS). Demyelination appears to result from interfering with the normal metabolism of oligodendrocytes (55). The subacute encephalopathy of AIDS is characterized clinically by impaired memory and concentration with psychomotor slowing (55a). The course is progressive and may be accompanied by motor or behavioral changes. On pathologic examination abnormalities are present in the white matter and in subcortical structures. They consist of white matter pallor, microglial nodules, and infiltrations of lymphocytes, macrophages and multinucleated giant cells (56,56a). The earliest pathologic feature of the subacute encephalopathy of AIDS is the white matter pallor and vacuolation (56). HIV has demonstrated in monocytes and multinucleated cells in the regions of demyelination, but not, so far, in oligodendrocytes (56). The mechanism of demyelination in subacute encephalopathy is, as yet, unknown. Subacute encephalopathy is a significant source of morbidity and mortality in AIDS. Establishment of the mechanism of demyelination is an important goal in the effort to design more effective therapies of AIDS.

D. Progressive Multifocal Leukoencephalopathy

Progressive multifocal leukoencephalopathy (PML) is a demyelinating disease due to human papovaviruses, the JC virus (JCV) (56,57) and the SV-40-like agent (58). PML occurs primarily in individuals with diseases which impair the immune system, such as leukemia, lymphoma, or AIDS (54,59-61). PML is seen in approximately 2% of patients with AIDS (62). If the present exponential increase in AIDS cases continues, PML will become as prevalent as Huntington's disease or myasthenia gravis by 1991 (62).

The clinical sign and symptoms of PML relate to the multifocal nature of the disease. PML is usually progressive and unrelenting, leading to death within six months of onset in immuno-incompetent individuals. PML may run a more protracted course in immuno-competent patients and, on rare occasion, may spontaneously resolve (63). Infection precedes without producing a fever or a pleocytosis in the CSF. PML can be diagnosed in the proper clinical setting with a characteristic appearance on CT scan: Multiple hypodense, nonenhancing lesions present in the white matter. These lesions do not respect vascular borders nor been demonstrated to be effective in the treatment of PML.

Pathologic examination of the brain of patients with PML reveals multifocal regions of demyelination, which become confluent as the disease progresses. Basophilic, enlarged oligodendrocytes, and bizarre, enlarged astrocytes with irregularly lobulated, hyperchromatic nuclei, are seen by light microscopy in conjunction with the demyelination (65). Large numbers of papovavirus particles, single or in crystalline arrays, can be visualized in the nuclei of oligodendrocytes by electron microscopy (66). Virus particles are not present in astrocytes or neurons. JCV nucleotide sequences are found in oligodendrocytes, occasionally in astrocytes, but not in vascular endothelial cells as detected by in situ hybridization techniques (67). PML probably arises as the result of reactivation of JCV in immunologically compromised patients (68). JCV virus is acquired subclinically during childhood (65). JCV virus can be recovered from spleen and bone marrow cells as well as mononuclear cells in the CSF. PML may occur as a result of JCV entering the perivascular space of the brain from tissues in which it has been dormant. Clinical signs develop when oligodendrocytes are infected (69).

The molecular basis of JCV-induced demyelination has recently been elucidated through application of the powerful techniques of modern molecular biology. Early attempts at identifying thee mechanism of demyelination were hampered by the restricted host range of JCV infection of oligodendrocytes (70). This barrier was overcome by creating transgenic mice containing the JCV early region (71). Transgenic mice which inherit the JCV early region develop "shaking", one phenotype seen in mice with defects in myelin synthesis. The JCV early region codes for the T-antigen. Expression of the T-antigen correlates with the severity of "shaking". The presence of the T-antigen in oligodendrocytes results in a decrease in the transcription, compared to the translation, of the major structural proteins of CNS myelin (72). The T-antigen shares a C-terminal subsequence with myelin basic protein (MBP). This sequence functions as a phosphate acceptor site in the latter. The T-antigen sequence appears to competitively inhibit the protein kinase phosphorylation of the Pro-Arg-Thr-Pro-Pro sequence of MBP (74). This blocks the production of myelin and arrests the maturation of oligodendrocytes. The T-antigen has been detected in the nuclei of oligodendrocytes of patients with PML by use of the immunoperoxidase staining technique (73). The T-antigen has not been demonstrated in oligodendrocytes of patients with MS (75). In conjunction with the rise in the number of cases of AIDS, PML promises to become an increasingly important clinical disease.

ANIMAL MODELS OF VIRAL-INDUCED DEMYELINATION

A variety of viruses which cause demyelination in animals have been studied as models of MS. These models have provided many insights into mechanisms of viral-induced demyelination. Martin and Nathanson (76) observed that these systems share the following characteristics: One, the diseases are biphasic with a stage of acute encephalitis followed by a stage of chronic demyelination. Two, virus persists in the white matter. Three, the lesions are multifocal, and are located primarily in the spinal cord. Recently, some of the models have been modified so that they more closely resemble MS.

A. JHM Strain of Mouse Hepatitis Virus

JHM virus (JHMV), the neurotropic strain of mouse hepatitis virus (MHV), is a coronavirus which produces an acute, diffuse encephalomyelitis with patchy demyelination in mice and rats (77,78,88). Lesions develop in the white matter five to seven days after intracerebral inoculation. Inflammation and necrotic lesions are present in gray and white matter. The degree of demyelination is dependent on the age and strain of the animal dose of virus and route of infection (79-82). Demyelination is due to the lytic effect of JHMV on oligodendrocytes (79). Demyelination occurs in conjunction with the presence of JHMV as demonstrated by fluorescent microscopy or immunoperoxidase techniques. JHMV can be visualized in oligodendrocytes by electron microscopy. There is no temporal or anatomic association with the occurrence of demyelination and thee presence of inflammatory cells; demyelination occurs even in the absence of perivascular inflammation aft treatment with cyclophosphamide (79). Animals which survive the acute encephalitis remain persistently infected and develop subclinical chronic recurrent demyelination (83,84). The study of viral-induced demyelination has been facilitated by the identification of temperature sensitive and antibody selected mutants of JHMV which cause chronic demyelination with minimal encephalitis (85-87) and a clinically relapsing disease in association with the recurrence of demyelination (89).

The immune system may play a role in the development of demyelination following JHMV infection of rats. Demyelination can be transferred from infected to naive rats by adoptive transfer of the lymphocytes, following in vitro stimulation with myelin basic protein (90). JHMV may cause demyelination by altering the regulation of cell mediated immunity in the brain. JHMV induces class II proteins on astrocytes (91), cells which do not ordinarily express class II (92). This may result in oligodendrocytes and/or myelin proteins becoming targets of the immune system, resulting in demyelination. JHMV remains a useful model for studying mechanisms of virus-induced demyelination.

B. Canine Distemper Virus

Canine distemper virus (CDV) is a paamyxovirus related to measles which produces either acute or chronic demyelinating disease in dogs, based on the strain of the virus (93-95). Demyelination is a result of a lytic infection of oligodendrocytes; myelin breakdown occurs in the absence of inflammatory cells. Demyelination occurs anatomically and temporally separate from inflammatory infiltrates. Mononuclear cells are present in the brain but occur around blood vessels, and represent a secondary response to demyelination. CDV is a useful model of PML because in both diseases demyelination is due to the oligodendrocidal effects of the virus.

C. Semliki Forest Virus

Semliki Forest Virus (SFC) is a non-human pathogenic alphavirus which was discovered in mosquitoes of the Semliki Forest of Uganda. SFV produces multifocal demyelination in the CNS when inoculated intracerebrally in mice (96). SFV-infected mice provide a very useful model to study the physiology of demyelination (97,98). SFV-induced demyelination is immune-mediated (99). Demyelination occurs in conjunction with inflammatory infiltrates. It does not occur in immuno-incompetent (100) or immunosuppressed mice, in spite of higher titer of virus in the brain tissue compared to control mice (101). Reconstitution of SFV-infected immuno-incompetent mice with normal spleen cells leads to demyelination (101). SFV is an excellent model of immune-mediated demyelination.

D. Herpes Simplex Virus

Herpes simplex virus type 1 (HSV-1) is a DNA virus which produces, on occasion, meningitis and encephalitis in man. HSV-1 has recently been found to produce demyelination in mice. Following oral-facial inoculation, HSV-1 induces lesions characterized by demyelination in association with an inflammatory mononuclear cell infiltrate in the brainstem adjacent to the trigeminal nerve root entry zone (102,103). Demyelination is immune-mediated; demyelination is prevented by immunosuppression with cyclophosphamide prior to infection (104). The extent of demyelination following infection with HSV-1 is under genetic control (105); certain strains of mice develop multifocal demyelination throughout the brain independent of the presence of virus. Demyelination in these latter strains is probably on an autoimmune basis. HSV-1 induced demyelination in mice is an important new model of MS.

E. Visna

Visna is a retrovirus which produces pneumonia and/or a chronic progressive, although occasionally relapsing-remitting, myelopathy in sheep (106). Visna persists at low levels for years, in part by evolving into antigenically distinct forms over time (107). Pathologically, demyelination occurs in two phases (108). During the initial phase, demyelination occurs in regions of inflammatory infiltrates with relatively little tissue necrosis. During the latter phase, demyelination occurs in conjunction with necrosis, of both gray and white matter. Immunosuppression inhibits early but not late demyelination (109). Visna may provide an excellent model for TSP and may be very useful as a means to test new therapies.

F. Theiler's Murine Encephalomyelitis Virus

Wild-type Theiler's murine encephalomyelitis virus (TMEV) usually produces an asymptomatic enteric infection in mice, and only rarely encephalomyelitis. One strain of TMEV, DA, has been isolated which reliably produces a biphasic neurologic disease in Swiss mice (110). Mice strains vary in their degree of susceptibility to TMEV (118,119). Nine to 20 days following intracerebral inoculation with the DA strain of TMEV, 80 percent of mice develop encephalomyelitis. Between one and five months post-infection, survivors develop a mild gait disturbance in conjunction with the occurrence of demyelination in areas of intense mononuclear inflammation in the spinal cord leptomeninges and white matter. During the acute phase, TMEV can be found in neurons and glial cells. During the late phase, TMEV is present only in glial cells (111,112). Immunosuppression prevents demyelination, although results in increased neuronal necrosis with a concomitant increase in mortality (113). Timing of the initiation of immunosuppression is critical in preventing demyelination (114). Immunosuppression initiated at the time of infection prevents early demyelination. Immunosuppression begun later is ineffective. MBP appears in the CSF and serum during chronic TMEV infection. The level of MBP parallels the clinical severity of demyelination (115). MBP appears to be a marker of demyelination rather than a target of attack by the immune system. Treatment with myelin components cannot prevent demyelination in TMEV (116) as is observed in experimental allergic encephalomyelitis, or even perhaps, in MS (117).

Demyelination occurs during TMEV infection as a result of a delayed type hypersensitivity (DTH) response against persistently infected

Table 1   Animal Models of Viral-Induced Demyelination

Virus	Host	Possible Mechanism
Theiler's	Mouse	Immunopathological in a persistent infection
Semliki Forest	Mouse	Immunopathological
JHM	Mouse	Oligodendrocidal in a persistent infection
	Rat	Immunopathological
Herpes Simplex	Mouse	Immunopathological
Canine Distemper	Dog	Oligodendrocidal
Visna	Sheep	Oligodendrocidal +/or immunopathologic

Table 2   Possible Mechanisms of Virus-Induced Demyelination (124)

Direct viral effects

- Viral infection of oligodendrocytes or Schwann cells causing demyelination through cell lysis or an alteration of cell metabolism

- Myelin membrane destruction by the virus or viral products

Virus-induced immune-mediated reactions

- Antibody and/or cell-mediated reactions to viral antigens on cell membrane

- Sensitization of host to myelin antigens

    - Breakdown or myelin by infection with introduction into the circulation

    - Incorporation of myelin antigens into the virus envelope

    - Modification of antigenicity of myelin membranes

- Cross-reacting antigens between virus and myelin proteins

- Bystander demyelination

Viral disruption of regulatory mechanisms of the immune system

oligodendrocytes (120). The development of demyelination correlates with the establishment of high levels of DTH against TMEV antigens (121). It does not appear to be due to an autoimmune response against CNS antigens. Examination of the fine specificity of class II restricted T cell responses reveals that the DTH is against viral antigens. Mice chronically infested with TMEV do not mount a DTH response against mouse spinal cord homogenate, myelin basic protein, or proteolipid protein (122). While demyelination during TMEV is not due to autoimmunity, procedures which increase inflammation, such as opening the blood brain barrier, lead to increased demyelination (123). TMEV provides an excellent system for studying viral-induced immune-mediated demyelination.

MECHANISMS OF VIRAL-INDUCED DEMYELINATION

Viral infections can induce demyelination through a variety of mechanisms (124). We have previously discussed new demyelination may result from lysis (JHMV), or interference with the normal metabolism (PML) of oligodendrocytes and immune-mediated destruction of virus infected oligodendrocytes (TMEV, SFV). Demyelination has recently been shown to arise as a consequence of molecular mimicry, where antibodies-synthesized against a viral protein inadvertently cross-react with a host protein. Experimental allergic encephalomyelitis (EAE) can be elicited by inoculation of MBP in Freund's adjuvant. MBP and the polymerase protein of hepatitis B are homologous for six amino acids. Inoculation of those six amino acids in Freund's adjuvant results in pathologic lesions which resemble EAE (125). Finally, demyelination may result as a consequence of a virus infection disrupting the normal regulatory mechanisms of the immune system which prevents autoimmunity. Chronic JHM infection in rats induces Class II antigens on glial cells, which may allow astrocytes to function as antigen-presenting cells and process an oligodendrocyte and/or myelin protein, such as MBP into an auto-antigen. Demyelination could result if a normal host protein becomes a target of the immune system. MS does not appear to arise as a consequence of the direct effect of a viral infection on oligodendrocytes and/or myelin. Instead, MS probably occurs as a result of either a virus-induced immune-mediated reaction or through alteration of the regulatory mechanisms of the immune system. Further studies into the pathogenesis of MS will be greatly aided by the availability of animal models of both mechanisms.

ACKNOWLEDGEMENTS

The authors wish to thank Angelina Morales for technical assistance in the preparation of this manuscript.

REFERENCES

1.   Waksman BH, Reingold SL: Viral etiology of multiple sclerosis. Where does the truth lie? Trends in Neuroscience 9:388, 1986
2.   Gonsdette RE, Delmotte P: Intensive immunosuppression with cyclophosphamide in multiple sclerosis: Follow-up of 116 patients for 2-6 years. J Neurol 173, 1977
3.   Weiner HL, Hauser SL, Hafler DA, Fallis RJ, Lehrich JR, Dawson DM: The use of cyclophosphamide in the treatment of multiple sclerosis. Ann NY Acad Sci 436:373, 1985
4.   Kappos L, Patzold V, Pommasch D, Poser S, Haas J, Krauseneck P, Malin JP, Fierz W, Graffensled B, Gugerli US: Cyclosporine versus azathioprine in the long-term treatment of multiple sclerosis. Results

of the German Multicenter Study. Ann Neurol 23:56, 1988

5. Aimard G, Confavrev XC, Devic M: Long-term immunosuppressive treatment with azathioprine in multiple sclerosis: A 10 year trial with 77 patients. IN: Bauer HJ, Poser C, Ritter G (eds.): Progress in Multiple Sclerosis, New York, Springer-Verlag, 371, 1980

6. Hafstein MP, Devereux C, Tronno R, Zito G, Vidaver R, Dowling PC, Lavenhar M, Cook SD: Total lymphoid irradiation in chronic progressive multiple sclerosis: A preliminary report. NY Acad Sci 436:397, 1984

7. Hafstein MP, Devereux C, Troiano R, Hafstein MP, Hernandez E, Lavenhar M, Vidaver R, Dowling PC: Effect of total lymphoid irradiation in chronic progressive multiple sclerosis. Lancet 1:1405, 1986

8. Weiner HL, Dawson DM: Plasmapheresis in multiple sclerosis: Preliminary study. Neurology 30:1029, 1980

9. Hohlfeld R, Michels M, Heininger K, Besinger U, Toyka KV: Azathioprine toxicity during long-term immunosuppression of generalized myasthenia gravis. Neurology 38:253, 1988

10. Visscher BR, Clark VA, Detels R, Malingren RM, Valdiviezo NL, Dudley JP: Two populations with multiple sclerosis. Clinical and demographic characteristics. J Neurol 225:237, 1981

11. Kurtzke JF: Epidemiology of multiple sclerosis. IN: Hallpike JF, Adams CWM, Tourtellotte WN (eds). Multiple Sclerosis: Pathology, Diagnosis and Management. Baltimore, Williams and Wilkins, 47-95, 1983

12. Matthews WB: The pattern of disease. IN: Matthews WB, Acheson ED, Batchelor JR, Weller RO (eds). McAlpine's Multiple Sclerosis. Edinburgh, Churchill Livingston, 3-26, 1985

13. Acheson ED: Epidemiology of multiple sclerosis. Br Med Bull 33:9, 1977

14. Limburg CC: The geographic distribution of multiple sclerosis and estimated prevalence for the United States. Proc Assoc Res Nerv Ment Dis 28:15, 1950

15. Alter M, Leibowitz V, Speen J: Risk of multiple sclerosis related to age at immigration to Israel. Arcch Neurol 15:234, 1966

16. Kurtzke JF: A method for estimating the age at which immigration of white immigrants to South Africa with an example of its importance. S Afr Med J 1:663, 1970

17. Jersild C, Svejgaard A, Fog T: HLA antigens and multiple sclerosis. Lancet 1:1240, 1972

18. Naito S, Namerow N, Mickey MR, Terasaki PI: Multiple sclerosis: Association with HLA-A3. Tissue Antigens 2:1, 1972

19. Jersild C, Dupont B, Fog T, Platz PJ, Svejgard A: Histocompatibility determinants in multiple sclerosis. Transplant Rev 22:148, 1975

20. Fog T, Schuller E, Jersild C, Engelfriet CF, Bertrams J: Neurology. Multiple sclerosis. IN: Dausset J, Sveggaard A (eds). HLA and Disease. Copenhagen, Munksgaard, 108-109, 1977

21. Eldridge R, McFarland H, Seyer J, Sadowsby D, Krebs V: Familial multiple sclerosis: Clinical, histocompatibility and viral serological studies. Ann Neurol 3:75, 1978

22. Roberts DF, Roberts MJ, Poskanzer DC: Genetic analysis of multiple sclerosis in Orkney. J Epidem Comm Health 33:229, 1979

23. Ebers GC: Genetic factors in multiple sclerosis. Neurologic Clinics 1:645, 1983

24. Compston DAS, Vakarelis BM, Paul E, McDonald WI, Batchelor JR, Mims CA: Viral infection in patients with multiple sclerosis and HLA-DR matched controls. Brain 109:325, 1986

25. Bobowick AR, Kurtzke JF, Brody JA, Hrubec Z, Gillespie M: Twin study of multiple sclerosis: an epidemiologic inquiry. Neurology 28:978, 1978

26. Kurtzke JF, Hyllested K: Multiple sclerosis in the Faroe

Islands : I. Clinical and epidemiological features. Ann Neurol 5:6, 1979

27. Kurtzke JF, Hyllested K: Multiple sclerosis in the Faroe Islands : III. An alternative assessment of the three epidemics. Acta Neurol Scan 76:317, 1987

28. Johnson RT: Viral infections of the nervous system. New York, Raven Press, 237-270, 1982

29. Haase AT, Ventura P, Gibbs CJ Jr, Tourtellotte WW: Measles virus nucleotide sequence: Detection by hybridization in situ. Science 212:672, 1981

30. Fraser NW, Lawrence WL, Wroblewska Z, Gilden DH, Koprowski H: Herpes simplex type I DNA in human brain tissue. Proc Natl Acad Sci 78:;6461, 1981

31. Tanaka R, Iwasaki Y, Koprowski H: Intracisternal virus-like particles in the brain of multiple sclerosis patients. J Neurol Sci 28:121, 1976

32. Adams JM, Imagawa DT: Measles antibodies in multiple sclerosis. Proc Soc Exp Bio Med 3:562, 1962

33. Norrby E: Viral antibodies in multiple sclerosis. Prog Med Virol 24:1, 1978

34. Haire M: Significance of virus antibodies. Br Med J 33:40, 1977

35. Cook SD, Dowling PC, Russell WC: Neutralizing antibodies to canine distemper virus and measles virus in multiple sclerosis. J Neurol Sci 41:61, 1979

36. Vandvic B, Degre M: Measles virus antibodies in serum and cerebrospinal fluid in patients with multiple sclerosis and other neurological disorders with special reference to measles antibody synthesis within the central nervous system. J Neurol Sci 24:201, 1975

37. Shirodaria PV, Haire M, Fleming E, Menett JD, Hawkins SA, Roberts SD: Viral antibody titers: Comparison in patients with multiple sclerosis and rheumatoid arthritis. Arch Neurol 44:1237, 1987

38. Koprowski H, DeFreitas EC, Harper ME, Sandberg-Wohlheim M, Sheremata WA, Robert-Guroff M, Saxinger CW, Feinberg MB, Wong-Staal F, Gallo RC: Multiple sclerosis and human T cell lymphotropic retroviruses. Nature 318:154, 1985

39. Ohta M, Ohta K, Mori F, Nishitani H, Saida T: Sera from patients with multiple sclerosis react with human T cell lymphotropic virus-1 gag proteins but not env proteins-Western blotting analysis. J Immunol 137:3440, 1986

40. Hauser SL, Aubert JS, Burks JS, Kerr C, Lyon-Caen O, de The G, Brahic M: Analysis of human T lymphotropic virus sequences in multiple sclerosis tissue. Nature 322:176, 1986

41. Karpas A, Kampf U, Siden A, Koch M, Poser M: Lack of evidence for involvement of known human retroviruses in multiple sclerosis. Nature 322:177, 1986

42. Rice GPA, Armstrong A, Bulman DE, Paty DW, Ebers GC: Absence of antibody to HTLV-I and III in sera of Canadian patients with multiple sclerosis and chronic myelopathy. Ann Neurol 20:533, 1986

43. Birnbaum G, Aubitz S, Kotilinek L: Search for autonomously proliferating spinal fluid lymphocytes in patients with multiple sclerosis. Neurology 38:28, 1988

44. Kuroda Y, Shibasaki H, Sato H, Okochi K: Incidence of antibody to HTLV-I is not increased in Japanese MS patients. Neurology 37:156, 1987

45. Madden DI, Mundon FK, Tzam NR, Fuccillo DA, Dalakas MC, Calabrese V, Elizan TS, Sever JL: Serologic studies of MS patients, controls and patients with other neurologic diseases: Antibodies to HTLV-I, II, III. Neurology 38:81, 1988

46. Gessain A, Barin F, Vernant JC, Gout O, Maurs L, Calender A, de The G: Antibodies to human T-lymphotropic virus type-1 in patients

with tropical spastic paraparesis. Lancet 2:407, 1985

47. Rodgers-Johnson P, Gajdusek DC, Morgano STC: HTLV-I and HTLV-III antibodies and tropical spastic paraparesis. Lancet 2:1247, 1985

48. Osame M, Usuko K, Izumo S. HTLV-I associated myelopathy, a new clinical entity: Lancet 1:1031, 1986

49. Osame M, Matsumoto M, Usuku K, Izumo S, Igichi N, Anitani H, Tara M, Igata A: Chronic progressive myelopathy associated with elevated antibodies to human T-lymphotropic virus type I and adult T cell leukemia-like cells. Ann Neurol 21:117, 1987

50. Vernant JC, Maurs L, Gessin A, Barin F, Gout O, Delaporte JM, Sanhadjis K, Buisson G, de The G: Endemic tropical spastic paraparesis associated with human T-lymphotropic virus type. A clinical and seroepidemiological study of 25 cases. Ann Neurol 21:123, 1987

51. Robertson WB, Cruickshank EK: Jamaican (tropical) myeloneuropathy. IN: Minckler J (ed). Pathology of the nervous system. New York, McGraw Hill, 2466-2476, 1972

52. Piccardo P, Ceroni M, Rodgers-Johnson P, Mora L, Asher DM, Char G, Gibbs CJ Jr, Gajdusek DC: Pathological and immunological observations on tropical spastic paraparesis in patients from Jamaica. Ann Neurol 23:S156, 1988

53. Petito CK, Navia BA, Cho ES, Jordan BD, George DS, Price RW: Vacuolar myelopathy pathologically resembling subacute combined degeneration in patients with the acquired immunodeficiency syndrome. N Eng J Med 312:874, 1985

54. Levy RM, Bredesen DE, Rosenblum ML: Neurological manifestations of the acquired immunodeficiency syndrome (AIDS): Experience at UCSF and review of the literature. J Neurosurg 62:475, 1985

55. Levy JA, Evans L, Chreg-Mayer C, Pan LZ, Lane A, Staben I, Dina D, Wiley C: The biologic and molecular properties of the AIDS-associated retrovirus on that affect antiviral therapy. Ann Inst Pasteur 138:101, 1987

55a. Navia BA, Jordan BD, Price RW. The AIDS dementia complex: I. Clinical features. Ann Neurol 19:517, 1986

56. Navia BA, Cho E-S, Petito CK, Price RW: The AIDS dementia complex: II. Neuropathology. Ann Neurol 19:525, 1986

56a. McArthur JC, Johnson RT: Primary infection with human immunodeficiency virus. IN: Rosenblum JL, Levy RM, Bredesen DE (eds). AIDS and the nervous system. New York, Raven Press, 183-201, 1988

57. Padgett BL, Walker DL, ZuRhein GM: Cultivation of papova-like virus from human brain with progressive multifocal leukoencephalopathy. Lancet 1:1257, 1971

58. Weiner LP, Herndon RM, Narayan O: Isolation of virus related to SV40 from patients with progressive multifocal leukoencephalopathy. N Eng J Med 286:385, 1972

59. Richardson EP Jr: Progressive multifocal leukoencephalopathy. IN: Vinken PJ, Bruyn GW (eds). Handbook of Clinical Neurology Vol 9. Multiple Sclerosis and Other Demyelinating Diseases. New York, Elsevier, 485-499, 1970

60. Miller JR, Barrett RE, Britton CB: Progressive multifocal leukoencephalopathy in a male homosexual with T cell immune deficiency. N Engl J Med 307:1436, 1982

61. Krupp LB, Lipton RB. Siverdlow ML, Leeds NE, Llena J: Progressive multifocal leukoencephalopathy: Clinical and radiographic features. Ann Neurol 17:344, 1985

62. Rosenblum ML, Levy RM, Bredesen DE: Overview of AIDS and the nervous system. IN: Rosenblum ML, Levy RM, Bredesen DE (eds). AIDS and the Nervous System. New York, Raven Press, 1-12, 1988

63. Fermaglich J, Hardman JM, Earle KM: Spontaneous progressive multifocal leukoencephalopathy. Neurology 20:479, 1970

64. Carroll BA, Lane B, Norman D: Diagnosis of progressive

multifocal leukoencephalopathy by computerized tomography. Radiology
122:137, 1977
    65. Brooks BR, Walker DL: Progressive multifocal
leukoencephalopathy. Neurologic Clinics 2:299, 1984
    66. ZuRhein GM: Polyoma-like virions in a human demyelinating
disease. Acta Neuropathol 8:57, 1967
    67. Aksamit AJ, Mourrain P, Sever JL, Major EO: Progressive
multifocal leukoencephalopathy investigation of three cases using in situ
hybridization with JC virus biotinylated DNA probe. Ann Neurol 18:490,
1985
    68. Chesters PM, Heritage J, McCance DJ: Persistence of DNA
sequences of BK virus and JC virus in normal human tissues and in
diseased tissues. J Inf Dis 147:676, 1983
    69. Houff SA, Major EO, Katz DA, Kufta CV, Sever JL, Pittaluga S,
Roberto JR, Gitt J, Sarni N, Lux W: Involvement of JC virus-infected
mononuclear cells from the bone marrow and spleen on the pathogenesis of
progressive multifocal leukoencephalopathy. N Eng J Med 318:301, 1988
    70. Walker DL, Padgett BL, ZuRhein GM: Human papovavirus (JC)
induction of brain tumors in hamsters. Science 181:674, 1973
    71. Small JA, Scangos GA, Cork L, Jay G, Khoury G: The early region
of human papovavirus: JC induces dysmyelination in transgenic mice.
Cell 46:13, 1986
    72. Trapp BD, Small JA, Pulley M, Khoury G, Scangos GA:
Dysmyelination in transgenic mice containing JC virus early region. Ann
Neurol 23:38, 1988
    73. Stoner GL, Ryschkewitsch CF, Walker DL, Webster H de F: JC
papovavirus large tumor (T)-antigen expression in brain tissue of
acquired immunodeficiency syndrome (AIDS) and non-AIDS patients with
progressive multifocal leukoencephalopathy. Proc NAtl Acad Sci 83:2271,
1986
    74. Chan K-FJ, Stoner GL, Hashim GA, Huang K-P: Substrate
specificity of rat brain calcium-activated and phospholipid-dependent
protein kinase. Biochem Biophys Res Commun 134:1358, 1986
    75. Stoner GL, Ryschkewitsch CF, Walker DL, Soffer D,
Webster H de F: Immunocytochemical search for JC papovavirus large T-
antigen in multiple sclerosis brain tissue. Acta Neuropath 70:345, 1986
    76. Martin JR, Nathanson N: Animal models of virus-induced
demyelination. Prog Neuropathol 4:27, 1979
    77. Bailey OT, Pappenheimer AM, Sargent F, Cheever FS, Daniels JB:
A murine virus (JHM) causing disseminated encephalomyelitis with
extensive destruction of myelin: II. Pathol J Exp Med 90:195, 1949
    78. Cheever FS, Daniels JB, Pappenheimer AM, Bailey OT: A murine
virus (JHM) causing disseminated encephalomyelitis with extensive
destruction of myelin: I. Isolation and biological properties of the
virus. J Exp Med 90:181, 1949
    79. Weiner LP: Pathogenesis of demyelination induced by mouse
hepatitis virus (JHM virus). Arch Neurol 28:298, 1973
    80. Stohlman SA, Frelinger JA: Resistance to fatal central nervous
system disease by mouse hepatitis virus, strain JHM. I. Genetic
analysis. Immunogenetics 6:277, 1978
    81. Stohlman SA, Frelinger JA, Weiner LP: Resistance to fatal
central nervous system disease by mouse hepatitis virus, strain JHM.
II. Adherent cell-mediated protection. J Immunol 124:1733, 1980
    82. Roos RP: Viruses and demyelination disease of the central
nervous system. Neurology Clinics 1681, 1983
    83. Herndon RM, Grifin DE, McCormick U, Weiner LP: Mouse hepatitis
virus-induced recurrent demyelination: A preliminary report. Arch
Neurol 32:32, 1975
    84. Stohlman SA, Weiner LP: Chronic central nervous system
demyelination in mice after JHM virus infection. Neurology 31:38, 1981
    85. Haspel MN, Lampert PW, Oldstone MBA: Temperature sensitive

mutants of mouse hepatitis virus produce a high incidence of demyelination. Proc Natl Acad Sci 75:403, 1978

86. Knobler RL, Lampert PW, Oldstone MBA: Virus persistance and recurring demyelination produced by a temperature-sensitive mutant of MHV-4. Nature 298:279, 1982

87. Fleming JO, Trousdale MD, El-Zaatari FAK, Stohlman SA, Weiner LP: Pathogenicity of antigenic variants of murine coronavirus JHM selected with monoclonal antibodies. J Virol 58:869, 1986

88. Nagashima K, Wege H, Meyermann R, ter Meulen V: Coronavirus induced subacute demyelinating encephalomyelitis in rats: a morphological analysis. Acta Neuropathol 44:63, 1978

89. Wege H, Watanabe R, ter Meulen V: Relapsing subacute demyelinating encephalomyelitis in rats during the course of coronavirus JHM infection. J Neuroimmunol 6:325, 1984

90. Watanabe R, Wege H, ter Meulen V: Adoptive transfer of EAE-like lesions from rats with coronavirus-induced demyelinating encephalomyelitis. Nature 305:150, 1983

91. Massa PT, Dorries R, Wege H, ter Meulen V: Analysis and pathogenetic significance of Class II MHC (Ia) antigen induction on astrocytes during JHM coronavirus infection in rats. IN: Lai MMC and Stohlman SA (eds). Coronaviruses, New York, Plenum Press, 203-218, 1987

92. Massa PT, Dorries R, ter Meulen V: Viral particles induce Ia antigen expression on astrocytes. Nature 320:543, 1987

93. Wisniewski H, Raine CS, Kay WJ: Observations on viral demyelinating encephalomyelitis. Canine distemper virus. Lab Invest 26:589, 1972

94. Raine CS: On the development of CNS lesions in natural canine distemper encephalomyelitis. J Neurol Sci 30:13, 1974

95. McCullough B, Krakowka S, Koestner A: Experimental canine distemper virus-induced demyelination. Lab Invest 31:216, 1974

96. Chew-Lim M, Cuckling AJ, Webb HE: Demyelination in mice after two or three infections with avirulent Semliki Forest virus. Vet Pathol 14:67, 1977

97. Tremain KE, Ikeda H: Physiological deficits in the visual system of mice infected with Semliki Forest virus and their correlation with those seen in patients with demyelinating disease. Brain 106:879, 1983

98. Tansey EM, Allen TGJ, Ikeda H: Enhanced retinal and optic nerve excitability associated with demyelination in mice infected with Semliki Forest virus. Brain 109:15, 1986

99. Fazakerley JK, Amor S, Webb HE: Reconstitution of Semliki Forest virus infected mice induces immune mediated pathological changes in the CNS. Clin Exp Immunol 52:15, 1983

100. Fazakerley JK, Webb HE: Semliki Forest virus-induced, immune mediated demyelination: The effect of radiation. Br J Exp Path 68:101, 1987

101. Fazakerley JK, Webb HE: Semliki Forest virus-induced, immune-mediated demyelination. Adoptive transfer studies and viral persistence in nude mice. J Gen Virology 67:377, 1987

102. Kristensson K, Suennerkelm B, Persson L: Latent herpes simplex virus trigeminal ganglionic infection in mice and demyelination in the central nervous system. J Neurol Sci 43:253, 1979

103. Townsend JJ: The demyelinating effect of cornial HSV infection in normal and nude (athymic) mice. J Neurol Sci 50:435, 1981

104. Townsend JJ, Baringer JR: Morphology of central nervous system disease in immunosuppressive mice after peripheral herpes simplex virus inoculation. Trigeminal root entry zone. Lab Invest 40:178, 1979

105. Kastrukoff LF, Lau AS, Kim SJ: Multifocal CNS demyelination following peripheral inoculation with herpes simplex virus Type I. Ann Neurol 22:52, 1987

106. Sigurdsson B: Rida, a chronic encephalitis of sheep. Br Vet J 110:341, 1954

107. Narayan O, Griffin DE, Chase J: Antigenic drift of visna virus in persistently infected sheep. Science 197:376, 1977

108. Petursson G, Nathanson N, Georgsson G, Panitch H, Palsson PA: Pathogenesis of visna: I. Sequential virologic serologic and pathologic studies. Lab Invest 35:402, 1976

109. Nathanson N, Panitch H, Palson PA, Petersson G, Georgsson G: Pathogenesis of visna: II. Effect of immunosuppression upon early central nervous system lesions. Lab Invest 35:444, 1976

110. Lipton HL: Theiler's virus infection in mice: An unusual biphasic disease leading to demyelination. Infect Immun 11:1147, 1975

111. Brahic M, Stroop WG: Theiler's virus persists in glial cells during demyelinating disease. Cell 26:123, 1981

112. Rodriguez M, Leibovitz JL, Lampert PW: Persistent infection of oligodendrocytes in Theiler's virus-induced encephalomyelitis. Ann Neurol 13:426, 1983

113. Lipton HL, Dal Canto MC: Theiler's virus-induced demyelination prevention by immunosuppression. Science 192:62, 1976

114. Roos RP, Firestone S, Wollmann R, Variakojes D, Arnason BGW: The effect of short-term and chronic immunosuppression on Theiler's virus demyelination. J Neuroimmunol 2:223, 1982

115. Rauch HC, Montgomery IN, Hinman CL, Harb W, Benjamins JA: Chronic Theiler's virus infection in mice appearance of myelin basic protein in the cerebrospinal fluid and serum antibody directed against MBP. J Neuroimmunol 14:35, 1987

116. Lang W, Wiley C, Lampert P: Theiler's virus encephalomyelitis is unaffected by treatment with myelin components. J Neuroimmunol 9:109, 1980

117. Bornstein MB, Miller A, Slagle S, Weitzman M, Crystal H, Drexler E, Keilson M, Mernam A, Wasser Theil-Smollers, Spada V, Weiss W, Arnon R, Jacobsohn I, Tertelbaum D, Sela M: A pilot trial of CoP-1 in exacerbating-remitting multiple sclerosis. N Eng J Med 317:408, 1977

118. Lipton HL: Persistent Theiler's murine encephalomyelitis versus infection in mice depends on plaque size. J Gen Viral 46:169, 1980

119. Lipton HL, Melvold R: Genetic analysis of susceptibility to Theiler's mice induced demyelinating disease in mice. J Immunol 132:1821, 1984

120. Clatch RJ,k Melvold R, Miller SD, Lipton HL: Theiler's murine encephalomyelitis virus (TMEV)-induced demyelinating disease in mice is influenced by the H-2D region. Correlation with TMEV-specific delayed-type hypersensitivity. J Immunol 135:1408, 1985

121. Clatch RJ, Melvold RW, Dal Canto MS, Miller SD, Lipton HL: The Theiler's murine encephalomyelitis virus (TMEV) model for multiple sclerosis shows a strong influence of murine equivalents of HLA-A, B, and C. J Neuroimmunol 15:121, 1987

122. Miller SD, Clatch RJ, Pevear DC, Trotter JL, Lipton HL: Class II restricted T cell responses in Theiler's murine encephalomyelitis virus (TMEV)-induced demyelinating disease: I. Cross specificity among TMEV substains and related picornaviruses, but not myelin proteins. J Immuno 138:3776, 1987

123. Love S, Wiley CA, Fujinami RS, Lampert PW: Effects of regional spinal x-radiation in demyelinating disease, caused by Theiler's virus, mouse hepatitis virus or experimental allergic encephalomyelitis. J Neuroimmunology 14:19, 1987

124. Weiner LP, Johnson RT, Herndon RM: Viral infections and demyelinating disease. N Eng J Med 288:1103, 1973

125. Fujinami RS, Oldstone MBA: Amino acid homology and immune response between the encephalitogenic site of myelin basic protein and virus: A mechanism for autoimmunity. Science 230:1043, 1985

11

RETROVIRAL DISEASE OF THE NERVOUS SYSTEM

B. Weinshenker and G.P.A. Rice

Department of Clinical Neurological Sciences
University of Western Ontario
London, Canada

INTRODUCTION

The recent explosion in the understanding of retrovirus biology and the recognition of viral neurotropism have shed new light on diseases of the nervous system of animals and man. In this chapter the biology of retroviruses and some diseases of the nervous sytem caused by retroviruses will be reviewed. Evidence that a novel human retrovirus might cause multiple sclerosis (MS) will also be considered.

Retroviruses

Retroviruses are enveloped RNA viruses (8). The name "retrovirus" derives from the fact that the RNA genome replicates through a DNA intermediate, which is synthesized by an RNA dependent DNA polymerase (reverse transcriptase). The virus genome codes for three principal classes of protein, in addition to those involved in gene regulation. The major proteins are the gag proteins which are the major structural proteins of the neucleocapsid, the pol protein which is the reverse transcriptase and the env proteins which are the envelope proteins (5).

Lentiviruses constitute a subfamily of retroviruses and derive their name from the slow time course of the diseases they cause in animals and man (8).

Visna

Visna is the prototypic lentivirus and the first one recognized to cause disease of the nervous system (18). In 1933, twenty Karakul rams were imported to Iceland from Halle, Germany as breeding stock. Within 6 years there were outbreaks of progressive pulmonary disease known as maedi (Icelandic for dyspnoea), and a neurological disorder, visna, in two of the farms which had received the Karakul rams. By the 1940's the disease had reached epidemic proportions and posed a serious threat to Icelandic sheep husbandry. The disease was eradicated by systematic slaughter of sheep in endemic areas.

Visna was first recognized in southwestern Iceland in flocks in which maedi had been recognized. A common agent was suspected to cause both conditions long before this observation was confirmed (8).

Visna had many clinical parallels to MS (18) (Table I). The disease occurred in only a few sheep in a given flock and never in sheep under 2 years of age. The clinical picture of visna was dominated by a subacute or chronic myelopathy. Early manifestations included hindlimb weakness and ataxia. The animals became less efficient at grazing and lost weight; the name visna derived from the Icelandic word for wasting. Relapses and remissions were occasionally observed. Although the hindlimbs became paralyzed over months to years, affected sheep maintained alertness and sphincteric control.

Although demyelination occurred, the pathology of this condition bore greater resemblance to allergic or postinfectious encephalomyelitis than to MS (4, 15, 29). In the central nervous system, sharply demarcated white matter plaques, typical of MS, were not seen. Microscopically, the principal features included meningeal and subependymal mononuclear cell infiltrates and microglial proliferation. The lesions were predominantly periventricular but could extend throughout the brain. Necrosis of confluent inflammatory lesions was a common feature. Severity of pathological lesions was not always associated with the duration of clinical disease.

It was Bjorn Sigurdson who recognized the infectious nature of the disease, and who demonstrated transmissibility by cell-free filtrates (27-29). He also recognized the long incubation period and the slow evolution of the clinical disease. It was his work with visna that led to the seminal concept of "slow virus infections".

The details of the virology of visna virus, its evolutionary relationship to other retroviruses, and other important aspects of its pathobiology have been reviewed recently (8).

Tropical Spastic Paraparesis (TSP)

A myelopathy occurring in tropical climates has been recognized for many years. In Colombia, the earliest case was described in 1952 (23). The clinical features are similar to the chronic idiopathic myelopathy that is commonly recognized in northern climates (19). The principal findings are those of a cervicothoracic myelopathy, with spasticity, weakness, hyperreflexia, pain, sphincteric disturbances, impotence, slight involvement of the dorsal columns and occasional involvement of the peripheral nervous system, (chiefly reduction in the ankle jerks and decrease in vibration sense).

TSP is the most common endemic paraplegia in the tropics. The prevalence appears to be approximately 1:1000 within endemic areas. Blacks are most frequently affected but no race is spared. Females appear to be affected more commonly than males (24).

Spinal fluid analysis reveals mild pleocytosis, a moderate increase in total protein and a relative increase in IgG. Pathologically, demyelination is confined to pyramidal tracts and dorsal columns with occasional involvement of the spinocerebellar and spinothalamic tracts. Optic atrophy and involvement of the eighth cranial nerve occasionally occur. The cerebrum and cerebellum are minimally involved. Perivascular mononuclear cell infiltrates, microglial and astrocytic gliosis and spongiform changes in myelin have been described. These latter changes are quite similar to the vacuolar myelopathy in AIDS (vide infra) (24).

The etiology is still uncertain. Genetic predisposition, exposure to toxins, such as cyanates, and infectious agents, including the spirochete

Table I   Similarities of MS and VISNA

	MS	VISNA
Predominantly myelopathic syndrome	+	+
Relapsing-remitting course	+	+
Age dependence	+	+
Genetically determined resistance	+	+
CSF - pleocytosis	+	+
increased IgG	+	+
Oligoclonal bands	+	+

Table II

	VISNA	TSP	HAM	AIDS	MS
DEMOGRAPHIC FEATURES					
Female predominance	-	+	+	-	+
Age dependent onset	+	?	?	?	+
Genetic resistance	+	?	?	?	+
CLINICAL INVOLVEMENT					
Myelopathy	+	+	+	+	+
Encephalopathy	-	-	-	+	+
Peripheral neuropathy	-	+	-	+	-
CSF FINDINGS					
Increased IgG	+	+	+	+	+
Oligoclonal bands	+	?	+	?	+
Pleocytosis	+	-	+	±	+
PATHOLOGY					
Demyelination	+	+	+	+	+
Vacuolar change	-	+	+	+	-
VIRAL DEMONSTRATION					
Isolation	+	+	+	+	-
Transmission	+	-	-	+	-
Retroviral seroreactivity	+	+	+	+	?

HTLV p19

6					11
SER	ARG	SER	ALA	SER	PRO
SER	ARG	SER	GLY	SER	PRO
161					166

Myelin Basic Protein

FIGURE 1   Homology of HTLV p19 with Myelin Basic Protein

toxins, such as cyanates, and infectious agents, including the spirochete which causes yaws, have been incriminated and these etiologic possibilities have been reviewed recently (24). Some cases appear to be associated with HTLV-1 exposure. This was first reported by Gessain et al in 60% of patients from Martinique (6, 30). HTLV-1 seropositivity has been reported in greater proportions of patients in more recent observations from Colombia, Jamaica, Trinidad and Tobago and the Seychelles (24, 25). Although reports of myelopathy and adult T-cell leukemia in endemic populations appear to be relatively common, only rarely do both diseases manifest in the same patient. There has been one report of an individual with TSP in whom acute T-cell leukemia developed (2). The disease has not yet been transmitted to primates.

HTLV-1 Associated Myelopathy (HAM)

A disorder similar to TSP has been described in Japan (17). The clinical features are again those of a progressive cervicothoracic myelopathy. Sensory changes and CSF pleocytosis appear to be more prominent than in TSP. Another major difference is the appearance of mononuclear cells with large multilobular nuclei in the spinal fluid. This has not yet been reported in patients with TSP from Colombia and the Seychelles (25).

Serological studies have revealed antibody to HTLV-1 in virtually all affected patients (17). This virus is endemic to the area of Japan in which most of the reported cases have been collected. The virus has been isolated from one patient with this myelopathy but transmission experiments have not yet been reported (24). Some have suggested that TSP and HAM are essentially the same disease (30). Further clinical, serological and imaging studies will likely resolve this issue (10).

Neurological Complications in Aids

It was the lymphotropism of human immunodeficiency virus I (HIV-1) that first drew attention to this virus in 1981 (reviewed 5). A wide spectrum of neurological disorders has now been recognized to be mediated directly by this virus (1).

The first complication identified was a progressive myelopathy. This observation was consolidated by Petito et al in 1985 (21). The clinical features were predominantly myelopathic in nature but dementia was reported in 70% of the patients. Pathological evidence of myelopathy was found in 20/89 consecutive AIDS autopsies. The pathology was a vacuolar myelopathy, chiefly affecting the dorsal and lateral columns. Morphologic similarity to the myelopathy of vitamin B12 deficiency has been suggested. The vacuolation is not accompanied by the intense demyelination recognized in the myelopathies that attend visna or TSP. Although HIV can be isolated from the brain and spinal fluid of such patients, the pathogenesis is unknown.

The encephalopathy associated with HIV infection appears to be an important part of the spectrum of neurological problems directly caused by HIV. Dementia is an occasional presenting feature of AIDS (1). In many cases, infection is attended by other opportunistic pathogens, especially by herpes viruses, but it is likely that HIV directly alters brain function. HIV gene expression appears to be concentrated in cells of microglial and endothelial cell origin (32). Occasionally neurons and astrocytes are involved. White matter changes exceed those in gray matter.

It has been shown recently that the virus glycoprotein (GP 120) binds to a CD4 epitope present largely in gray matter. This suggests, at least theoretically, an interesting means by which the virus might gain access to the nervous system (20).

The Role of a Retrovirus in the Cause of Multiple Sclerosis

The theory that a virus might participate in the pathogenesis of MS has been popular for several years, but concrete evidence for this suspicion has been lacking. Evidence has derived largely from epidemiological studies working at the limit of their power. Studies of the occurrence of MS in migrant populations have suggested that it might be an acquired disease, as migrants from high to low risk areas carry their higher risk only if migration occurs after the age of 15. This suggests that exposure to an agent in early adolescence might be critical. Rare epidemics, such as that which occurred in the Faroe islands after World War II, have suggested that the MS agent might be transmissible. However, direct attempts to isolate a virus, to show a unique antibody, or to demonstrate virus gene products in brain have failed to show findings unique to MS patients. The repeated failure to corroborate various reports of virus isolation has engendered a healthy skepticism among followers of this literature.

The report of antibody to the p24 protein of HTLV-I was naturally met with cautious enthusiasm. Koprowski and his colleagues demonstrated antibody to HTLV-I in 7 of 17 MS patients from Key West, Florida (12). The mean antibody titer of 35 Swedish MS patients to the p24 protein was significantly higher in the spinal fluid and serum of these patients than in those of normal individuals and patients with other neurological diseases. Three patients with other neurological disorders also had significantly reactive antibody titers. Competition assays with purified virions appeared to show that the antibodies were specific to HTLV viruses although HTLV-I, II and III all inhibited binding in some cases. In some patients, the antibody titers fluctuated over time. The authors also demonstrated that an HTLV-I genomic sequence (the probe used was a 3' HTLV-I riboprove specific for long terminal repeat sequences) could be demonstrated in 1/10 000 leukocytes derived from interleukin 2 expanded lymphocytes from 4 of 8 MS spinal fluid samples. The hybridization stringency conditions, however, were low.

The authors were cautious in the interpretation of their studies. They did not incriminate HTLV-I directly, but rather suggested that this antibody might represent a cross reactive antibody, perhaps reacting with a novel human retrovirus. A Japanese group has also found antibody in low titer to this virus in 25% of Japanese patients with clinically definite MS (16). The antibody was directed largely to the p15, 19 and 25 proteins which are all products of the GAG gene. Other investigators have been unable to demonstrate seroreactivity to HTLV-I using standard commercial or cell based ELISA technology, in approximately 400 patients from Canada, France, USA, Sweden, Germany, Japan, and the United Kingdom (3, 7, 9, 11, 14, 22).

Some criticism of the negative studies has been raised. The principal objection has been that not enough of the p24 antigen was available in the immunoassays to detect a weakly cross reactive antibody (13). We have identified an antibody that is weakly reactive to HTLV-I preparations by immunoblot technique, (although antibody cannot be shown by conventional ELISA), in a small proportion of patients with MS and occasionally in those with other neurological diseases. Although much of the reactivity appears to be directed to higher molecular weight proteins

(probably unrelated to the virus), in occasional patients antibody
appears to have reacted to proteins of 19 or 24 kilodalton molecular
weight (31).

What might this antibody represent?

1. Molecular mimicry. Perhaps the p24 reactive antibody is simply a
cross reactive antibody. In patients with MS, autoantibodies to a
variety of CNS antigens can be demonstrated. We have compared the amino
acid sequence of HTLV-I to the sequence of several brain proteins to
which MS patients occasionally make antibodies. We found an interesting
homology between myelin basic protein and the p19 of HTLV (Figure 1)
(31). This potential cross reactivity is under study.

2. Simple cross reactivity. Some of the bands that we have
demonstrated can be blocked or competed by incubation with a lysate
derived from an uninfected cell line similar to the T-cell lines from
which HTLV-I is commonly prepared. Precedent for this observation exists
and others have demonstrated false seropositivity of patients to HIV by
immunoblot analysis due to cross reactivity with a similar molecular
weight cellular protein (26).

3. Nonspecificc activation of the immune system. It is well
recognized that MS patients appear to generate slightly higher antibody
titers to a variety of viruses. The reason for this is unknown. One
might speculate that MS susceptibility genes are linked to a variety of
immune response genes relevant to those which are targeted against
viruses.

4. The antibody might react with a novel human retrovirus. This is
a realistic possibility. A retrovirus was first incriminated as the
cause of AIDS by the demonstration that 30% of patients with AIDS had
antibody reactive in immunofluorescence assays to cells infected with
HTLV-I. Only later was the closely related retrovirus HTLV-III (or HIV-
I) identified (5).

Strategy for Identification of a Novel Human Retrovirus

1. Serological. Although the serological data which have been
collected to date have been conflicting, further studies might resolve
this. Many investigators are now studying p24 reactivity with different
p24 preparations, different virus isolates and different MS populations.

2. Virus isolation. The first human retrovirus was identified in
1981 by Poiesz and Gallo after years of development of viral isolation
techniques (5). It should be remembered that in many retroviral
infections, virus gene expression is incomplete and that latency can be
broken only by the addition of the appropriate stimulus or by cultivation
of latently infected cells with permissive cell lines. As our
understanding of retroviral culture techniques improves, we will be in a
better position to isolate the putative MS agent or to gauge the
significance of the negative isolation attempts which have been reported
to date (11).

3. Hibrydization studies. The earliest attempts to look for HTLV-I
genes in MS leukocyte lines or in MS brains have been negative (9) or
unconvincing (12). This kind of technology might be more revealing when
some of the more neurotropic retroviruses are sequenced. Probes derived
from such strains would then be used to survey MS tissues or leukocyte
lines derived from MS patients.

CONCLUSION

The features of diseases suspected or established to be caused by retroviruses are summarized in Table II as is the evidence that supports such an etiology. The incrimination of retroviruses as the cause of MS would have a drastic impact on MS research. The focus of clinical trials would suddenly shift. New strategies of MS prevention would emerge. Work that examines this important question should proceed with the utmost enthusiasm.

ACKNOWLEDGEMENTS

The authors thank Donna Greer for expert secretarial assistance with the manuscript. Brian Weinshenker is a fellow of the MS Society of Canada, and George P.A. Rice is a recipient of a Career Scientist award from the Ministry of Health of Ontario.

REFERENCES

1. Anders KH, Guerra WF, Tomiyasu U, Verity MA, Vinters HV: Neuropathology of AIDS. Am J Path 124:537-558, 1987
2. Bartholomew C, Cleghorn F, Charles W, Ratan P, Roberts L, Maharay K, Jankey N, Daisley H, Hanchard B, Blattner W: HTLV-I and tropical spastic paraparesis. Lancet II:99-100, 1986
3. DeRossi A, Gallo P, Tavolato B, Callegaro L, Chieco-Bianchi L: Search of HTLV-I and LAV/HTLV-III antibodies in serum and CSF of MS patients. Acta Neurol Scand 74:161-164, 1986
4. Georgsson G, Martin JR, Klein J, Palsson PA, Nathanson N, Petursson G: Primary demyelination in visna. Acta Neuropath 57:171-178, 1982
5. Gall RC, Wong-Staal F: Human T lymphotrophic retroviruses. Nature 317:395-403, 1985
6. Gessain A, Barin F, Vernant JC, Gout O, Maurs L, Calendar A, de The G: Antibodies to human T-lymphotropic virus type 1 in patients with tropical spastic paraparesis. Lancet II:407-411, 1985
7. Gessain A, Abel L, de The G, Vernant JC, Raverdy P, Guillard A: Lack of antibody to HTLV-I and HIV in patients with MS from France and French West Indies. Brit Med J 293:424-425, 1986
8. Haase AT: Pathogenesis of lentivirus infections. Nature 322:130-136, 1986
9. Hauser SL, Aubert C, Burks JS, Kerr C, Lyon-Caen O, de The G, Brahic M: Analysis of human T-lymphotropic virus sequences in multiple sclerosis tissue. Nature 322:176-177, 1986
10. Johnson RT, McArthur JC: Myelopathies and retroviral infections. Ann Neurol 21:113-116, 1987
11. Karpas A, Kampf U, Siden A, Koch M, Poser S: Lack of evidence of involvement for known human retroviruses in MS. Nature 322:177-178, 1986
12. Koprowski H, DeFreitas EC, Harper ME, Sandberg-Wollheim M, Sheremata WA, Robert-Guroff M, Saxinger CW, Feinberg MA, Wong-Staal F, Gallo RC: Multiple sclerosis and human T-cell lymphotropic viruses. Nature 318:154-160, 1985
13. Koprowski H, DeFreitas EC, Harper ME, Sandberg-Wollheim M, Sheremata WA, Robert-Guroff M, Saxinger CW, Feinberg MA, Wong-Staal F, Gallo RC: Letter. Nature 322:178, 1986
14. Kuroda Y, Shibasaki H, Sato H, Okochi K: Incidence of antibody to HTLV-I is not increased in Japanese MS patients. Neurology 37:156-158, 1987

15. Nathanson N, Palsson PA, Petursson G: The pathology of visna and maedi in sheep. In R.H. Kimberlin (ed): Slow virus diseases of animals and man. North Holland Publishing, Amsterdam 1976

16. Ohta M, Ohta K, Mori F, Nishitani H, Saida T: Sera from patients with MS react with human T-cell lymphotropic virus I GAG proteins but not ENV proteins - Western Blotting analysis. J Immunol 137:3440-3443, 1986

17. Osame M, Matsumoto M, Usuku K, Izumu S, Ijichi N, Amitani H, Tara M, Igata A: Chronic progressive myelopathy associated with elevated antibodies to HTLV-I and adult T-cell leukemia-like cells. Ann Neurol 21:117-122, 1987

18. Palsson PA: Maedi and visna in sheep. In R.H. Kimberlin (ed): Slow virus diseases of animals and man. North Holland Publishing, Amsterdam 1976

19. Paty DW, Blume WT, Brown WF, Jaatoul N, Kertesz A, McInnis W: Chronic progressive myelopathy. Investigation with CSF electrophoresis, evoked potentials and CT scan. Ann Neurol 6:419-424, 1979

20. Pert C, Hill JM, Ruff MR, Berman RM, Robey WG, Arthur LD, Ruscetti FW, Farrar WL: Octapeptides derived from the neuropeptide receptor-like pattern of antigen T4 in brain potently inhibit HIV receptor binding and T-cell infectivity. Proc Nat Acad Sci 83:9254-9258, 1986

21. Petito CK, Navia BA, Cho ES, Jordan BD, George DC, Price RW: Vacuolar myelopathy pathologically resembling subacute combined degeneration in patients with AIDS. New Engl J Med 312:874-879, 1985

22. Rice GPA, Armstrong HA, Bulman DE, Paty DW, Ebers GC: Absence of antibody to HTLV-I and III in sera of Canadian patients with MS and chronic myelopathy. Ann Neurol 20:533-534, 1986

23. Roman GC, Roman LN, Spencer PS, Scchoenberg BS: Tropical spastic paraparesis: A neuroepidemiological study in Columbia. 17:361-365, 1985

24. Roman GC: Retrovirus associated myelopathies. Arch Neurol 44:659-663, 1987

25. Roman GC, Schoenberg BS, Madden DL, Sever JL, Hugon J, Ludolph A, Spencer PS: HTLV-I antibodies in the serum of patients with tropical spastic paraparesis in the Seychelles. Arch Neurol 44:605-607, 1987

26. Roy S, Fitz-Gibbon L, Spira B, Portnoy J, Wainberg MA: False-positive results of confirmatory testing for antibody to HIV-I. Can Med Assoc J 136:612-615, 1986

27. Sigurdsson B, Palsson PA: Visna of sheep - a slow demyelinating infection. Brit J Exp Path 39:519-528, 1958

28. Sigurdsson B: Maedi, a slow progressive pneumonia in sheep. Brit Vet J 110:255-270, 1954

29. Sigurdsson B, Palsson PA, Van Bogaert L: Pathology of visna. Acta Neuropath 1:343-363, 1962

30. Vernant JC, Maurs MD, Gessain A, Barin F, Gout O, Delaporte JM, Sanhadji K, Buisson G, de The G: Endemic tropical spastic paraparesis associated with HTLV-I: A clinical and seroepidemiological study of 25 cases. Ann Neurol 21:123-130, 1987

31. Weinshenker B, Rice GPA: Antibodies to HTLV-I in Canadian patients with multiple sclerosis. Can J Neurol Sci (submitted 1987)

32. Wiley CA, Schrier RD, Nelson JA, Lampert PW, Oldstone MBA: Cellular localization of HIV infection within the brains of AIDS patients. Proc Nat Acad Sci 83:7089-7093, 1986

12

HERPES SIMPLEX VIRUS TYPE 1 (HSV1):   A MURINE MODEL OF VIRUS INDUCED

CENTRAL NERVOUS SYSTEM DEMYELINATION

L. Kastrukoff, A. Lau, D. Osborne, and S.U. Kim

Division of Neurology
University of British Columbia
Vancouver, Canada

INTRODUCTION

Although demyelination may be observed in a wide variety of pathological conditions affecting both the peripheral and central nervous systems (PNS and CNS), it is the major feature of multiple sclerosis (MS), a disease affecting the human CNS.

The characteristics of this disease including its epidemiology, genetics, clinical, and pathological features have been reviewed elsewhere (8). The understanding of the etiology and pathogenesis of the disease is of utmost importance, as it is unlikely that any significant advancements in the treatment or prevention of this illness will occur without it. Despite considerable work by numerous investigators, the cause of MS remains unknown.

A number of theories have been proposed, based on available information, which attempt to explain the development of the disease. Among them, two have achieved prominence. One theory suggests that MS represents a disorder of the immune system, resulting in immune dysregulation and the development of an autoimmune disease. The other theory invokes an infectious agent as playing a causative role. Experimental evidence is available which would support both of these theories (24, 25) but has not as yet defined the fundamental immune defect(s) or infectious agent(s) involved. It is entirely possible that these two theories are not mutually exclusive and that both may be operational. In a genetically susceptible population (9), MS may result from defective immune regulation but be triggered by infectious agent(s).

Infectious Agents and Multiple Sclerosis

Classically, the evidence offered in support of infectious agents playing a role in MS comes from studies of epidemiology, serology, and virology (26, 27). Additional indirect support comes from the recognition that a number of viruses are capable of inducing chronic neurological disease in man. Such diseases include subacute sclerosing panencephalitis (SSPE), progressive multifocal leuco-encephalitis (PML), and progressive congenital rubella. Critical evaluation of these results, however, reveals that they are largely circumstantial in nature or as in the case of viral isolation studies, suffer from a lack of reproducibility. The reasons for the latter are not entirely clear but may reflect a lack of sensitivity on the part of the methodology to identify infectious agents, the inability to identify infectious agents which may play a role only during the early phases of the disease, involvement of a number of different infectious agents which could potentially trigger the disease, or the possibility that infectious agents do not play a role and that there is some other explanation for the epidemiology and serologic results.

We are interested in examining the possibility that a number of different viruses may play a role in MS by acting as "triggers" to a potentially defective immune system. Furthermore, we are interested in defining mechanisms by which viruses can induce central nervous system demyelination. One approach to this problem involves the use of experimental animal models of virus induced CNS demyelination.

Experimental Animal Models of Virus Induced CNS Demyelination

Theoretically, viruses could induce CNS demyelination by a number of different mechanisms (Table 1). The existence of some of these mechanisms has been confirmed in experimental animal models and are discussed in greater detail by Dr. L. Weiner, elsewhere in this monograph.

Various animal models and viruses have been studied in an effort to identify a model system which reproduces all of the features of multiple sclerosis in man. The various models and their relevance to MS has been the subject of review (7). Although none of the model systems reproduce all aspects of the human disease, much information has been gained on various mechanisms of CNS demyelination induced by a number of viruses including Theiler murine encephalomyelitis virus, JHM virus, Visna, and canine distemper virus, among others.

Herpes Simplex Virus

Herpes simplex virus (HSV) has also been studied as an experimental animal model of CNS demyelination (7).

HSV, a member of the family Herpesviridae, is a complex virus consisting of a double stranded DNA core and surrounded by a capsid of 162 capsomeres (Figure 1). Together they form the nucleocapsid which is approximately 100 nm in diameter. This in turn is surrounded by an envelope giving a final diameter of 150-200 nm (figure 2). The epidemiology, virology, and clinical characteristics of this virus have been extensively reviewed (11).

HSV infection may have relevance to MS for a number of reasons. It is ubiquitous in man; with neutralizing antibodies present in >50% of the population by age 5 and 80% by adult life (28). Furthermore, the virus has a predilection for establishing a latent infection of the nervous system. It is present in ganglia of the peripheral nervous system (PNS) of up to 40-60% of random autopsies in which co-cultivation techniques are employed (34) and up to 80% if more sensitive complementation techniques (3) are used. Recently, HSV complementary nucleic acid sequences have also been reported to be present in the central nervous system (CNS) of some controls and patients with MS (10).

Table 1   Possible Mechanisms Mediating Virus Induced CNS
Demyelination

A. Viral Infection of Glial Cells
    - Acute Infection
    - Latent Infection
    - Chronic or Slow Infection

B. Interaction of Viral and Immune Systems
    - Humoral and/or cellular response to viral antigens
    - "Bystander effect"
    - Post-infectious encephalomyelitis
    - Immune response to neoantigens

C. Demyelination Without Glial Cell Infection
    - Molecular mimicry
    - Direct effect of virus on components of the immune
      system resulting in altered immune function

D. Combined Cytotoxic Effect of Virus on Glial Cells
    and Immune Mediated Response

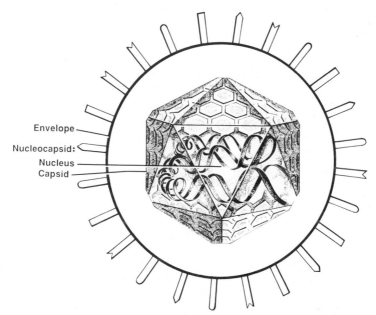

Figure 1   Herpes simplex virus. The double stranded DNA core (100
million daltons) is surrounded by an icosahedral capsid of 162 capsomers.
The capsomers in cross section appear to be hexagonal or pentagonal (at
the vertices). The capsid is assembled in the nucleus with maturation
occuring as the nucleocapsid acquires an envelope at the nuclear
membrane. Intact virions that have been recently released from cells
contain envelopes that form a tight sheath around the capsid but in older
preparations, the envelope often becomes large, irregular, and separates
from the capsid.

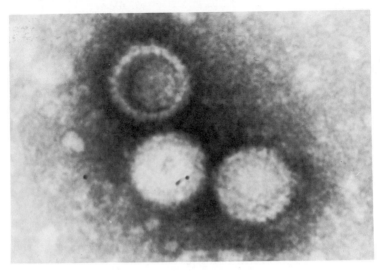

Figure 2   EM photomicrograph of Herpes simplex viruses. In this
negatively stained preparation, the naked nucleocapsid is identifiable.
x 132,00

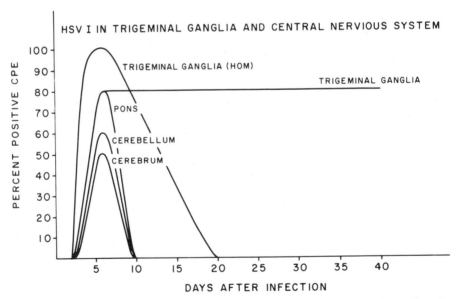

Figure 3   Male Balb/c mice, 8-10 weeks of age, were lip inoculated
with HSV 1 laboratory strain 2. Groups of 10-20 mice were sacrificed
each day in serial fashion. The trigeminal ganglia and brains were
removed, sectioned and either tissue fragments or homogenates were
incubated with indicator cells (CV_1 cells). Cultures were considered
positive for infectious virus if a typical spreading cytopathic effect
was observed. Two phases of infection occur in the peripheral and two in
the central nervous systems.

A Murine Model of HSV 1 Infection of the Nervous System

HSV 1 infection in experimental animals has allowed the study of a number of biological phenomena including the spread of virus to the nervous system, the development of CNS demyelination, the development of viral latency, and interactions between virus and the host immune system (1). This chapter will focus on the first two.

A variety of factors can influence the nature of the model system which is obtained. The route of administration of virus (intra-cerebral, intravenous, intraperitoneal), the strain and size of the viral inoculum, and the age and strain of experimental animals will significantly alter the outcome. In this system, adult mice (8-10 weeks of age) are peripherally (lip) inoculated with HSV 1, laboratory strain 2. This route was chosen to more closely approximate the route of infection in man and the strain of virus chosen to produce low mortality and high latency rates. Following lip inoculation, virus can be detected in the nervous system of mice using either tissue homogenates or co-cultivation techniques (6, 14). Two phases of infection occur in the PNS and two phases in the CNS (Figure 3). In the PNS, infectious virus can be isolated from the trigeminal ganglia (TG) up to twenty days after infection (acute infection) while latent virus can be observed for prolonged periods (up to ten months in our studies). In the CNS, infectious virus can be isolated from different areas of the brain for up to ten days post-infection (acute infection). Thereafter, virus can not be isolated using either homogenate or co-cultivation techniques but complementary viral nucleic acid sequences can be identified in the brain using hybridization techniques (chronic infection) (22,4,29). These studies indicate that basic differences exist between the interaction of virus and cells of the peripheral and central nervous systems.

Spread of virus from the site of inoculation to the peripheral ganglia is likely by retrograde intra-axonal transport (30,2,12) but cell to cell spread by fibroblasts and schwann cells may also contribute (13,21). Hematogenous routes have generally not been found to be of primary importance in viral spread with HSV (6,13) and these results have been confirmed in this model (Kastrukoff, unpublished). Viral spread to the CNS likely represents an extension of the process in the PNS. Knotts et al (18) concluded from their studies in mice, that HSV 1 spreads from the peripheral ganglia to the CNS along nerve fibers and tracts by cell to cell extension and our studies are in agreement with these observations (Kastrukoff, unpublished). Additional studies were performed to determine if cerebrospinal fluid (CSF) could act as a route of viral spread in the CNS but no evidence was found which would support this possibility.

Following the peripheral inoculation of HSV 1 in mice, demyelinating lesions are observed at the trigeminal root entry zone (TREZ) of the CNS (Figure 4). In both Swiss and Balb/c mice, early demyelination can be observed 5-8 days post infection (pi) and are associated with an inflammatory cell infiltrate including myelin laden macrophages (33,19). Somewhat later, naked axons are observed and this may be followed by remyelination (19,32). The mechanisms mediating the development of this viral induced CNS demyelination remain unclear. Townsend and Baringer observed a reduction in demyelination using nude mice or immuno-suppression with cyclophosphamide (33,31). They interpreted these results as implicating a role for the host immune system in the development of the lesions. In contrast, Kristenson et al (20) performed similar experiments with cyclophosphamide but observed increased demyelination. They interpreted their results as indicating an increased spread of virus from the periphery and in support of a direct cytocidal

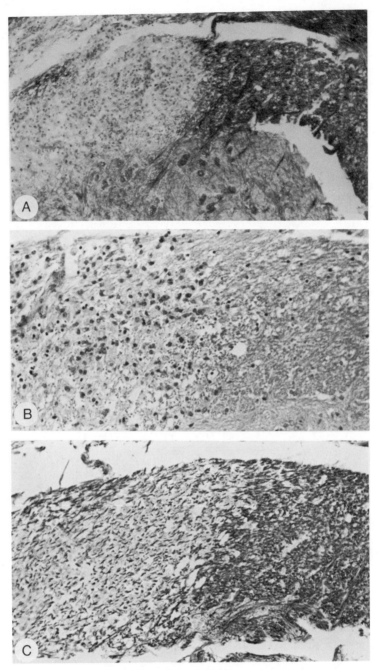

Figure 4   Male Balb/cByJ mice, 8-10 weeks of age, were lip inoculated
with HSV 1 laboratory strain 2. Mice were sacrificed on Day 6 pi by
in-vivo perfusion with 80% methanol, 10% acetic acid, and 10%
formaldehyde. CNS tissue was dehydrated in alcohol and toluol, embedded
in paraffin, and serial six-micron-thick coronal sections were obtained.
(A) A focal area of demyelination on the CNS side of the trigeminal root
entry zone is observed (Luxol fast blue-cresyl fast violet stain, x 100)
(B) The demyelination is associated with a mononuclear cell infiltrate
(Hematoxylin-Eosin stain, x 200)   (C) The demyelinating lesions are
associated with a relative preservation of axons (Luxol fast blue-Holmes
silver nitrate stain, x 200).

effect of virus on oligodendrocytes. The differences in results likely reflect differences in experimental methodology and most likely both mechanisms play a role in the development of these lesions. Confirmation of this hypothesis, however will require further study.

HSV 1 Induced Multifocal Demyelination of the CNS in Mice

Although HSV 1 induced demyelination of the CNS is of interest and relevant to MS, it can be criticized because of its unifocal and uniphasic nature; features unlike the human disease (7). We began to investigate the possibility that both a multifocal and multiphasic disease model might exist. In an earlier study, viral involvement of the CNS following lip inoculation with HSV 1 was found to vary depending on the murine strain examined (15). Furthermore, the extent of viral involvement of the CNS appeared to correlate with the degree of mortality. The latter developing mainly from Day 9 to 12 pi and resulting from viral encephalitis.

It was reasoned that since mortality might reflect the extent of CNS involvement by virus, it might also reflect differences in the pathological appearance of the CNS. Studies were undertaken examining mortality following lip inoculation with HSV 1 in a number of inbred, congenic, and F1 hybrid strains of mice (16). The results of these studies, outlined in Table 2, indicate that a continuum of resistance exists among the different strains examined. Among the inbred strains, C57BL/10J is most resistant while PL/J is most susceptible.

Inbred strains of mice were further categorized into resistant, moderately resistant, and susceptible groups (16). Representative stains from each group were then selected and examined histologically (17). A resistant stain (C57BL/6) was found to have focal mononuclear cell infiltrates at TREZ associated with little if any demyelination (Figure 5). This was not associated with any other lesions in the brain. A

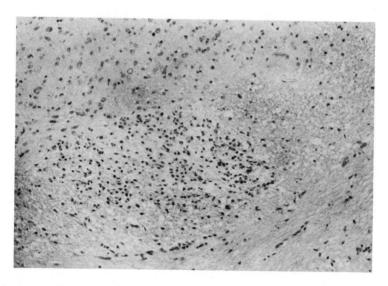

Figure 5   Male C57BL/6J mice (resistant strain) 8-10 weeks of age, were lip inoculated with HSV 1 laboratory strain 2 and sacrificed in serial fashion every 3 days pi. Mononuclear cell infiltrates are observed on the CNS side of the trigeminal root entry zone associated with minimal demyelination. (Hematoxylin-Eosin stain; x 200).

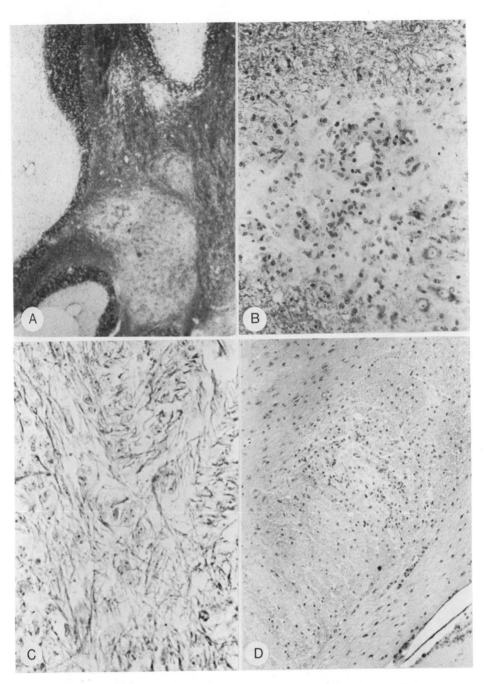

Figure 6   Male A/J mice (susceptible strain) 8-10 weeks of age, were lip inoculated with HSV 1 laboratory strain 2 and sacrificed in serial fashion every 3 days pi.   (A)   Multiple areas of demyelination are present in the cerebellum, 12 days pi (Luxol fast blue-cresyl fast violet; x 40)   (B)   Mononuclear cell infiltrates (often perivascular in location) are associated with the demyelinating lesions (Hematoxylin-Eosin stain, x 200)   (C)   The lesions are associated with relative preservation of axons (Luxol fast blue-Holmes silver nitrate stain, x 400)   (D)   Similar lesions are observed in the cerebral white matter, 15 days pi (Luxol fast blue-cresyl fast violet, x 100).

moderately resistant strain (Balb/cByJ) was found to have a mononuclear cell infiltrate at TREZ (Figure 4). This was associated with demyelination and relative preservation of axons in the area but was not associated with a mononuclear cell infiltrate and relative preservation of axons were found not only at TREZ but throughout the brain (Figure 6). The histological appearance of the CNS therefore appears to correlate with mortality. It is possible that the genetics determining natural resistance to mortality (16) may also apply to the pathological appearance but will require examination of additional inbred strains of mice to confirm. Strain PL/J (a susceptible strain) has recently been examined histologically and found to exhibit multifocal demyelinating lesions in the brain as well. This would suggest that the observations made in strain A/J are not restricted to one strain alone but are part of a more general phenomena.

Although, lesions appearing at the TREZ in all strains examined develop at about Day 6 pi, lesions in other parts of the brain in strains A/J and PL/J appear to develop sequentially. In the latter two strains, lesions first appear to developing the cerebellum on Day 12 pi while in the cerebral hemispheres, on Day 15 pi. The sequential appearance of these lesions is reminiscent of the viral spread through the CNS (15) and is not consistent with a post-infectious encephalomyelitis. Furthermore, since the only variable in these studies is murine strain, the results suggest that host factors, likely genetically determined, play an important role in determining the pathological appearance of the CNS.

Electron microscopy studies were performed to further characterize the nature of the multifocal demyelinating lesions in A/J mice. Cerebellar lesions, developing of Day 12 pi, consisted of abnormally myelinated or demyelinated fibers but with relative preservation of axons (Figure 7). A mononuclear cell infiltrate, abnormally myelinated or demyelinated axons, loosening of the interstitial space, and thinly myelinated axons suggesting the possibility of remyelination were characteristic of a number of lesions examined. The mononuclear cell infiltrate consisted primarily of macrophages although occasional lymphocytes were also observed (Figure 8). In some cases, macrophages were found to extend cytoplasmic processes around abnormally myelinated axons. Careful examination of both neurons and oligodendrocytes however failed to reveal any evidence of virus (Figure 9). The absence of virus distinguishes the multifocal lesions from the unifocal lesions previously described. In the latter case, virus could be observed in neurons, astrocyte, and oligodendrocytes (18,32). These results suggest that the mechanisms involved in the development of multifocal lesions are different from those mediating the development of unifocal lesions.

Since virus was not observed in the EM studies of the multifocal lesions, viral titration studies were performed to further examine the presence of virus in the CNS (17). The results of the titrations in all three murine strains examined are given in Figure 10. The results confirm the sequential spread of virus through the CNS. In the resistant strain (C57BL/6), virus could only be isolated from the trigeminal ganglia and pons. The presence of virus correlates with the appearance of TREZ lesions. A similar correlation is observed for the moderately resistant (Balb/cByJ) and susceptible (A/J) strains. The development of multifocal lesions in the cerebellum and cerebral hemispheres of A/J mice however, occurs at a time when infectious virus could not be isolated from the CNS. These results are consistent with those from the EM studies and suggest that mechanisms other than a direct effect of virus may play a role in the development of these lesions. It must be remembered that infectious virus could be isolated from the cerebellum and cerebral hemispheres of A/J mice during the acute stage of the infection and it is possible that multifocal demyelination represents a late effect of direct damage by the virus. Against this would be the titration studies in Balb/cByJ mice, where infectious virus could also be identified in the cerebellum and cerebral hemispheres during the acute stage but where multifocal demyelination is not observed. It is possible that the development of the multifocal lesions is immune mediated but "triggered" by the acute infection of the CNS in genetically susceptible strains of mice.

As it can be argued that both EM and viral titration studies may not be sensitive enough to detect low levels of virus, PAP studies to detect viral antigen and in-situ hybridization studies with viral probes are currently in progress.

Variable Resistance to HSV 1 Among Oligodendrocytes Derived From Different Strains of Mice

Although the mechanisms mediating the development of both unifocal and multifocal demyelinating lesions in the CNS remain unclear, emphasis has been placed on the immune system and the direct cytocidal effect of

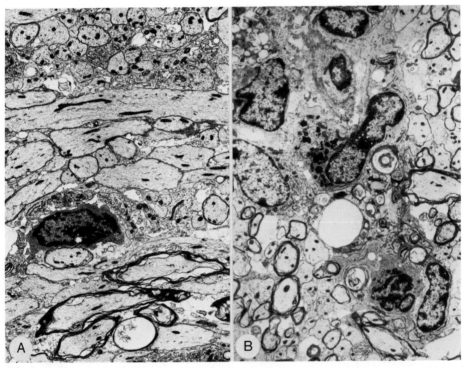

Figure 7   EM photomicrographs of demyelinating lesions in the cerebellum of A/J mice, 12 days post-infection. Mice were perfused in-vivo with 1.5% paraformaldehyde and 2% gluteraldehyde in PBS. CNS tissue was fixed in 3.1% gluteraldehyde at 4 degrees C for 24 hours, sectioned on a vibrotome, post-fixed, stained with 1% osmium tetroxide for 60 minutes and dehydrated in alcohol. Sections were embedded in Epon, sectioned and examined with a Phillips 300 electron microscope.   (A) Both demyelinating and demyelinated axons are observed in the cerebellar lesions but with relative preservation of axons x 1350   (B) In addition to the demyelinating axons, a mononuclear cell infiltrate, loosening of the interstitial space, and occasionally thinly remyelination, are features of the lesions x 1350.

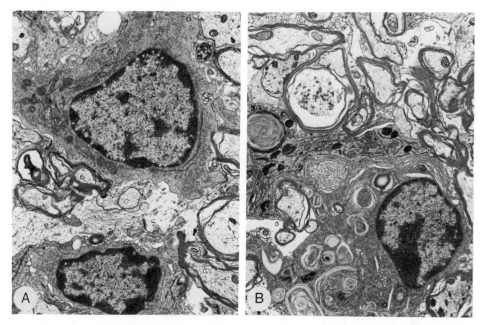

Figure 8   EM photomicrographs of demyelinating lesions in the
cerebellum of A/J mice, 12 days post-infection.   (A)   The mononuclear
cell infiltrate in the lesions consisted primarily of macrophages but
with occasional lymphocytes interspersed x 1600   (B)   Macrophages laden
with myelin debris are observed to extend cytoplasmic processes around
demyelinating axons x 2350.

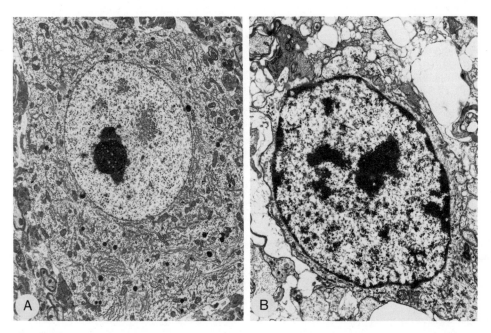

Figure 9   EM photomicrographs of cerebellum of A/J mice, 12 days pi.
(A)   Neurons in close proximity to the demyelinating lesions show no
evidence of viral infection x 950   (B)   Oligodendrocytes in the
demyelinating lesions also show no evidence of viral infection x 1350

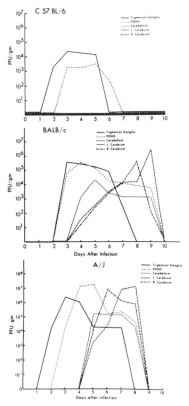

Figure 10   Viral titrations in the trigeminal ganglia and central
nervous system of three strains of mice. Following lip inoculation with
HSV 1, groups of 5 mice from each strain were sacrificed daily for 10
days. The trigeminal ganglia were removed and the brain sectioned.
Homogenates were obtained by freeze-thawing of tissue 3 times followed by
grinding in a Ten Broeck homogenizer. Serial dilutions of tissue
homogenates were plaque-assayed.

virus. This concept is supported by the studies of Lopez on the genetics
of natural resistance to mortality (23). The results of his study were
interpreted as indicating an important role for the immune system while
no differences in resistance were observed at the level of structural
cells (fibroblasts). Recently Collier et al (5) presented results which
suggested that differences in resistance to HSV 1 might also exist at the
level of non-neural structural cells. Using primary oligodendrocyte (OD)
cultures derived from different strains of mice (Figure 11), we have
examined the possibility that differences in resistance to HSV 1 might
also exist at this level and contribute to the pathological appearance
observed among the different strains of mice (17). OD cultures were
examined using $TCID_{50}$, immunofluorescence, and EM techniques. Results of
the $TCID_{50}$ studies (Figure 12) indicate that differences in resistance to
HSV 1 exist among OD derived from different inbreed, congenic, and Fl
hybrid strains of mice. Higher titers of virus are required to infect
cells derived from A/J and PL/J strains. It is of interest that results
of these in-vitro studies do correlate to a great extent with in-vivo
studies of mortality (Table 2). Differences between these two sets of
data likely reflect other factors contributing to the final outcome.
Results of immunofluorescence studies are given in Table 3. Cultures
infected with HSV 1 were stained with anti-HSV 1 monoclonal to
glycoprotein-C in serial fashion and the time at which 50% of the cells

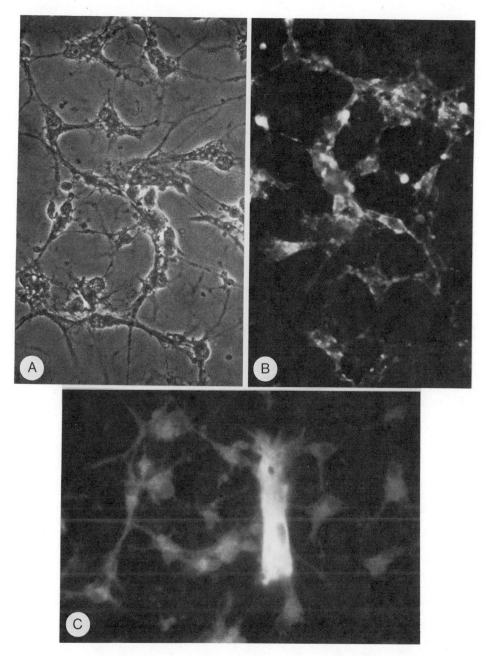

Figure 11   Primary murine oligodendrocyte (OD) cultures derived from
different stains of mice. CNS tissue from mice 10-12 weeks of age were
dissected into 3 mm pieces, incubated in 0.25% trypsin plus 20 mg/ml
DNAse. Digested tissue was passed through a 150 um nylon mesh filter and
centrifuged at 1500 rpm for 10 minutes. Tissue was then rewashed and
suspended in Percoll-Hanks media. This was centrifuged at 15,000 rpm at
4 degrees for 30 minutes. The OD layer was removed, washed repeatedly,
and the cells were plated on polylysine-coated Aclar coverslips. (A)
Phase contrast photomicrograph of cultures 14 days after being
established x 400. The same culture in (A) was doubly stained with anti-
galactocerebroside (B) and anti-glial fibrillary acidic protein (C).
Cultures are 95-98% pure anti-Gal-C positive oligodendrocytes.

were positive were recorded. Differences in resistance could again be
identified among the OD derived from different strains. OD from C57BL/6
became positive at 30 hours while cells from Balb/cByJ at 21 hours, and
those from A/J mice at 15 hours. Overall, the results of the IF studies
in-vitro were similar to mortality studies in-vivo (Table 2). In EM
studies, OD cultures were infected with HSV 1, fixed at various times
post infection, and examined. At 24 hours, major differences in the
appearance of OD derived from the three inbred strains of mice were
observed (Figure 13). In A/J derived cultures, virus was observed in
both the cytoplasm and nuclei of virtually all cells examined. In
Balb/cByJ cultures, less virus was identified in the cytoplasm as
compared to A/J cultures and none in the nuclei. In C57BL/6 cultures,
few cells were found to be infected by virus.

Table 2   Resistance to mortality of inbred strains, F1 hybrids, a
congenic strains lip inoculated with HSV 1

Inbred Strains	H-2	LD_{50} (HSV1)[1]	Strain Effect Coefficient and S.E. (Sample 1)[2]
C57BL/10J	b	$>10^{9.59}$	<-2.33[3]
C57BL/6J	b	$>10^{9.59}$	<-2.33[3]
DBA/1J	q	$>10^{9.59}$	<-2.33[3]
LP/J	b	$10^{8.77}$	$-2.33 \pm .42$
CBA/J	k	$10^{8.12}$	$-1.36 \pm .39$
Balb/cByJ	d	$10^{7.80}$	$-.89 \pm .38$
SWR/J	q	$10^{7.06}$	$.20 \pm .38$
AKR/J	k	$10^{6.96}$	$.35 \pm .38$
DBA/2J	d	$10^{6.85}$	$.51 \pm .38$
A/J	a	$10^{6.52}$	$1.00 \pm .39$
PL/J	u	$10^{6.06}$	$1.69 \pm .42$
C3H/HeJ	k	NR	NR
F1 Hybrids			
B6AF1/J	bxa	$10^{9.59}$	<-2.33[3]
B6D2F1/J	bxd	$10^{9.59}$	<-2.33[3]
(C57BL/6xCBA/J)F1	bxk	$10^{9.59}$	<-2.33[3]
(C57BL/6xLP/J)F1	bxb	$10^{9.59}$	<-2.33[3]
Congenic Strains			
B10.A/SgSnJ	a-b	$>10^{9.59}$	<-2.33[3]
B6.C-JH2^{d}/ByJ	d-b	$>10^{9.59}$	<-2.33[3]
B10.D2/nSnJ	d-b	$>10^{9.59}$	<-2.33[3]
B10.BR/SgSnJ	k-b	$>10^{9.59}$	<-2.33[3]
A.BY/SnJ	b-a	$>10^{6.59}$	$.84 \pm .39$
C3H.SW/SnJ	b-k	NR	NR

[1] Spearman - Karber estimate
[2] Strain effect coefficient from logistic regression analysis of
mortality
[3] All mice survived at all dilutions. Coefficient less than smallest
estimated coefficient; no SE available.
NR   Dose response pattern could not be adequately summarized by LD_{50} or
logistic model.

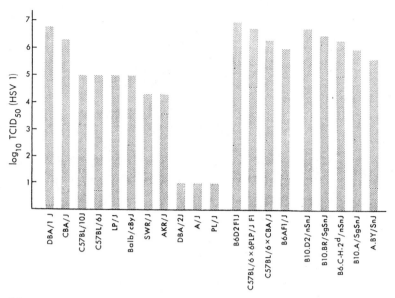

Figure 12   Determination of TCID$_{50}$ for primary oligodendrocyte cultures derived from different strains of mice. Poly-lysine coated Aclar coverslips with OD (5×10^3 cells) were placed in a 24 well Linbro flat-bottomed plate. Media was removed and replaced by tenfold dilutions of virus, eight wells/dilution. Cultures were maintained at 37 degrees for 90 minutes, then washed repeatedly and resuspended in media. Titers were calculated in terms of 50% tissue culture infective dose by the method of Spearman-Karber.

The results of these preliminary studies indicate that resistance to HSV 1 can also exist at the level of neural structural cells. At this time it remains unclear if differences at this level contribute to the different pathological appearances observed in the CNS of the different inbred strains of mice. However, the pattern of results observed in-vitro does appear to correlate with both the in-vivo mortality results and CNS pathological appearance to a great extent and merits further evaluation.

SUMMARY

The etiology and pathogenesis of multiple sclerosis remains unclear although available evidence would indicate that a genetic predisposition, an "abnormal" immune system, and environmental factors (possibly infectious in origin) may play a role. Experimental animal models offer one approach to the study of how viruses might induce central nervous system (CNS) demyelination.

Herpes simplex virus type 1 (HSV 1) is a ubiquitous infection in man and capable of inducing latent infection of the nervous system. Following peripheral inoculation (lip) in mice, virus will spread to the trigeminal ganglia, likely by retro-grade intra-axonal transport, and then to the CNS. Two phases of infection (acute and latent/chronic) can be identified in both the peripheral and central nervous systems. Approximately six days after inoculation, demyelinating lesions can be observed at the trigeminal root entry zone (TREZ) of the CNS. The mechanisms mediating the development of these lesions are unclear but

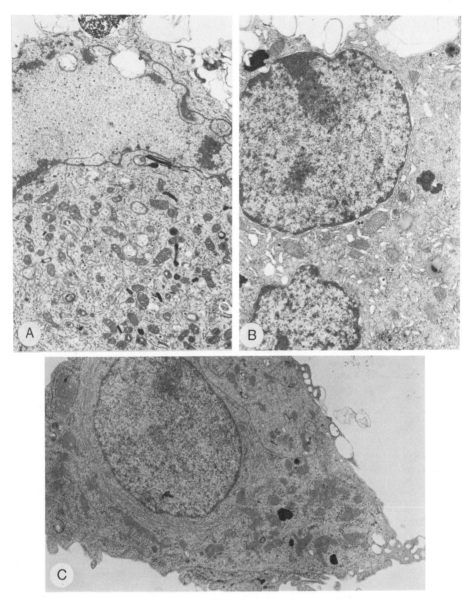

Figure 13   EM photomicrographs of HSV 1 infected oligodendrocytes
derived from three stains of mice.   Cultures were infected at an MOI of 5
for 60 minutes at 37 degrees.   Incubation was continued at 37 degrees,
and at appropriate intervals the coverslips were removed, washed, and
fixed in 3% glutaraldehyde in PBS.   They were post-fixed in 1% osmium
tetroxide for 60 minutes and dehydrated in alcohol.   Cells were
infiltrated with EM Bed 812 over 36 hours, sectioned on an ultra
microtome, and examined with a Phillips 300 electron microscope.
(A) Oligodendrocytes derived from A/J mice contain numerous virions in
both the cytoplasm and nucleus at 24 hours post-infection x 9500   (B)
Oligodendrocytes derived from Balb/cByJ mice are found to have virions in
the cytoplasm (generally fewer than in A/J) but none in the nucleus (24
hours post-infection, x 9500) (C) Oligodendrocytes derived from
C57BL/6J mice are generally not infected at 24 hours post-infection
although less than 5% of cells did show occasional virions in the
cytoplasm x 9500.

Table 3   Resistance of Primary Oligodendrocyte Cultures Derived From Inbred, Fl Hybrids, and Congenic Strains of Mice to HSV1 as Determined by Appearance of Membrane Glycoprotein (gCl)

Group	Strain	Appearance of gCl + Cells (50%)(HRS)[1,2]
Inbred	C57BL/10J	42
	C57BL/6J	30
	DBA1/J	39
	LP/J	33
	CBA/J	36
	Balb/cByJ	21
	SWR/J	27
	AKR/J	21
	DBA2/J	27
	A/J	15
	PL/J	15
Fl Hybrids	(C57BL/6xLP/J)Fl	39
	B6D2Fl/J	33
	(C57BL.6xCBA/J)Fl	42
	B6AFl/J	33
Congenic	B10.A/SgSnJ	39
	A.BY/SnJ	36
	B6.C-H-2/ByJ	42
	B10.D2/nSnJ	30
	B10.BR/SgSnJ	39

1. All cultures were infected at a MOI of 10^8 PFU/ml.
2. Cultures were stained with anti-gC_1 every 3 hours.

probably involve both the immune system and a direct cytocidal effect of virus.

Although this is an interesting model of demyelination and relevant to MS, it can be criticized for its unifocal and uniphasic nature; features unlike the human disease. Studies are being performed to identify a multifocal and multiphasic disease model. Various inbred, congenic, and Fl hybrid strains of mice have been categorized on the basis of resistance to mortality following lip inoculation with HSV 1. As mortality likely reflects CNS involvement by virus, it was reasoned that the pathological appearance of the CNS might also correlate with mortality. Representative strains from each group: resistant (C57BL/6J), moderately resistant (Balb/cByJ), and susceptible (A/J) were examined histologically. The appearance of the CNS was variable and strain dependent. In the susceptible strain, not only were demyelinating lesions observed at TREZ but also in the cerebellum and cerebral hemispheres. These lesions appeared sequentially and were characterized by demyelination, relative preservation of axons, and a mononuclear cell infiltrate. EM studies confirmed the characteristics of the multifocal lesions observed by light microscopy but failed to identify the presence of virus; a characteristic feature of the unifocal lesions observed at the TREZ. The appearance of the multifocal lesions in the cerebellum and cerebral hemispheres, at a time when infectious virus could not be isolated from the CNS, was confirmed by viral titration studies. These studies have been interpreted as indicating that multifocal demyelinating lesions in the CNS are likely immune mediated but "triggered" by an acute HSV 1 infection of the CNS in genetically susceptible mice.

Although the host immune system and a direct cytocidal effect of virus likely represent the major mechanisms involved in mediating the development of HSV 1 induced CNS demyelination, differences in resistance to virus at the level of the oligodendrocyte (OD) may also contribute. Preliminary studies were performed using primary cultures of OD derived from different strains of mice. Immunofluorescence, $TCID_{50}$, and EM studies indicate that differences in resistance to HSV 1 do exist at this level and correlate with in-vivo mortality results and pathological appearances of the CNS.

These studies have identified a multifocal model of HSV 1 induced CNS demyelination and additional mechanisms which may play a role in their development. Further studies with this model are likely to increase our understanding of how viruses can induce CNS demyelination and possibly our understanding of MS.

REFERENCES

1. Baringer JR: Herpes simplex virus infection of nervous tissue in animals and man. Prog Med Virol 20:1, 1975
2. Baringer JR, Swoveland P: Persistent herpes simplex virus infection in rabbit trigeminal ganglia. Lab Invest 30:230, 1974
3. Brown SM, Subak-Sharpe J, Warren KG, Wroblewska Z, Koprowski H: Detection by complementation of defective or uninducible (herpes simplex type 1) virus genomes latent in human ganglia. Proc Natl Acad Sci 6:2634, 1979
4. Cabrera C, Wohlenberg C, Openshaw H, Rey-Mendez M, Puga A, Notkins AL: Herpes simplex virus DNA sequences in the CNS of latently infected mice. Nature 228:288, 1980
5. Collier LH, Scott QJ, Pani A: Variation in resistance of cells from inbred strains of mice to herpes simplex virus type 1. J gen Virol 64:1483, 1983
6. Cook ML, Stevens JG: Pathogenesis of herpetic neuritis and ganglionitis in mice: evidence for intra-axonal transport of infection. Infect Immun 7:272, 1973
7. Dal Canto MC, Rabinowitz GS: Experimental models of virus-induced demyelination of the central nervous system. Ann Neurol 11:109, 1982
8. Demyelinating Disease. Handbook of Neurology, Vinken PJ, Bruyen GW (Eds) North Holland Press, Amsterdam, 1980
9. Ebers GC: Genetic factors in Multiple Sclerosis. In: Neurologic Clinics 1:645, JP Antel (ed), 1983
10. Fraser NW, Lawrence WC, Wroblewska Z, Gilden DH, Koprowski H: Herpes simplex type 1 DNA in human brain tissue. Proc Natl Acad Sci 78:6461, 1981
11. The Herpesviruses. Vol. 1-4. Roizman B. (Ed.), Plenum Press, New York, 1985
12. Hill TJ, Field MJ, Roome APC: Intra-axonal location of herpes simplex virus particles. J gen Virol 15:253, 1974
13. Johnson RT: The pathogenesis of herpes virus encephalit, Part 1. (Virus pathways to the nervous system of suckling mice demonstrated by fluorescent antibody staining). J Exp Med 119:343, 1964

14. Kastrukoff LF, Long C, Koprowski H: Herpes simplex virus immune system interaction in a murine model. In: The Human Herpes viruses. Nahmias AJ, Dowdle WR, Schinazi RF (Eds.) Elsevier Biomedical. NY, Amsterdam, Oxford, 1981

15. Kastrukoff LF, Hamada T, Schumacher U, Long C, Doherty PC, Koprowski H: Central nervous system infection and immune response in mice inoculated into the lip with Herpes simplex virus type 1. J Neuroimmunol 2:295, 1982

16. Kastrukoff LF, Lau AS, Puterman ML: Genetics of natural resistance of Herpes simplex virus type 1 latent infection of the peripheral nervous system in mice. J gen Virol 67:613, 1986

17. Kastrukoff LF, Lau AS, Kim SU: Multifocal CNS demyelination following peripheral inoculation with Herpes simplex virus type 1. Ann Neurol 22:52, 1987

18. Knotts FB, Cook ML, Stevens JG: Pathogenesis of herpetic encephalitis in mice after ophthalmic inoculation. J Inf Dis 130:16, 1974

19. Kristensson K, Svenerholm B, Vahlne A, Lycke E: Latent Herpes simplex virus trigeminal ganglionic infection in mice and demyelination in the central nervous system. J Neurol Sci 43:253, 1979

20. Kristensson K, Svenerholm B, Vahlne A, Nilheden E, Persson L, Lycke E: Virus-induced demyelination in Herpes simplex virus-infected mice. J Neurol Sci 53:205, 1982

21. Lascano EF, Berria MK: Histological study of the progression of Herpes simplex virus in mice. Arch Virol 64:67, 1980

22. Lawrence W, Kastrukoff L, Gilden D, Fraser N: 5th Cold Spring Harbour Meeting on Herpes virus, Abstract 21, 1980

23. Lopez C: Genetics of natural resistance to Herpes virus infections in mice. Nature 258:152, 1975

24. McFarlin DE, McFarland HF: Multiple Sclerosis (first of two parts). N Engl J Med 307:1183, 1982

25. McFarlin DE, McFarland HF: Multiple Sclerosis (second of two parts). N Engl J Med 307:1246, 1982

26. Multiple Sclerosis. Hallpike JF, Adams CWM, Tourtellotte WW (Eds.) Williams and Wilkines, Baltimore, 1983

27. Multiple Sclerosis. McDonald WI, Silberberg DH (Eds.) Butterworths, 1986

28. Nahmias AJ, Roziman B: Infection with herpes-simplex viruses 1 and 2. N Engl J Med 289:781, 1973

29. Rock DL, Fraser NW: Detection of HSV 1 genome in central nervous system of latently infected mice. Nature 302:523, 1983

30. Stevens JG, Cook ML: Latent herpes simplex virus in spinal ganglia in mice. Science 173:843, 1971.

31. Townsend JJ: The demyelinating effect of corneal HSV infections in normal and nude (athymic) mice. J Neurol Sci 50:435, 1981

32. Townsend JJ: Schwann cell remyelination in experimental Herpes simplex encephalitis at the trigeminal root entry zone. J Neuropath Exp Neur 42:529, 1983

33. Townsend JJ, Baringer JR: Morphology of central nervous system disease in immunosuppressed mice after peripheral herpes simplex virus inoculation. Lab Invest 40:178, 1979

34. Warren KG, Moira-Brown S, Wroblewska Z, Gilden D, Koprowski H, Subak-Sharpe J: Isolation of latent Herpes simplex virus from the superior cervical and vagus ganglions of human beings. N Engl J Med 298:1068, 1978

13

EXPERIMENTAL ALLERGIC ENCEPHALOMYELITIS

AS A MODEL OF MULTIPLE SCLEROSIS

Ellsworth C. Alvord Jr.

Department of Pathology
University of Washington
Seattle, WA

Opinions have changed many times during the past several decades as to whether experimental allergic encephalomyelitis (EAE) is a good model of multiple sclerosis (MS) or not. It is not likely today that everyone's thoughts are synchronized, indeed probably quite likely that each of us has retained our first or last impression based on thoughts of several or many years ago. Since the situation has changed rapidly in the last few years, I welcome this opportunity to try to bring everyone up to speed. I hope that I can convince most of you that EAE is a good, perhaps even a perfect, model of MS (Fig. 1) - but we must be careful in defining just which form of EAE we are talking about.

It is somewhat ironic that EAE was originally thought to be one disease, specifically only monophasic and demyelinating, even though each of those early investigators described several histology and clinical patterns of EAE following the use of whole central nervous system (CNS) tissues for sensitization. However, they had little or no control over the production of these patterns and did not know that this apparently single model, eventually identified as "EAE", contained a mixture of reactions that we have come to recognize as just that, a mixture largely related to different antigens. The irony is compounded by the fact that one of those antigens, myelin basic protein (BP), is so dominant that the effects of the others can hardly be recognized without it. Having spent decades studying the dominant one, I have to admit that it is probably not sufficient to explain the whole mixture of reactions - but I also have to report that studying the others in isolation did not, in retrospect could not, lead very far beyond the observation in tissue cultures of demyelinating or myelination-inhibiting antibodies.

The birth of the idea that EAE and MS might be related goes back over 50 years to the time when Rivers et al (38, 39) showed that sensitization of monkeys to whole CNS tissues could produce an inflammatory and demyelinating reaction confined to the CNS. Although confirmed within the next decade (21), the original experimental model was very cumbersome, requiring months of repeated injections, and it was not until the development of Freund's adjuvants that the technique for producing EAE became very simple, requiring only a single injection and a few weeks for the response to occur. From those first experiments in the last half

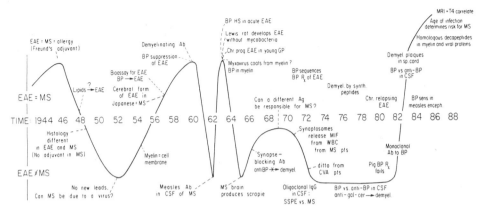

Figure 1   Fluctuations over the past 40 years in the belief that EAE is a good (EAE = MS) or a not-so-good (EAE = MS) model for MS.

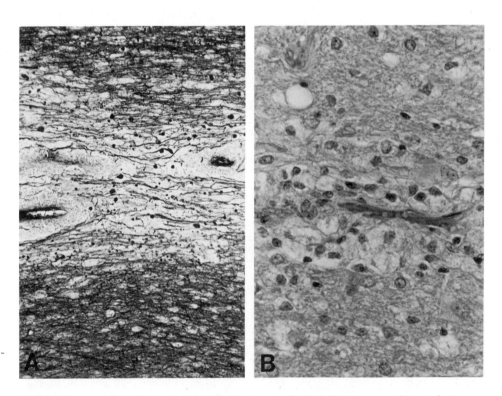

Figure 2   Subclinical healed lesions of EAE showing perivascular demyelination and foamy macrophages in monkey 84255 whose CD4+ lymphocytes decreased markedly 20 days after sensitization to BP but who never showed any clinical signs and was sacrificed 57 days later. Data courtesy of LM Rose and S Hruby. A, LFB-Holmes stain for myelin and axons, x 32; B, H and E, x 130.

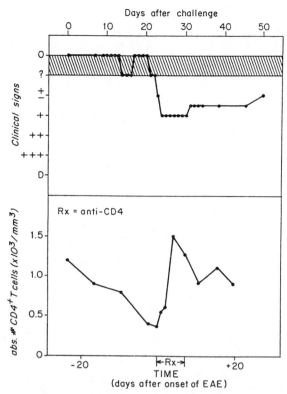

Figure 3   Correlation of clinical signs and circulating CD4+
lymphocytes in monkey 86213 developing EAE and responding to treatment
with a monoclonal antibody to CD4 antigen.   Data courtesy of LM Rose and
S Hruby.

of the 1940's (1, 12, 22, 27, 28, 31, 35, 36) the model has become
increasingly popular and subjected to extensive biochemical,
immunological, genetic, phyiological, histopathological and
electronmicroscopic analyses.   The pace has increased so rapidly in the
last few years that I have had trouble keeping up even with the results
being generated by my own associates!

     As shown in Figure 1, as recently as the last half of the 1970's I
believed that EAE was important only to study for itself, that it was not
likely to provide any meaningful leads for the study of MS.   What is it
that has caused me to change my mind beginning in the early 1980's and
increasing to the point that I would now suggest that EAE may even be a
perfect model of MS?

     This new evidence consists of five major elements: 1) the
development of a 2-cell model of chronic relapsing demyelinating EAE in
guinea pigs (5, 16, 19, 20) and rats (33, 34); 2) the development of a
relapsing model of EAE in non-human primates by the simple device of
inadequately prolonged treatment of acute attacks (10, 45, 47); 3) the
definition of different regions of BP as being encephalitogenic for
different species and strains of animals (3); 4) the discovery of
"homologous" decapeptides (4, 7, 8, 26, 48) in many micro-organisms that
may be biochemically similar enough to "mimic" corresponding sequences in
myelin proteins and to evoke immunological cross-reactions with myelin,

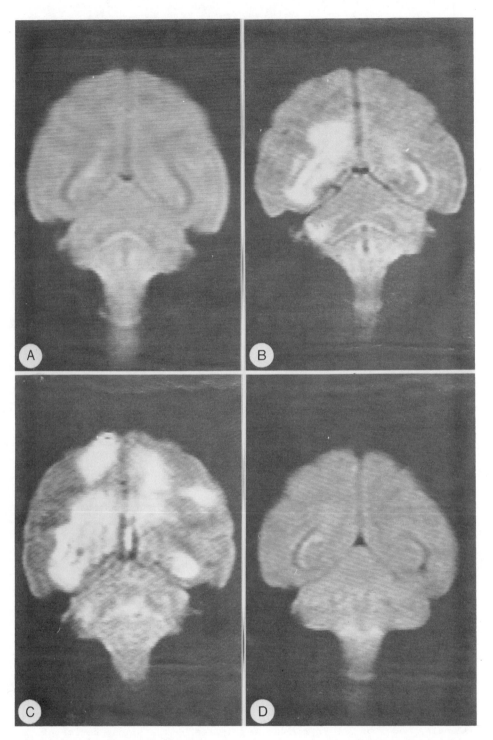

Figure 4   Serial MRI's at 17, 22, 25 and 46 days after sensitization of the monkey shown in Fig. 3. Data courtesy of TL Richards. Note the drop in lymphocytes (Fig. 3) and the first MRI-visible lesion (Fig 4B), both occurring 3 days before the appearance of clinical signs (Fig. 3).

thereby inducing EAE not only in animals (23, 24) but also in humans, where it appears as either acute disseminated encephalomyelitis (DEM) or its peripheral equivalent, the Guillain-Barre-Strohl (GBS) syndrome (4, 7, 8, 26); and 5) the recognition that increasing age of infection with certain common world-wide infections (specifically measles, mumps and rubella) increases the risk that the individual will develop demyelinating diseases, especially MS, years to decades later (18).

The last factor may also account not only for the geographic distribution of MS but also for the development of MS in migrants (32). The combination of the last two factors in the human makes possible the extrapolation of the first three factors from animals to humans and the recognition of the virtual identity of one particular form of EAE and MS (4, 8). The combination of all five elements makes a very real possibility of the prediction that MS is already being prevented and that 1987 is the last year that we will see unmodified MS. Within the next decade we should have clear-cut evidence that MS is disappearing.

Let me consider each of these five elements in detail:

1) The 2-cell model of chronic relapsing demyelinating EAE is based on the observation in strain 13 guinea pigs that passive transfer of suboptimal numbers of BP-sensitized lymph node cells creates a state of partial resistance to subsequent active challenge (sensitization) to produce EAE (19). Such guinea pigs are actually completely resistant to challenge with BP (in Freund's complete adjuvants) and half of them are equally completely resistant to challenge with whole CNS (also in adjuvants). The other half of the guinea pigs, however, develop chronic remitting-relapsing or progressive neurologic signs lasting weeks to months (19) and show extensive demyelination at autopsy (5). This demyelination correlates well with the presence of demyelinating (myelination-inhibiting) antibodies in their sera (16). These antibodies and the demyelination result from the presence of non-encephalitogenic components of myelin in the whole CNS challenge (20) but no disease occurs unless the BP-sensitized T cells are also present.

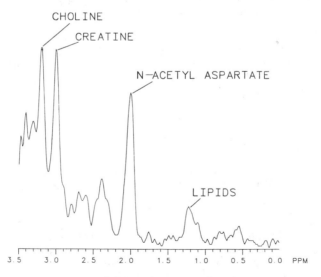

Figure 5   Proton spectral signal indicating lipids that appeared transiently on day 25 in the lesion shown in Fig. 4B just dorso-medial to the ventricle.   Data courtesy of TL Richards.

In Lewis rats comparable antibodies have been identified as being directed against a minor glycoprotein in CNS myelin (33, 34) but the clinical sequence has not yet been converted to the chronic patterns seen in the guinea pigs. In addition, however, antibodies to galactocerebroside have been shown to produce necrosis _in vivo_ when superimposed on the inflammation induced by the passive transfer of BP-sensitized T lymphocytes (33). Whether necrosis and demyelination are merely quantitatively different so that each can be produced by appropriately different concentrations of these antibodies remains to be proven but we have shown that synthetic galactocerebroside, free of any possible natural contaminant, can evoke demyelinating antibodies, not necrosis, detectable in tissue culture (25). In any event, these results already show that there is a spectrum of inflammation-demyelination-necrosis that can be produced at will by appropriate combinations of specific T and B cells.

Since two antigens and two immunologic effector systems are involved in these new guinea pig and rat models, it is easy to see that two cells, specifically but differently sensitized T and B cells, are involved in these species. In rabbits and monkeys, however, the necessity for two cells is not so clear since demyelinating EAE can be produced by sensitization only to BP or BP fragments (30, 47). Since we have not seen EAE develop in monkeys in the absence of anti-BP antibodies (42), we suspect that two cells, T and B cells, are still involved but that the two antigens may be different epitopes (antigenic determinants) on BP itself. The possibility of some minor contaminant of BP has not been completely ruled out, however. Whether humans require two antigens, such as BP and non-BP myelin components, or only one with two epitopes, such as BP, remains to be seen but it seems likely that both T and B cells are involved in MS as well as in demylinating EAE.

Even without this proof of two cells being involved, the monkey model has been very useful in providing many direct comparisons between EAE and MS, most recently in the areas of lymphocyte subsets (40-43) and magnetic resonance imaging or MRI (37, 50). In some cases changes in circulating lymphocytes have occurred without any clinically apparent EAE ever developing but with subclinical healed histologic lesions detectable at post mortem examination (Fig 2). In other cases the correlation of lymphocyte changes and MRI has shown the actual onset to be days ahead of what we could detect clinically (Figs. 3 and 4). As in MS, the EAE is simply still subclinical, apparent by MRI or fluorescence-activated cell-sorting (FACS) early in the course, just as we had observed when recording evoked potentials or analyzing cerebrospinal fluid (6, 11). The newest observations of Richards et al (37) concern proton nuclear magnetic resonance spectroscopy (a technique which uses similar equipment as MRI), which shows the development (Fig. 5) and resolution of lipids _in vivo_ within the acute EAE lesion in monkeys.

2) The treatment of EAE in two strains of monkeys has become remarkably successful with BP and an antibiotic in M. mulatta and either BP and steroid (9, 44, 46) or anti-CD4 monoclonal antibodies (40) in M. fascicularis. While we were looking for the shortest effective treatment, we found that inadequately prolonged treatment of EAE resulted in the reappearance of clinical and histological signs of EAE in about half of the monkeys (10, 47). Such a relapsing course has many similarities with MS, not only clinically but also histologically, since less hyperacute necrotizing EAE and more chronic demyelinating EAE become obvious. Since one of the treatments in monkeys includes steroids, another analogy with MS is apparent with either endogenous or exogenous steroids helping MS patients to improve or go into remission.

3) The definition of encephalitogenic sequences in BP has progressed far beyond the initial belief that there might be only a single encephalitogenic determinant on a single encephalitogenic molecule. While BP remains the dominant encephalitogenic molecule, with many different regions being specifically encephalitogenic for each species and even for each strain so far studied (3), at least one other molecule, proteolipid protein (PLP), has been shown to have some encephalitogenic activity (17). Extrapolation to humans has so far been impossible, largely because of the inability to develop a strategy that would permit identification in BP or PLP of the encephalitogenic determinant(s) for humans. From the answers available in different strains and species of animals it appears likely that humans will resemble non-inbred rabbits and monkeys in being susceptible to several different epitopes in BP (and probably also PLP) that can evoke specifically sensitized T cells. From quantitative comparisons it does not seem likely that other myelin antigens will be found to have encephalitogenic activity, but at least two other myelin antigens are already known to be responsible for demyelinating or necrotizing antibodies (25, 33, 34). Possibly others (probably including lipids and carbohydrates as well as other proteins) will be found to have similar effects. Just as with BP and PLP (4, 8, 48), it seems likely that homologies with decapeptides or other epitopic structures with sufficient molecular mimicry will be found in many micro-organisms to evoke these B cells and their antibodies which augment the inflammatory effects of the encephalitogenic T cells.

4) "Homologous" decapeptides (4, 8, 26), analogous sequences (48) or molecular "mimicry" (23, 24) are concepts that may explain immunologic cross-reactions between exogenous infecting micro-organisms and endogenous targets in a wide variety of disease, specifically MS in the CNS (8) and the GBS syndrome in the peripheral nervous system (7). The most relevant that so far have appeared relate measles C and nucleocapsid proteins to myelin BP as follows:

Measles nucleocapsid: (142)SRFGWFENKE.......(429)LPRLGGKEDR
Myelin BP: (111)LSRFSWGAEGQ......(153)IFKLGGRDSR
Measles C: (1)MSKTEWNASQ

The number of the first aminoacid in each decapeptide is in parentheses and the single letter code is used for each aminoacid. The identical ones are indicated by underlining, but many of the others are similar enough to give total scores that are far beyond chance occurrence (26).

Computer searches turn up a surprisingly large number of biochemically similar decapeptides (4, 7, 8, 26), so many that one can only wonder why auto-immune diseases are not much more common than already known or suspected. Probably most of these biochemical similarities are not really similar enough to be immunologically acceptable as cross-reacting epitopes. Also, probably most are only minor potential epitopes consisting of continuous sequences of amino acids buried in the depths of the protein molecule and revealed only after the protein is degraded by macrophages. Thus, they may be destroyed and be completely non-antigenic or they may evoke relatively few T or B cells. Since they appear to have no relation to neutralization, hemagglutination or other classical techniques for classifying sub-strains of micro-organisms, they have up to now been generally ignored. I hope that this neglect changes with the recognition of the disease-inducing potential of at least some of them (4, 7, 8, 23, 24, 26, 48).

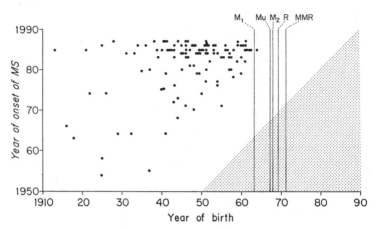

Figure 6   Year of birth and year of onset of MS in patients presenting to neurologists for the first time in Seattle in 1986-1987. For comparison with Figs. 7-9 the dates of introduction of certain vaccines are indicated as follows: M_1 = killed measles, M_2 = live measles, Mu = live mumps, R = live rubella and MMR - live triple vaccine.

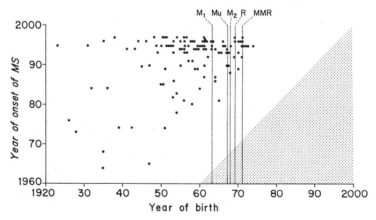

Figure 7   A similar survey as in Fig. 6, if conducted in 1996-1997, should show a similar distribution of new cases of MS if nothing has been introduced to prevent MS.

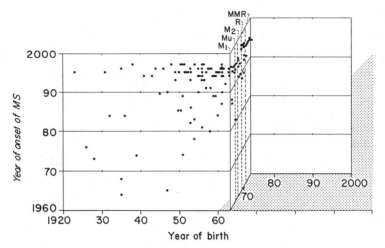

Figure 8   The same theoretical survey as in Fig. 7 but, by contrast with Fig. 7, if the increasingly effective vaccines against measles, mumps and rubella (c.f., Fig. 9) have prevented MS, then those cases shown climbing the "wall of resistance" to MS should not exist.

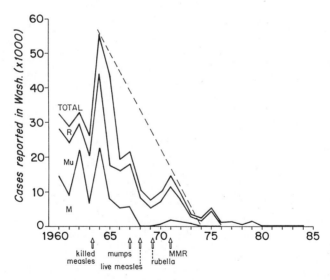

Figure 9   The progressive elimination of measles, mumps and rubella infections in the state of Washington during the decade 1964-1974.   The converse of this curve is shown in Fig. 8 as the steeply inclined "wall of resistance" to developing MS.

181

5)   The importance of age has been long recognized in determining susceptibility to EAE in animals (51) and to acute DEM in humans (2, 13, 14, 49) but only recently to play a role in determining susceptibility to MS (18).   The observations of Compston et al (18) concerning single infections have been extended to combinations of two or three infections and the same relationship demonstrated (8).   There is an exponentially linear relationship between the average age of infection with measles, mumps and rubella and susceptibility to demyelinating diseases, especially MS.   The neutral age is about 5 to 7 years, with progressively younger children increasingly resistant and progressively older children increasingly susceptible.

Since children in the tropics are regularly infected with measles before 1 year of age (15, 29), it seems likely that both the low incidence of MS in the tropics and the resistance to MS of migrants from the tropics may be directly related to this early age of infection with such world-wide agents.   Such an explanation seems much more likely than any based on a single cause of MS, since how would persons become resistant if the organism was not present in their environment?   Also, such common infections as measles, mumps and rubella would seem more likely to have been introduced into the Faeroes by relatively small numbers of otherwise healthy British soldiers in World War II than the relatively rare MS-specific virus which has been postulated to explain the mini-epidemic of MS occurring in the Faeroes a few years later (32).

In conclusion then, even if all the above is true, so what?   Well, for starters I suggest that there will be an important by-product of the routine immunizing of young children against measles, mumps and rubella (MMR).   Since these children are less than 5 years old, it seems likely that they will be also immunized specifically against MS or at least made non-specifically resistant to MS.   How soon will this effect be noticed? If the present suggestion had not been made, such a disappearance in the youngest candidates for developing MS would be so gradual that it would probably not be noticed for a decade or two after it had begun and then only because neurologists began to realize that they had not seen a case of MS in a 20-year-old person - and a decade later, no case of MS in a 30-year-old person.

With this in mind I asked my Neurology friends in the Seattle area to provide a survey of all cases of MS seen by them for the first time in 1986-1987.   As you may have guessed, there are no new cases yet in persons born after the introduction of the triple live MMR vaccine in 1971 (Fig. 6), but then MS is not very common in 16-year-old persons. However, within the next decade, if similar surveys are continued (Figs. 7 and 8), the lack of such cases should become increasingly apparent and by 1997 it should be so obvious that every one would recognize that no new cases had developed in persons born since 1971 and very few persons immunized in the period 1964-1974 as the effects of the improving vaccines became increasingly detectable (Fig. 9), as though there is a rising "wall of resistance" to MS through which it is difficult for MMR-immunized persons to pass (Fig. 8).

I hope we all live long enough to see this prediction come to pass.

SUMMARY

The convergence of five independent lines of research strongly suggests that EAE is a good, perhaps even a perfect, model of MS.   If so, the elimination of MS appears to be a real possibility that should become obvious within the next decade.

ACKNOWLEDGEMENTS

With many thanks to many collaborators: Marian W. Kies, Cheng-Mei Shaw, Sarka Hruby, Rosemarie Petersen, Bernard F. Driscoll, Russell E. Martenson, Gladys E. Deibler, Lynn M. Rose, Edward A. Clark, Ulrike Jahnke, Edmond H. Fischer, Lowell H. Ericsson, Todd L. Richards, Nigel P. Groome, Wendy A. Stewart, Donald W. Paty, and D. Alastair S. Compston.

Supported in part by research grants RG-805-E-20 and 1708-A-21 from the National Multiple Sclerosis Society and RR-00166 from the National Institutes of Health.

REFERENCES

1. Alvord EC Jr: Distribution and nature of the "antigen" responsible for experimental meningoencephalitis in the guinea pig. Proc Soc Exp Biol Med 67:459-461, 1948
2. Alvord ED Jr: Acute dissseminated encephalomyelitis and 'allergic' neuro-encephalopathies. In Vinken PJ, Bruyn GW (eds). Handbook of Clinical Neurology, Vol. 9, Amsterdam, North-Holland Publ Co, 1970, pp 500-571
3. Alvord EC Jr: Species-restricted encephalitogenic determinants. Prog Clin Biol Res 146:523-537, 1984
4. Alvord EC Jr: Disseminated encephalomyelitis: Its variations in form and their relationships to other diseases of the nervous system. In Vinken PJ, Bruyn GW, Klawans HL (eds). Handbook of Clinical Neurology. Amsterdam, Elsevier Science Publ, 1985, vol 3(47), pp 467-502
5. Alvord EC Jr, Driscoll BF, Kies MW: Large subpial plaques of demyelination in a new form of chronic experimental allergic encephalomyelitis in the guinea pig. Neurochem Path 3:195-214, 1985
6. Alvord EC Jr, Hruby S, Shaw CM, Slimp JC: Myelin basic protein and its antibodies in the cerebrospinal fluid in experimental allergic encephalomyelitis, multiple sclerosis, and other disease. Prog Clin Biol Res 146:359-363, 1984
7. Alvord EC Jr, Jahnke U, Fischer EH: The causes of the syndromes of Landry (1859) and of Guillain, Barre and Strohl (1916). Rev Neurol in press
8. Alvord EC Jr, Jahnke U, Fischer EH, Kies MW, Driscoll BF, Compston DAS: The multiple causes of multiple sclerosis: The importance of age of infections in childhood. J Chil Neurol in press
9. Alvord EC Jr, Shaw CM, Hruby S: Myelin basic protein treatment of experimental allergic encephalomyelitis in monkeys. Ann Neurol 6:469-473, 1979
10. Alvord EC Jr, Shaw CM, Hruby S, Sires LR: Chronic relapsing experimental allergic encephalomyelitis induced in monkeys with myelin basic protein. J Neuropath Exp Neurol 39:338, 1980
11. Alvord EC Jr, Shaw CM, Hruby S, Sires LR, Slimp JC: The onset of experimental allergic encephalomyelitis as defined by clinical, electro-physiological and immuno-chemical changes. Prog Clin Biol Res 146:461-466, 1984
12. Alvord EC Jr, Stevenson LD: Experimental production of encephalomyelitis in guinea pigs. Res Publ Assoc Res Nerv Ment Dis 28:99-12, 1950
13. Berger K, Puntigam F: Uber das Vorkommen von postvakzinalen Enzephalitiden in Osterreich bei alteren Personen. Wien med Wschr 108:59-61, 1958
14. Berger K, Puntigam F: Die Alterdisposition bei postvakzinalen Enzephalitis. Dtsch med Wschr 85:1520-1524, 1960

15.   Black FL: Measles.   In Evans AS (ed).   Viral infections of humans.   Epidemiology and control.   New York, Plenum Medical Book Co, 1982 2nd ed, pp 397-418

16.   Bourdette DN, Driscoll BF, Seil FJ, Kies MW, Alvord EC Jr: Severity of demyelination in vivo correlates with serum myelination inhibition activity in guinea pigs having a new form of experimental allergic encephalomyelitis.   Neurochem Path 4:1-9, 1986

17.   Cambi F, Lees MB, Williams RM, Macklin WB: Chronic experimental encephalomyelitis produced by bovine proteolipid apoprotein: Immunological studies in rabbits.   Ann Neurol 13:303-308, 1983

18.   Compston DAS, Vakarelis BN, Paul E, McDonald WI, Batchelor JR,Mims CA: Viral infection in patients with multiple sclerosis and HLA-DR matched controls.   Brain 109:325-334, 1986

19.   Driscoll BF, Kies MW, Alvord EC Jr: Suppression of acute experimental allergic encephalomyelitis in guinea pigs by prior transfer of suboptimal numbers of EAE-effector cells: Induction of chronic EAE in whole tissue-sensitized guinea pigs.   J Immunol 128:635-638, 1982

20.   Driscoll BF, Kira J, Kies MW, Alvord EC Jr: Mechanism of demyelination in the guinea pig.   Separate sensitization with encephalitogenic myelin basic protein and nonencephalitogenic brain components.   Neurochem Path 4:11-22, 1986

21.   Ferraro A, Jervis GA: Experimental disseminated encephalopathy in the monkey.   Arch Neurol Psychiat 43:195-209, 1940

22.   Freund J, Stern ER, Pisani TM: Isoallergic encephalomyelitis and radiculitis in guinea pigs after one injection of brain and mycobacteria in water-in-oil emulsion.   J Immunol 57:179-194, 1947

23.   Fujinami RS: Molecular mimicry: Mechanism of virus-induced auto-immunity.   This volume

24.   Fujinami RS, Oldstone MBA: Amino acid homology between the encephalitogenic site of myelin basic protein and virus: Mechanisms for auto-immunity.   Science 230:1042-1045, 1985

25.   Hruby S, Alvord EC Jr, Seil FJ: Synthetic galactocerebrosides evoke myelination-inhibiting antibodies.   Science 195:173-175, 1977

26.   Jahnke U, Fischer EH, Alvord EC Jr: Sequence homology between certain viral proteins and proteins related to encephalomyelitis and neuritis.   Science 229:282-284, 1985

27.   Kabat EA, Wolf A, Bezer AE: Rapid production of acute disseminated encephalomyelitis in rhesus monkeys by injection of brain tissue with adjuvants.   Science 104:362-363, 1946

28.   Kabat EA, Wolf A, Bezer AE: The rapid production of acute disseminated encephalomyelitis in rhesus monkeys by injection of heterologous and homologous brain tissue with adjuvants.   J Exp Med 85:117-130, 1947

29.   Kenny MT, Jackson JE, Medler EM, Miller SA, Osborn R: Age-related immunity to measles, mumps and rubella in Middle American and . United States children.   Am J Epidemiol 103-174-180, 1976

30.   Kira J, Bacon ML, Martenson RE, Deibler GE, Kies MW, Alvord EC Jr: Experimental allergic encephalomyelitis in rabbits: A major encephalitogenic determinant within residues 1-44 of myelin basic protein.   J Neuroimmunol 12:183-193, 1986

31.   Kopeloff LM, Kopeloff N: Neurologic manifestations in laboratory animals produced by organ (adjuvant) emulsion.   J Immunol 57:229-237, 1947

32.   Kurtzke JF: Epidemiology of multiple sclerosis.   In Vinken PJ, Bruyn GW, Klawans HL (eds).   Handbook of Clinical Neurology, Amsterdam, Elsevier Science Publ, vol 3(47), 1985, pp 259-287

33.   Lassman H: Principal mechanisms of demyelination in experimental allergic encephalomyelitis and multiple sclerosis.   Northern Lights Neuroscience Symposium on Myelin and Demyelination, Bergen, Norway, May 1987, Acta Neurol Scand (in press)

34. Lassman H, Linington C: Demyelination in vivo mediated by a monoclonal antibody specific for a minor glycoprotein. Neuroimmunology Symposium, London, Ontario, July 1986

35. Morgan IM: Allergic encephalomyelitis in monkeys in response to injection of normal monkey nervous tissue. J Exp Med 85:131-140, 1947

36. Morrison LR: Disseminated encephalomyelitis experimentally produced by the use of homologous antigen. Arch Neurol Psychiat 58:391-416, 1947

37. Richards TL, Rosse LM, Golden RN, Alvord EC Jr: In vivo proton spectroscopy of the primate brain after induction of experimental allergic encephalomyelitis. Soc Mag Reson Med 1987

38. Rivers TM, Schwentker FF: Encephalomyelitis accompanied by myelin destruction experimentally produced in monkeys. J Exp Med 61:689-702, 1935

39. Rivers TM, Sprunt DH, Berry GP: Observations on attempts to produce acute disseminated encephalomyelitis in monkeys. J Exp Med 58:39-53, 1933

40. Rose LM, Alvord EC Jr, Hruby S, Jackevicius S, Petersen R, Warner N, Clark EA: In Vivo administration of anti-CD4 monoclonal antibody prolongs survival in long-tailed macaques with experimental allergic encephalomyelitis. Clin Immunol Immunopath in press

41. Rose LM, Clark EA, Ginsberg A, Ledbetter JA, Hruby S, Alvord EC Jr: Modulation of a subset of CD4+ (T4) "T helper" cells in active multiple sclerosis (MS) and experimental allergic encephalomyelitis (EAE) in macaques. Ann NY Acad Sci 475:418-419, 1986

42. Rose LM, Clark EA, Hruby S, Alvord EC Jr: Fluctuations of T- and B-cell subsets in basic protein-induced experimental allergic encephalomyelitis (EAE) in long-tailed macaques. Clin Immunol Immunopath 44:in press

43. Rose LM, Ginsberg AH, Rothstein TL, Ledbetter JA, Clark EA: Selective loss of a subset of T helper cells in active multiple sclerosis. Proc Natl Acad Sci USA 82:7389-7393, 1985

44. Shaw CM, Alvord EC Jr: Treatment of experimental allergic encephalomyelitis in monkeys. II. Histopathological studies. In Shiraki H, Yonezawa T, Kuroiwa Y (eds). The Etiology and Pathogenesis of the Demyelinating Diseases. Tokyo, Japan Science Press, 1976, pp 377-395

45. Shaw CM, Alvord EC Jr: A morphologic comparison of three experimental models of experimental allergic encephalomyelitis with multiple sclerosis. Prog Clin Biol Res 146:61-66, 1984

46. Shaw CM, Alvord EC Jr, Hruby S: Treatment of experimental allergic encephalomyelitis in monkeys. I. Clinical studies. In Shiraki H, Yonezawa T, Kuroiwa Y (eds). The Etiology and Pathogenesis of the Demyelinating Diseases. Tokyo, Japan Science Press, 1976, pp 367-376

47. Shaw CM, Alvord EC Jr, Hruby S: Chronic remitting-relapsing experimental allergic encephalomyelitis induced in monkeys with homologous myelin basic protein. Submitted

48. Shaw SY, Laursen RA, Lees MB: Analogous amino acid sequences in myelin proteolipid and viral proteins. FEBS Letters 207:266-270, 1986

49. Shiraki H, Otani S: Clinical and pathological features of rabies post-vaccinal encephalomyelitis in man. In Kies MW, Alvord EC Jr (eds). "Allergic" Encephalomyelitis. Springfield, Ill., Charles C. Thomas Publ, 1959, pp 58-129

50. Stewart WA, Alvord EC Jr, Hruby S, Hall LD, Paty DW: Early detection of experimental allergic encephalomyelitis by magnetic resonance imaging. Lancet 2:898, 1985

51. Stone SH, Lerner EM II: Chronic disseminated allergic encephalomyelitis in guinea pigs. Ann NY Acad Sci 122:227-241, 1965

14

IMMUNE REGULATION IN MULTIPLE SCLEROSIS

Jack P. Antel[1] and Rolf Loertscher[2]

Departments of Neurology[1] and Medicine[2]
Montreal Neurological Institute, McGill University
Montreal, Canada

INTRODUCTION

The capacity of the immune system to recognize antigens, to
distinguish self from non-self, and to generate an effective yet
controlled response against a specific target is dependent on a series of
ordered interactions between the cellular and soluble components which
comprise the immune system. Immunoregulatory mechanisms refer to those
components and properties of the immune system which contribute to the
initiation, magnitude, and termination of the immune response.
Derangements in the above mechanisms have been speculated to contribute
to the development of autoimmune disease. In this presentation we will
review studies which (A) implicate involvement of immunopathogenic
mechanisms in MS; (B) describe some of the specific immunoregulatory
mechanisms which may be relevant to the disease process multiple
sclerosis (MS), and present data directly indicating perturbed
immunoregulatory mechanisms in MS; (C) postulate about specific
properties of the central nervous system (CNS) which may contribute to a
state of immune reactivity developing within the CNS; and (D) indicate
potential therapeutic means to effect immunoregulatory function.

Immunopathogenic Mechanisms in MS

The postulate that immune mechanisms are involved in the MS disease
process is derived from pathologic observations and from direct in vitro
studies of the immune status of MS patients. The pathologic process in
MS is confined to the CNS, raising the need to define the mechanisms
whereby an autoimmune disease can be restricted to this anatomic
substrate. The hallmark lesions are multifocal areas of demyelination
with relative or total preservation of axons. The active MS lesions are
characterized by an inflammatory infiltrate comprised of T cells and
macrophages. Immunohistochemical studies indicate that both major T cell
subsets, CD4[+] and CD8[+], are present in and around MS lesions (17,40,90).
Several groups have found that CD8[+] cells are the predominant cell
(17,40), whereas Traugott et al found mainly CD4[+] cells extending into
the adjacent white matter (90). This inflammatory response provides a
major basis for implicating immune-mediated events contributing to MS.
Expression of major histocompatibility complex (MHC) antigens on glial
and endothelial cells in the region of MS lesions are now reported (91).
Whether the cellular infiltrate found within the CNS of MS patients has
been recruited as a result of immune sensitization to a specific

endogenous (myelin, oligodendrocyte) or exogenous (viral) antigen remains to be established. Hafler et al found no evidence of myelin basic protein (MBP) or proteolipid protein (PLP) antigen-specific T cells amongst T cells cloned from MS plaque regions (36). In contrast, MBP reactive cells could be derived from CNS tissue of cases of acute disseminated encephalomyelitis (ADEM). T cells sensitized against MBP can be derived from both MS and control donors' cerebrospinal fluid (CSF) (18). Whether these cells are directed against the same or different MBP epitopes is unknown as yet. The demonstration that "normal" individuals have autoreactive T cells, indicates that regulatory mechanisms likely exist to prevent these autoreactive clones from inducing autoimmune disease.

Both B and T cells appear to demonstrate ongoing reactivity in MS. A disease hallmark of MS is intrathecal production of Ig, which when subjected to electrophoresis is found to be "oligoclonal". The oligoclonal band patterns of IgG eluted from different CNS lesions of individual MS cases seem to be rather constant according to Walsh et al (94), although Mattson et al (57) report variability. To date the Ig has not been shown to be directed against specific antigen. The levels of Ig secretion in the CSF do not appear to correlate with changes in disease activity; the number of bands tends to increase over time (58). The basis for this Ig production is unknown with speculation including the presence within the CNS of B cell activators which could be products released by lymphocytes, glial cells, or disrupted myelin; the effects of persistent virus; or failure of normally existing regulatory mechanisms to actively shut off the response. The high intrathecal antibody levels found in cases of subacute sclerosing panencephalitis (SSPE) provide an example of heightened antibody response as a consequence of inefficient clearing of antigen.

Evaluation of results of in vitro studies of the immune status of MS patients must take into account the clinical status of the patients. A wide range of such studies indicate activation of the immune system, particularly at times of clinical disease activity. In our experience, mononuclear cells (MNCs) derived from patients with progressive MS and stimulated in vitro with the polyclonal T-cell dependent activator, pokeweed mitogen (pwm), secrete increased levels of IgG secretion seemingly reflect altered T cell regulatory effects. Heightened T cell activation in patients with active clinical disease is evidenced by detection of an increased proportion of T cells bearing specific cell surface proteins referred to as activation antigens as they are expressed only by T cells which have progressed to the G_1B stage of the cell cycle (31,37). Further evidence of T cell activation in such patients was provided by Noronha et al (68), who demonstrated that the RNA/DNA content of MS-T cells from CSF were increased, indicating entry into the cell cycle. Thus, although one has yet to demonstrate that the observed T and B cell reactivity is induced by a specific antigen, the evidence does suggest that MS is characterized by a state of heightened immune reactivity. The role of immunoregulatory mechanisms in contributing to these findings will be discussed.

Immunoregulatory Mechanisms relevant to MS

i) Immunogenetic factors. The magnitude of the immune response that individuals generate after exposure to specific antigens is greatly determined by genetic factors. Classic studies of inbred animal strains indicated that the gene loci contributing to the magnitude of immune response are located within a chromosomal region termed the major histocompatibility complex (MHC) which in man is located on the short arm of chromosome 6 (HLA-locus), and on chromosome 17 in mice (h-2 locus). The organization of the MHC locus in humans, as well as mice, has been studied in great detail, and the contribution of specific gene products within this region to immune reactivity defined (Reviewed in 30,54). The different gene loci of the MHC region and their products have been divided into two major classes, class I and II. Class I-MHC antigens (HLA-A,B,C in man and H-2K,D in mouse) are expressed on all cells of the lymphoid system as well as on virtually all nucleated cells of the body (see section on MHC antigens on neural cells) and were initially thought to be the major antigens recognized by the immune system of tissue allograft recipients. Subsequently, class I-MHC antigens were demonstrated to be the favored restriction element for CD8[+]-cytotoxic T cells. Class II-MHC antigens (HLA-DR,DQ,DP in man, and H-2IA,IE in mouse) are expressed only on certain lymphoid cells - ie. on macrophages, B cells, and activated but not resting T cells. Bone-marrow derived cells, such as dendritic cells and Langerhans' cells in the skin also express these antigens and at high density (84). Under certain conditions endothelial cells and glial cells may also express class II MHC antigens. The class II MHC antigens are the restriction element for CD4[+]-T cells. Class II MHC antigens are critical elements involved in cell-cell interactions within the lymphoid system during the generation of an immune response.

The high degree of polymorphism which characterizes the MHC system contributes greatly to individual variability in the magnitude of reactivity to a specific antigen. Classic studies with simple synthetic antigens indicate that actual non-response to an antigen can be a function of class II MHC phenotype determinants. As shown in studies of mouse T cell reactivity to specific MBP fragments, the degree of reactivity is dependent on specific epitopes encoded by either IA or Ie region MHC genes (100). In MS populations, the genetic predisposition to disease is well illustrated by the relatively high, although not absolute concordance rate for disease between identical twins (27). Results from most epidemiologic studies of MS populations indicate that specific MHC-II antigens, particularly HLA-DR2, are over-represented, indicating that the disease is occurring in a population with characteristic genetically-determined immune response traits which likely greatly influence their immune capability or repertoire. Jacobsen et al demonstrated that class II MHC-dependent CD4[+] cell cytotoxic responses to measles-infected cells were impaired in MS patients bearing the DR2 phenotype (45). Ongoing studies in MS include sequencing the MHC-II gene regions to look for nucleotide sequences characteristic of affected individuals, as well as transfection experiments evaluating how specific MHC class II antigen products influence immune reactivity (12). Note that studies of the classic animal "autoimmune" disorders experimental allergic encephalomyelitis (EAE) and experimental myasthenia gravis indicate that non-MHC genes also contribute to the magnitude of immune reactivity to the specific antigens and to overall levels of susceptibility to disease.

Genetic factors also determine the capacity of individual T cells to recognize specific antigens, presented to them by accessory cells. Antigen recognition by T cells occurs via a specific site on the T cell surface termed the T cell receptor (TcR) (Reviewed in 51). The TcR is comprised of two polypeptide chains (alpha, beta) each of which contains a variable and constant region. The structural similarities which exist between the TcR, the Ig molecule and MHC antigens as well as neural cell adhesion molecules, suggest these all were originally descended from a common ancestral gene giving rise to the term, Ig supergene family. The polymorphism within each gene region encoding for the TcR and the various combinations of rearrangements of individual TcR component genes during the process of forming the complete gene, result in T cells with indifferent repertoires with regard to antigen recognition. Thus T cells bearing specific TcR gene rearrangement patterns will respond to specific antigens. Recognition of antigen by T cells also requires the presence of MHC antigens, a feature which distinguishes these cells from B cells which recognize antigen without MHC requirement.

The precise contribution of the above genetic factors in determining susceptibility to MS requires further definition.

    ii) _Immunoregulatory Mechanisms and MS_. a) T cell mediators of immune regulation. Within the T cell population, there are cell subsets possessing the functional properties of either (a) inducing or amplifying an immune response, or (b) suppressing the response. The involvement of inducer/amplifier T cells (T helper cells) in generation of cytotoxic T cell responses and T-cell dependent B cell responses are well established, particularly by studies of cloned T helper cell lines (76). In humans, the T helper cells are contained within the $CD4^+$ cell population, and more precisely within the $CD4^+$ - $CDW29^+$ ($4B4^+$) population (63). The T helper ($CD4^+$)cells mediate their effects via secretion of soluble factors, termed lymphokines. Defined lymphokines include those required for maintaining T cell proliferation (IL-2), for recruiting and effecting a delayed type hypersensitivity response, and for inducing B cell activation and maturation (Reviewed in 69). To date, no claims of excess T helper activity are reported to occur in MS.

Suppressor T cells have been defined as those which can down regulate the immune response, using non-cytotoxic mechanisms (61). The mechanism whereby T suppressor (T_S) cells are activated and mediate their effects are not entirely defined. Generation of antigen-specific T_S cloned cell lines has not been as reproducible as the generation of T_{CTL} and T helper lines, perhaps because of requirement for different growth factors (29). A consensus is emerging, however, regarding the existence of specific cell circuits which result in generation of suppressor cells. Mohaghepour et al (62) demonstrated that generation of antigen-specific T_S clones requires an initial MHC-II restricted interaction between a T helper cell and an antigen-presenting cell. A soluble factor released by this T_S cell, (termed a T_S1 cell) directly activates a T_S effector cell (T_{S2}) which exerts its action in a non-cytotoxic class I MHC-restricted manner. Thus, this model implies that T_S cell generation occurs via interaction of an intermediary inducer cell and not via direct interaction with antigen, in contrast to T helper cells which promote T and B cell effector responses.

The identification of the cell types and soluble factors involved in the postulated human suppressor network continues to advance. Putative T suppressor/inducer cells are contained within the $CD4^+$ cell subset and are defined by the presence of the epitope CD45R recognized by mAbs 2h4+ and Leu18 (24,26). The suppressor/inducer (CD45R) and helper cell (CDW29+) $CD4^+$ populations are mutually exclusive. Precursors of suppressor effector cells are defined by the presence of the CD8 protein and the absence of the CD28 epitope which is recognized by the mAb 9.3 (98). Leu 15+ antibody is reported to selectively detect $CD8^+$ suppressor rather than cytotoxic T cells. Fox et al have recently demonstrated that the cellular mediators of the suppressor network ($CD45R^+$ - $CD4^+$ and $CD28^-$;$CD8^+$ cells) may have requirements for cell proliferation distinct from cell types which augment or effect T cell responses (29). They have identified an 8 Kd T suppressor cell growth factor released by pokeweed mitogen-stimulated $CD4^+$ cells that induces IL-2 receptor expression only on the suppressor network cells. Conversely, the cells in the suppressor cascade are not activated by binding of anti-CD3 mAb, which does induce IL-2 receptors on helper and cytotoxic cells.

The above emerging data on T_S circuits in humans appears to reasonably parallel the suppressor cell cascade observed in murine systems in which generation of T suppressor factor secreting hybridomas have allowed partial characterization of soluble mediators included in the suppressor network (21,66,85,92). In the murine system, studies of simple antigens (e.g. GAT) have provided insight regarding soluble factors mediating interaction between T cell subsets. Sorensen and Pierce (85) have identified two suppressor factors, termed T_SF_1 and T_SF_2. T_SF_1 is produced by an antigen-specific T_S inducer cell (Lyt 1^+2^-, IJ^+, Qa 1^+) and is composed of a single polypeptide chain bearing determinants encoded by the I-J region of the MHC, and which can bind antigen. T_SF_1-mediated suppression is not MHC restricted. T_SF_2 is produced by the T_S effector cell (Lyt 2^+1^-, IJ^+, Qa 1^+) and is composed of two polypeptide chains, one bearing I-J determinants and the other capable of binding antigen. T_SF_2-mediated suppression is MHC-restricted. Generation of T_XF_2 requires subimmunogenic dosages of antigen and T_SF_1. The nature of the antigen receptor on the T_SF molecules remain to be defined, as does the gene loci encoding for these factors. Exposure of naive spleen cells to T_SF_1 or T_SF_2 in the absence of antigen, can result in generation of idiotype-specific T_S cells - ie. these T_S cells release factors which bind antigen-specific idiotypes. The factors are comprised of 2 polypeptide chains, one bearing I-J and the other id-binding activity. The factors are restricted by MHC and IgH genes. T_SF_2 interacts with T helper cells and blocks the latters' effect on B cells (92). The T_SFs may act not only directly on T cells, but also via accessory cells (66). Hapten-specific T_SF may bind antigen on macrophages and induce the macrophages to secrete non-specific immunosuppressive factors.

Although the above cited data have begun to define specific soluble factors involved in the suppressor network, the field is further complicated by description of a wide array of soluble factors mainly isolated from cell lines maintained in vitro, which are capable of suppressing immune reactivity but whose in-vivo biologic role is less clear. A partial list of such factors is present in Table 1 below.

Table 1

T-SUPPRESSOR LYMPHOKINES - soluble factors (glycoproteins, polypeptides) produced by T cells and which inhibit cell proliferation

Soluble Factor	Reference
- IFN-$_\gamma$	(99)
- lymphotoxin	(32)
- soluble immune suppressor supernatants (SISS) - B, T	(33)
- suppressor activating factor (SAF)	(52)
- soluble immune response suppressor (SIRS)	(82)
- suppressor induction factor (SIF)	(46)
- B-cell growth inhibitory factor (BIF)	(48)
- inhibitor of DNA synthesis (IDS)	(67)
- T-leukemia-derived suppressor lymphokine (TLSL)	(81)

b) Suppressor cell mechanisms in MS. The approach to evaluating suppressor cell activity in MS has been to devise in-vitro assays which demonstrate functional suppressor cell activity and then attempt to determine the cellular basis for the observed functional effects. As previously mentioned, the failure as yet to identify a putative specific antigen in MS, has required that studies of suppressor cell function be conducted using either non-specific cell activators - e.g. mitogens or non-CNS antigens, particularly MHC antigens, either in allo-activating (mixed lymphocyte reaction (MLR) or auto-activating (autologous MLR) systems. Functional suppressor assays might be considered in two categories; either resting or non-activated suppressor cell assays, or activated suppressor cell assays.

Resting suppressor cell assays: In these assays, a putative suppressor cell, usually a T cell, is either added without prior manipulation to immune cells which are being induced to respond to a mitogen or antigen, or depleted from an immune cell population which is about to be induced to respond. Suppressor function is calculated based on the extent of the dampening or enhancement of the observed responses, as compared to a control culture without regulatory cells. The suppressor properties of $CD8^+$-cells were initially defined in such systems, particularly in assays in which the mitogen, pokeweed, was used to induced IgG secretion.

Data from early studies using T-cell dependent polyclonal B cell stimulation (ie. pokeweed mitogen (pwm) stimulation) indicated that in normal young individuals, levels of IgG secretion by MNCs depended predominantly on T cell influences (49,79). The normal population could be divided into those who were "high" and "low" responders to pwm (41). Radiation of T cells resulted in augmented IgG levels, implicating the presence of a specific radiation-sensitive suppressor T cell. By use of cross-over experimental paradigms between T cells (i.e. rosette with sheep red blood cells) (E^+) and B cells (E^-), one could further demonstrate that T cells determined the levels of observed IgG secretion. With the availability of $CD8^+$ mAbs, this cell subset was shown to mediate the suppressor function. However, the levels of pwm-induced IgG secretion by MNCs did not correlate with the absolute number of $CD8^+$-T cells within the MNC population. We demonstrated that high and low levels of pwm-induced IgG secretion by MNCs were, however, correlated with the functional suppressor activity of the $CD8^+$ cell subset. Our experimental paradigm involved adding a constant number of $CD8^+$ T cells obtained from either high or low IgG secretors to a standard pool of $CD4^+$ T helper cells + B cells to which pwm was added (2). IgG levels were higher in cultures containing $CD8^+$ cells from high secretors compared to levels in cultures containing cells from low secretors. Our data in chronic progressive MS indicated that these patients were as a group high secretors, and this finding could not be accounted for by a depletion of the $CD8^+$-T cell subset (11). We did find that adding increasing numbers of $CD8^+$ cells from MS patients did result in suppression, indicating both that the suppressor defect is not an absolute one in MS and that $CD8^+$ cell numbers are an important factor. In this latter regard, reduction in the proportion of $CD8^+$ cells within the overall MNC population is variably observed in patients with active MS (22,73,77). Our data favor the postulate that high IgG secretion in the MS population reflects a fundamental derangement in the $CD8^+$-T suppressor function mediating cell population or in the circuitry of cells involved in inducing the final suppressor effector function. More detailed studies of IgG secretion in MS as a function of clinical disease are presented in the chapter by Oger.

The definition of the cell circuits involved in the suppressor pathway, as described in a previous section, have permitted more detailed analysis of this regulatory function in MS patients. The studies of Rose et al (51) and Hafler et al (65) have implicated a numerical reduction of the $CD4^+$-$CD45R^+$ suppressor/inducer subset in patients with active MS.

One notes that a reduction in T cell subsets as defined by mAbs can reflect either an actual deletion of the putative cell subset or deletion of the epitope recognized by the mAb from the cell surface. If the former were the correct explanation, one would expect an overall reduction in CD4[+] cells in MS and an increase in the proportion of CD4 [+]-CDW29[+] cells. Taken together with the putative reduction in CD8[+]-T cells reported by some in MS, one would have expected a prominent reduction in total T cell number in MS. To date the data is incomplete in resolving these issues. One need further note that activation of CD4[+]-CD45R cells with mitogens results in reduced expression of the CD45R antigen (53,55); heightened T cell activation is a feature of active MS (31,37). One anticipates even more precise definition of cell subsets involved in the immune regulatory circuits, with use of new panels and combinations of mAbs and increasingly sophisticated cell analysis techniques, such as the multiparameter techniques described by Kastrukoff et al (47).

"Activated" suppressor cell assays: The activated suppressor cell assays as applied to MS have involved either (a) preactivation of MNCs or T cells with mitogenic lectins (ConA) or mitogenic mAbs (OKT3) and assessing these cells' effects on a subsequent immune response, or (b) use of an AMLR to generate a putative suppressor population which is then again added to a subsequent immune response. In our initial studies, we preactivated MNCs, isolated from peripheral blood of MS patients and controls, in vitro with ConA for 48-96 hours then treated these cells with mitomycin C to prevent further proliferations, and assessed their suppressor influences on a second cell population, namely either autologous or allogeneic MNCs which were themselves stimulated with ConA. In this assay system we found that patients with active disease, either early in the relapse phase or in progressive phase of the disease had reduced suppressor levels compared to controls (3). More recent serial studies of patients with progressive disease conducted over periods of > 1 year indicated that the suppressor defect is a persistent one (11). Patients with relapsing-remitting disease were found to have more variable levels of suppressor function, with levels suggestively increased during the several weeks following an acute relapse (recovery phase) (3). Patients with clinically stable disease were found to have suppressor levels which did not differ statistically from controls, although individual low values were observed (4). One suggest that further studies correlating suppressor values with magnetic resonance imaging-defined disease relapses will determine how well this in-vitro assay correlates with disease activity. Suggestive data indicating such correlations have been presented by Oger et al (70).

We have attempted to define some of the cellular factors accounting for this activated suppressor cell defect in MS by expanding our studies of patients with progressive disease - i.e. the group in which we find the most consistent functional suppressor defect. To establish the principal cell mediator of mitogen activated suppressor activity, we attempted to prepare isolated T cells or subsets thereof, activate these cells, and measure their suppressor capabilities. Activation of T cells or subsets with either ConA or anti-CD3 mAb (OKT3) requires the presence of accessory cells, supplied either as non-T cells (E[-]) or as enriched macrophage populations. We found that suppressor activity was contained within the T cell (E[+]) population and predominantly within the CD8[+] population (5).

The properties of the CD8[+] population which are associated with reduced suppressor function mediating capability continue to be studied. FACS analysis of the CD8[+] population indicate a wide variability in fluorescence intensity and by inference CD8[+] antigen density on CD8[+]-T cells. Whether CD8[+] cells bearing different antigen densities comprise different functional cell subtypes remains problematic. High density CD8[+] cells have been reported to be mediators of cytotoxicity (89). The function of low CD8 density cells remains unclear, particularly with regard as to whether all these cells are truly T cells or whether some are best classified as NK cells. CD3[-], CD8[+] bearing the NK marker CD11 (Leu 15) are now well recognized. Some of these latter cells do bear the CD2 (sheep red blood cell receptor) T cell marker, making the definition of a T cell based on surface phenotype markers rather arbitrary.

The precise role of the CD8 antigen in mediating either suppressor function or the other function attributes to CD8[+] cells, namely MHC-class I restricted cytotoxicity is not totally established. By reducing CD8 antigen density on MNCs derived from normal individuals by means of modulating the antigen from the cell surface using OKT8 mAb, we did find that their activated suppressor mediating function was reduced (6). Reder et al (74) and Thompson et al (88) reported that CD8 antigen density was reduced in MS patients with progressive disease; Hirsch et al (42) did not find this in relapsing patients. Reder et al (74) found no such decrease in stable MS patients. The factors contributing to the reduced antigen density are not defined, but could include soluble factors such as macrophage-derived products, particularly prostaglandins, other lymphokines, and perhaps neurotransmitters and neurohormones. Using 3 different restriction enzymes and Southern blot analysis, we found no polymorphism in the CD8 gene in MS patients, including those with low suppressor function (75).

To address whether all CD8[+] functions were perturbed in MS, we compared progressive MS patients and controls with regard to class I allo-antigen directed cytotoxic function (7). We sensitized MNCs from test donors in vitro with HLA-bearing Epstein-Barr virus (EBV) transformed B cells and after 5 days assessed the capacity of these sensitized cells to lyse the same EBV cells in chromium release assays. We found no difference between MS patients and controls, suggesting that the suppressor defect was a selective one. The cell targets we used in our assays were not susceptible NK cell targets. As yet, we have not noted significant differences between MS patients and controls with respect to relative proportions of CD8[+] suppressor (Leu15+) and cytotoxic (9.3+) cell subsets present within the systemic MNC population.

To further characterize the mediators of activated suppressor function, we generated IL-2 dependent CD8[+] T-cell lines by isolating these cells from peripheral blood MNCs using a panning technique, inducing proliferation with anti-CD3 mAb (OKT3) and maintaining the cells in the presence of IL-2 (ie. MLA cell line supernatant) and autologous feeder cells (radiated MNCs) (8). The proliferative rate of these cells, as evaluated by ^{3}H-thymidine uptake after 2 weeks in culture was dependent on the presence of IL-2 in the media. In studies conducted on CD8[+] cells maintained in culture for 14 days, then given fresh anti-CD3 mAb and feeders for a subsequent 2 days, followed by treatment with mitomycin, we found that progressive MS patient-derived CD8[+] cells mediated lower levels of suppressor activity than did control donor-derived cells. These data suggest that the reduced suppressor function observed in MS patients could not be attributed to transient effects of factors exogenous to the T cells, such as serum factors, antibodies, proteolytic enzymes, immune complexes, or monokines. Whether such exogenous factors can induce long acting effects on CD8 cells cannot be excluded. We continued to find defective suppressor function mediated by MS CD8[+] cells up to 4 weeks of cultures (last time point examined). These CD8[+] cell lines mediated significant levels of cytotoxicity, comparable to those mediated by control donors' cells.   Further studies using cloned CD8[+] cell lines should help delineate whether quantitative differences in CD8[+] cell subsets or intrinsic cellular properties account for the observed findings.   In CD8[+]-cell line studies, similar to those described above, conducted on patients with stable MS, we found that suppressor cell activity in these patients were significantly greater than those observed in progressive patients, although suggestively lower than control values (4).

The above studies have been conducted using systemic blood-derived cells.   An important issue regarding the immunopathogenic mechanisms which may be operative in MS, is whether the T-cell accumulation in the CNS represents a recruitment of a selective subpopulation of T cells to this tissue compartment.   Hafler et al showed by labelling systemic T cells with mÅb, that these cells could be detected in the CSF within several days (38).   One postulate suggests that during active disease there is a shift of specific subsets of T cells from the peripheral circulation into the CNS.   The studies of Cashman et al (17) indicate a reduction rather than over-representation of CD8[+] T cells in the CSF. Data from studies of EAE indicate that most of the T cells recruited to the CNS in this disease are non-antigen specific, and thus do not set a precedent for selective recruitment of a given subset to the brain. Selective subset recruitment does, however, occur in some viral infections, particularly lymphocytic choriomeningitis (LCM) virus infection in animals (10).

Studies of CSF T cells face the technical problem of retrieving sufficient cell numbers from the CSF to permit most functional in vitro assays without prior expansion of the cells in vitro.   This need to expand the T cell population in vitro warrants caution in interpreting results regarding a selective accumulation of particular ccells in the CSF since the apparent selectivity may reflect the in vitro growth properties of the cells rather than their in-vivo biologic properties. Data regarding the restricted heterogeneity of TcR gene rearrangement patterns found in studies of CSF T cells must be interpreted in this context (39,80).   We have compared activated suppressor mediating capacities of T cell lines derived from blood and CSF of single donors, most of whom did not have inflammatory CNS diseases.   Although we usually found rather comparable levels of suppression mediated by identical numbers of activated T cells, we occasionally noted discrepancies.   Thus, we have not yet demonstrated a consistent difference in suppressor-mediating function in cells recruited to the CNS (CSF) compared to controls.

Defective suppressor cell function in active MS has been reported when the autologous mixed lymphocyte reaction (AMLR) has been used to induce suppressor cells (20,34).   The AMLR is an in vitro paradigm in which T cells respond to autologous non-T cells.   This response results in generation of MHC-antigen specific helper and suppressor T cells and thus can be used as an in vitro model of immune regulation.   Hafler et al (20) and Hirsch et al (43) have reported reduced AMLR responses in MS patients.   In the former study, aberrant responses were found in "active" MS patients, mainly those with progressive disease, but not "stable" patients.   In the latter studies, aberrant responses were found in "stable" patients.   Reduced AMLR is also reported in other diseases of suspected autoimmune etiology such as SLE.   The predominant proliferating cell in the AMLR is reported to be contained in the CD4[+] population, and seemingly is the suppressor-inducer network.   In this regard, Hafler et al (20) have observed a correlation between decreased AMLR-generated T cell suppressor activity in MS patients and reduced numbers of circulating suppressor-inducer cells. In contrast, Birnbaum (13) observed heightened AMLR responses in the MS population, using systemic MNCs. He then further extended his observations to CSF-derived T cell cloned cell lines and again observed heightened frequency of autologous T cell reactivity in the MS population (14).

The mechanisms accounting for the observed disease activity correlated alterations in immuno-regulatory circuits in MS remain unknown. Inconsistent reports exist of serum factors arising in MS which selectively recognize or alter function of specific T cell subsets. The wide array of soluble products released by activated lymphocytes and monocytes may themselves induce changes in immunoregulatory T cell numbers and function. An additional attractive hypothesis is that CNS-derived factors arising as a consequence of tissue injury, either neurotransmitters, neurochormones, or inflammatory mediators feed back on the immune system, thus creating a circuitry which could contribute to a self-perpetuating disease process.

c) Lymphokine production and MS. The array of soluble factors produced by T cells (lymphokines) play an essential role in cell-cell interactions both within the immune system and for immune cell interaction with non-lymphoid tissues, including endothelial cells and glial cells. Some of these factors are produced by one T cell subset and act upon a second cell type within the immune system, such as B cell growth and differentiation factors produced by T cells; other factors fulfil criteria for autocrine factors in that the same cells produce and respond to the factor. The response of T cells to IL-2, the specific factor required for T cell growth, is such an example. Defraitis et al have reported that T cell clones from MS patients can become independent of IL-2 with regard to ability to survive and proliferate in vitro (25). This data has not, however, yet been confirmed. The number of identified lymphokines continue to expand, and specific abnormalities of their production and effects will continually be examined.

d) Non-T cell immunoregulatory mechanisms. In addition to T cell regulatory circuits, one need consider how non-T cells contribute to overall regulation of immune reactivity. NK cells, whose lineage seemingly is distinct from T cells, were initially defined functionally by their ability to lyse specific cell targets without need for prior sensitization. NK cells, as defined phenotypically and morphologically, are now shown to function as immunoregulatory cells, particularly with regard to modulating antigen presentation by dendritic cells to T cells (1). NK function has variably been reported to be decreased or normal in MS; NK cell numbers as defined by monoclonal antibodies are usually reported as normal (44). NK cells are also a source of IFN-γ, when administered to MS patients results in exacerbation of the disease (72). IFN-γ production in MS by peripheral blood lymphocytes is normal or reduced compared to controls (93).

B cells also contribute to multiple aspects of immune regulation. B cells are shown to be important antigen-presenting cells. B-cell depleted animals do not develop EAE (96), even though T-cells are essential for passive transfer of the disease. The presence of heightened levels of intra-blood brain barrier IgG in MS has led to studies to determine whether defects in B cell regulatory mechanisms exist. The classic B-cell feedback control mechanism is termed the idiotype-anti-idiotype network hypothesis, in which initial production of antibody (Ab) results in generation of a second series of antibodies (anti-idiotype (anti-id) antibodies) directed against unique determinants on the variable regions of the initial (first order) antibody (idiotypes). Antibody to the second order Abs influencing the activity of the B cells bearing the Ig molecule recognized by the "anti-id" Abs. Anti-id Abs also can act back upon regulatory T cells. Whether a similar network of interacting T cells exist is an attractive postulate. To date, little evidence exists for the presence of common idiotypic antibodies mongst MS patients.

The rule of accessory cells has received considerable attention in MS. Monocytes/macrophages are required as antigen-presenting cells for both B and T cells. With regard to the latter, MHC antigen are essential requirements. The role of accessory cells in MS takes on further importance in MS with the demonstration that glial cells can subserve this function (see later section). Monocytes from MS patients are reported to secrete excess levels of prostaglandins, specifically PGE_2, which can suppress immune reactivity, as well a possibly modulate expression of surface antigens on T cells (26).

CNS-Immune Cell Interactions

As mentioned previously in this review, in postulating a role for immune mechanisms in mediating the lesions of MS, one needs to define those factors which lead to a CNS-specific disease. One hypothesis which forms the basis of considerable current research efforts in MS, is that the presence of a unique endogenous or exogenous antigen within the CNS initiates the sensitization and recruitment of effector T cells to the brain. One need also consider, however, that immunocytes could be recruited to the CNS in response to multiple antigens, including non-myelin antigens and that myelin is damaged as an innocent bystander. Whatever the mechanisms accounting for recruitment of immune cells to the CNS, one need consider how the unique properties of glial cells could contribute to promoting immune reactivity. In this regard the following possibilities must be considered:
   a)   the influence of glial cells on immune cells;
   b)   the influence of immune cells on glial cell function.
Glial cells, both astrocytes and oligodendrocytes, have classically been considered as not expressing MHC antigens. However, under select in vitro conditions, both glial cell types can express class I and possibly class II MHC antigens, as determined by immunohistochemical methods. Interestingly, mouse and rat strains with the highest susceptibility to EAE are characterized by a hyperinducibility of class II MHC antigens (56).

Astrocytes have been demonstrated to be capable of acting as antigen-presenting cells in vitro (86). Cashman and Noronha demonstrated that sheep oligodendrocytes can serve as accessory cells to support activation of human T cells with ConA (18). Our recent data, derived from studies of dissociated human adult glial cell cultures, indicate that human astrocytes and oligodendrocytes can also serve accessory cell functions for T cell response to mitogen although, in these studies, a possible contribution of the microglial cell population cannot be totally excluded (19). Glial cells can also serve as stimulators in allogeneic mixed lymphocyte-astrocyte reactions, implying the functional presence of class II MHC antigens (9). Thus, in considering mechanisms whereby an immune response may be selective for CNS, one need consider these specific properties of the glial cells. Cerebral endothelial cells also can fulfil similar functions in contributing to immune reactivity, and these cells are likely to play a critical role in the process of lymphocyte trafficking to the CNS.

Glial cells also may contribute to immune reactivity within the CNS by production of soluble factors, initially considered to be unique to lymphoid cells, including IL-1, prostaglandins, and suppressor factors (28,87). Gurney et al demonstrated that the mRNA for neuroleukin, a potent inducer of Ig secretion by B cells, is contained within glial cells (35). This array of soluble factors likely influences the function of lymphocytes entering the CNS. One need also consider that lymphocyte products also influence glial reactivity. As mentioned,, IFN-γ induces expression of MHC antigens on glial cells. One can demonstrate that induction of MHC antigens on glial cells makes them susceptible targets of cytotoxic T cells (83). Specific lymphokines are described which induce glial cell proliferation and maturation (59). Given that MS is characterized by the presence of activated immune cells and extensive glial proliferation, one might anticipate that the glial-immune reactivity would occur at an increased level in this setting.

Pharmacologic Therapy Affecting Regulatory Mechanisms

Interest in immunoregulatory mechanisms in autoimmune disorders has been generated in the past partly by the opportunities which exist to pharmacologically manipulate these processes. Although, as mentioned, derangement of immunoregulatory mechanisms in MS show a correlation with disease activity, to date these correlations are insufficient to utilize the assays as predictive tests of pending alterations in clinical disease activity. Nor is the data in hand to establish whether the observed abnormalities are primary or secondary events in the disease course.

The effects of multiple pharmacologic agents on regulatory immune functions have been evaluated in animal models and in some human studies. One can utilize a strategy of either ablating helper cell function or augmenting suppressor activity. Anti-helper cell antibody administration is effective in treatment of EAE (94), although no long term trials have yet been carried out in MS.

Amongst agents under trial in MS, cyclosporin A (CsA) is shown to inhibit helper T cell activity and in particular IL-2 secretion, while sparing suppressor function. In our experience, CsA did not augment suppressor activity in progressive MS patients over a 6-12 month treatment period (11); Kerman et al have found such increases in patients followed for a longer duration (50). Azathioprine in humans exerts its influence initially by ablating B cell response; at higher dosage, the drug impairs suppressor function (71). T helper cells are the most resistant to the drug. Cyclophosphamide at low doses can ablate suppressor function, a property thought to explain why pre-treatment with low dosage of the drug can result in augmented disease severity in autoimmune models such as EAE (60). Total lymphoid irradiation, a therapy of suggestive benefit for treatment of progressive MS, may act at least partially by generation of "natural" suppressor cells (23).

More selective regulatory cell directed therapy remains under active investigation in autoimmune disease. Neuropharmacologic agents can selectively augment suppressor function, particularly H_2-agonists. The role of other neurotransmitters and neurohormones are under investigation, since a large number of these agents are shown to have receptors on lymphocytes and to influence some functional property of these cells. A further understanding of neural influences on lymphoid cells may thus provide not only insight into mechanisms of autoimmune neurologic diseases, but also upon potential avenues of therapy.

REFERENCES

1.   Abruzzo LV, Mullen CA and Rowley DA: Immunoregulation by natural killer cells. Cell Immunol 98:266-278, 1986

2.   Antel JP, Reder AT and Noronha AB: Cellular immunity and immune regulation in multiple sclerosis. Seminars in Neurology 5:117-126, 1985

3.   Antel JP, Arnason BGW and Medof ME: Suppressor cell function in multiple sclerosis - correlation with clinical disease activity. Ann Neurol 5:338-342, 1979

4.   Antel JP, Brown M, Nicholas MK, Blain M, Noronha A and Reder A: Activated suppressor cell function in multiple sclerosis - clinical correlations. J Neuroimmunol (in press)

5.   Antel JP, Bania MB, Reder AT and Cashman NR: Activated suppressor cell dysfunction in progressive multiple sclerosis. J Immunol 137:137-141, 1986

6.   Antel J, Oger JJF, Jackevicius S, Kuo HH and Arnason BGW: Modulation of T-lymphocyte differentiation antigens: Potential relevance for multiple sclerosis. Proc Natl Acad Sci USA 79:3330-3334, 1982

7.   Antel JP, Nicholas MK, Bania MB, Reder AT, Arnason BGW and Joseph L: Comparison of T8[+] cell-mediated suppressor and cytotoxic functions in multiple sclerosis. J Neuroimmunol 12:215-224, 1986

8.   Antel J, Bania M, Noronha A and Neely S: Defective suppressor cell function mediated by T8[+] cell lines from patients with progressive multiples sclerosis. J Immunol 137:3436-3439, 1986

9.   Antel JP. Submitted for publication

10.   Baezinger J, Hengartner H, Zinkernagel RM and Cole GA: Induction or prevention of immunopathological disease by cloned cytotoxic T cell lines specific for lymphocytic choriomeningitis virus. Eur J Immunol 16:387, 1986

11.   Bania MB, Antel JP, Reder AT, Nichlas MK and Arnason BGW: Suppressor and cytolytic cell function in multiple sclerosis - Effects of Cyclosporine A and interleukin-2 (IL-2). J Clin Invest 78:582-586, 1986

12.   Bell JI, Denney D, MacMurray A, Foster L, Watling D and McDevitt HO: Molecular mapping of class II polymorphisms in the human major histocompatibility complex. J Immunol 139:562-573, 1987

13.   Birnbaum G and Kotilinek L: Autologous lymphocyte proliferation in multiple sclerosis and the effect of intravenous ACTH. Ann Neurol 9:439-446, 1981

14.   Birnbaum G, Kotilinek L, Schwartz M and Sternard M: Spinal fluid lymphocytes responsive to autologous and allogenic cells in patients with multiple sclerosis. J Clin Invest 74:1307-1317, 1984

15.   Booss J, Esiri MM, Tourtelotte WW, Mason DY: Immunohistological analysis of T-lymphocyte subsets in the central nervous system in chronic progressive multiple sclerosis. J Neurol Sci 62:219-232, 1983

16.   Burns J, Rosenzweig A, Zweiman B and Lisak RP: Isolation of myelin basic protein-reactive T-cell lines from normal human blood. Cell Immunol 81:435-440, 1983

17.   Cashman N, Martin C, Eizenbaum J-F, Degos J-D, Bach M-A: Monoclonal antibody defined immunoregulatory cells in multiple sclerosis cerebrospinal fluid. J Clin Invest 70:387-392, 1982

18.   Cashman NR, Noronha ABC: Accessory-cell competence of ovine oligodendrocytes in mitogenic activation of human peripheral T cells. J Immunol 136:4460-4463, 1986

19. Cashman N, Boulet S, Hamel S, Blain M, Antel J: Accessory cell capabilities of human glial cells in mitogenic activation of allogenic T cells. J Neuroimmunol 16:30, 1987

20.   Chofflon M, Weiner HL and Hafler DA: Loss of functional suppression is linked to decreases in circulating suppressor-inducer (CD4[+]2H4[+]) T cells in multiple sclerosis. J Neuroimmunol 16:33, 1987

21. Chu WS and Rich S: Suppressor T cell growth and differentiation: Evidence for induced receptors in suppressor T cells that bind a suppressor cell differentiation factor. J Immunol 138:504-512, 1987

22. Compston A: Lymphocyte subpopulations in patients with multiple sclerosis. J Neurol Neurosurg Psychiatry 46:105-114, 1983

23. Cook SD, Troiano R, Zito G et al: Effect of total lymphoid irradiation in chronic progressive multiple sclerosis. Lancet II: 405-409, 1986

24. Dalchau R, Fabre JW: Identification with a monoclonal antibody of a predominantly B lymphocyte-specific determinant of the human leukocyte common antigen: evidence for structural and possible functional diversity of the human leukocyte common molecule. J Exp Med 1981:152

25. DeFreitas EC, Sandberg-Wollheim M, Schonely K, Boufal M and Koprowski H: Regulation of interleukin 2 receptors on T cells from multiple sclerosis patients. Proc Natl Acad Sci USA 83:2637-2641, 1986

26. Dore-Duffy P, Zurier RB, Donaldson J: Lymphocyte adherence in multiple sclerosis. Neurology (NY) 29:232-235, 1979

27. Ebers GC, Bulman DE, Sadovnick: A population-based study of multiple sclerosis in twins. New Engl J Med 315:1638-1642, 1979

28. Fontana A, Kristensen F, Dubs R, Gemsa D and Weber E: Production of prostaglandin E and interleukin-1 like factor by cultured astrocytes and C_6 glioma cells. J Immunol 129:2413, 1982

29. Fox EJ, Lewis DE, Deemer KP, ElMasry MN and Rich RR: T suppressor cell growth factor and anti-CD3 antibodies stimulate reciprocal subset of T lymphocytes. J Exp Med 166:404-418, 1987

30. Giles RC and Capra JD: Biochemistry of MHC class II molecules. Tissue Antigens 25:57-68, 1985

31. Golaz J, Steck A, Moretta L: Activated T lymphocytes in patients with multiple sclerosis. Neurology 33:1371-1373, 1983

32. Gray PW, Aggarwal BB, Benton CV et al: Cloning and expression of cDNA for human lymphotoxin, a lymphokine with tumor necrosis activity. Nature 312:721, 1984

33. Greene WC, Fleisher TA and Waldmann TA: Soluble suppressor supernatants elaborated by Concanavalin A-activated human mononuclear cells. I. Characterization of a soluble suppressor of T ccell proliferation. J Immunol 126:1185-1191, 1981

34. Greenstein JI and Crisp DT: Cellular regulation of the autologous mixed lymphocyte response in multiple sclerosis. Ann Neurol 20:P184, 1986

35. Gurney ME, Apatoff BR, Spear GT, Baumel MJ, Antel JP, Bania MB and Reder AT: Neuroleukin: A lymphokine product of lectin-stimulated T cells. Science 234:574-581, 1986

36. Hafler DA, Benjamin DS, Burks J and Weiner HL: Myelin basic protein and proteolipid protein reactivity of brain- and cerebrospinal fluid-derived T cell clones in multiple sclerosis and postinfectious encephalomyelitis. J Immunol 139:68-72, 1987

37. Hafler DA, Fox DA, Manning ME, Schlossman SF, Reinherz EL and Weiner HL: In vivo activated T lymphocytes in the peripheral blood and cerebrospinal fluid of patients with multiple sclerosis. New Engl J Med 312:1405, 1985

38. Hafler DA and Weiner HL: In vivo labelling of peripheral blood T-cells using monoclonal antibodies: rapid traffic into cerebrospinal fluid in multiple sclerosis. J Immunol 139:68-72, 1987

39. Hafler DA, Duby A, Lee SJ, Seidman DBJ and Weiner HL: Oligoclonal T lymphocytes in the cerebrospinal fluid (CSF) of patients with chronic progressive multiple sclerosis. J Neuroimmunol 16:69, 1987

40. Hauser SL, Bhan AK, Gilles F, Kemp M, Kerr C and Weiner HL: Immunohistochemical analysis of the cellular infiltrate in multiple sclerosis lesions. Ann Neurol 19:578-587, 1986

41. Haynes BF and Fauci AS: Activation of human B lymphocytes. XIII. Characterization of multiple populations of naturally occurring immunoregulatory cells of polyclonally-induced in vitro human B cell function. J Immunol 123:1289, 1979

42. Hirsch RL, Ordonez J, Panich HS and Johnson KP: T8 antigen density on peripheral blood lymphocytes remains unchanged during exacerbations of multiple sclerosis. J Neuroimmunol 9:391-398, 1985

43. Hirsch RL: Defective autologous mixed lymphocyte reactivity in multiple sclerosis. Clin Exp Immunol 64:107-113, 1986

44. Hirsch RL and Johnson KP: Natural killer cell activity in multiple sclerosis patients treated with recombinant interferon. Clin Immunol and Immunopathol 37:363-344, 1985

45. Jacobson S, Flerlage ML and McFraland HF: Impaired measles virus-specific cytotoxic T cell responses in multiple sclerosis. J Exp Med 162:839-850, 1985

46. Kasakura M, Taguchi M, Murachi T, Uchino H and Hanakoa M: A new mediator (suppressor cell induction factor) activating T cell-mediated suppression: Characterization of suppressor cells, kinetics of their generation, and mechanism of their action. J Immunol 131:2307-2315, 1983

47. Kastrukoff LF and Buican TN: Multiplex labelling analysis of peripheral blood lymphocytes in multiple sclerosis. Ann Neurol 22:138, 1987

48. Kawano M, Iwato K and Kuramoto A: Identification and characterization of a B cell growth inhibitory factor (BIF) on SCGF-dependent B cell proliferation. J Immunol 134:375-381, 1985

49. Keightley RG, Cooper MD and Lawton AR: The T cell dependence of B cell differentiation induced by pokeweed mitogen. J Immunol 117:1538, 1976

50. Kerman RH, Wolinsky JS, Nath A, Sears ES, Franklin GM and Nelson LM: Immunoregulation in MS patients treated with cyclosporine. J Neuroimm 16:90, 1987

51. Kronenberg M, Siu G, Hood LE, Shastri N: The molecular genetics of the T-cell antigen receptor and T-cell antigen recognition. Ann Rev Immunol 4:529-591, 1986

52. Lau CY, Wang EY, Li D: Mechanism of action of a suppressor-activating factor (SAF) produced by a human T cell line. J Immunol 134:3155-3162, 1985

53. Ledbetter JA, Rose LM, Spooner CE, Beatty PG, Martin PJ, Clark EA: Antibodies to common leukocyte antigen p220 influence human T cell proliferation by modifying IL 2 receptor expression. J Immunol 135:1819-1825, 1985

54. Lew AM, Lillehoj JP, Cowan EP, Maloy WL, Van Schravendijk MR and Coligan JE: Class I genes and molecules: an update. Immunol 57:3-18, 1986

55. Loertscher R, Antel JP, Cashman NR and Duquette P: Expression of T cell surface antigens as a function of cell activation. J Neuroimmunol 16:110, 1987

56. Massa PT, Ter Meulen V and Fontana A: Hyperinducibility of Ia antigen on astrocytes correlates with strain-specific susceptibility to experimental autoimmune encephalomyelitis. Proc Natl Acad Sci USA 84:4219-4223, 1987

57. Mattson DH, Roos RP and Arnason BGW: Isoelectric focusing of IgG eluted from multiple sclerosis and subacute sclerosing panencephalitis brains. Nature 287:335, 1980

58. Mattson DH, Roos RP and Arnason BGW: Oligoclonal IgG in multiple sclerosis and subacute sclerosing panencephalitis. J Neuroimmunol 2:261-276, 1982

59. Merrill JE, Kutsinai S, Mohlstrom C, Hofman F, Groopman J and Golde DW: Proliferation of astroglia and oligodendroglia in response to human T-cell derived factors. Science 224:1428, 1984

60. Miyazaki C, Nakamura T, Kaneko K, More R and Shibasaki H: Reinduction of experimental allergic encephalomyelitis in convalescent Lewis rats with cyclophosphamide. J Neurol Sci 67:277-284, 1985

61. Mohaghehpour N, Danile NK, Takasa S and Engelman EG: Generation of antigen receptor-specific T cell clones in man. J Exp Med 164:950-955, 1986

62. Mohaghehpour N, Damile NK, Takasa S and Engelman EG: Generation of antigen receptor-specific suppressor T cell clones in man. J Exp Med 164:950-955, 1986

63. Morimoto C, Letvin N, Boyd AW, Hagan M, Broiwn HM, Kornacki MM and Schlossman SF: The isolation and characterization of the human helper inducer T cell subset. J Immunol 134:3762-3769, 1985

64. Morimoto C, Letvin NL, Distaso JA, Aldrich WR, Schlossman SF: The isolation and characterization of the human suppressor inducer T cell subset. J Immunol 134:1508-1515, 1985

65. Morimoto C, Hafler DA, Weiner HL: Selective loss of the suppressor-inducer T-cell subset in progressive multiple sclerosis: analysis with anti-2H4 monoclonal antibody. New Engl J Med 316:67-72, 1987

66. Moser G, Cheng G and Abbas AK: Accessory cells in immune suppression. III. Evidence for two distinct accessory cell-dependent mechanisms of T lymphocyte-mediated suppression. J Immunol 137:3074-3079, 1986

67. Namba Y, Jegasothy BV and Waksman BH: Regulatory substances produced by lymphocytes. V. Production of inhibitor of DNA synthesis (IDS) by proliferating T lymphocytes. J Immunol 118:1379-1384, 1977

68. Noronha AB, Otten GR, Richman DP and Arnason BGW: Multiple sclerosis: Flow cytometry with simultaneous detection of surface antigen and cell cycle phase reveals that activated cells in cerebrospinal fluid in acute attacks are not OKT8 positive. Ann Neurol 12:104, 1983

69. Nossal GJV: Current Concepts: Immunology: The basic components of the immune system. New Engl J Med 316:1320-1325, 1987

70. Oger, Willoughby E, Paty DW: Immune function in MS: Correlation with disease activity as revealed by MRI. J Neuroimmunol 16:134, 1987

71. Oger J, Antel JP, Kuo HH and Arnason BGW: Influence of azathioprine (Imuran) on in vitro immune function in multiple sclerosis. Ann Neurol 11:177-181, 1982

72. Panitch HS, Haley AS, Hirsch RL, Johnson KP: A trial of gamma interferon in multiple sclerosis: clinical results. Neurology 36:285, 1986

73. Paty DW, Kastrukoff L, Morgan N, Hiob L: Suppressor T lymphocytes in multiple sclerosis: Analysis of patients with acute relapsing and chronic progressive disease.

74. Reder AT, Antel JO, Oger JJF, MacFarland TA, Rosenkoetter M and Arnason BGW: Low T8 antigen density on lymphocytes in active multiple sclerosis. Ann Neurol 16:242-249, 1984

75. Reder AT, Antel JP, Singh S, Stoklosa C and Sukhatme V: Molecular analysis of the CD8 gene in multiple sclerosis. J Neuroimmunol 14:183-187, 1986

76. Reinherz EL, Kung PC, Goldstein G and Schlossman SF: Separation of functional subsets of human T cells by a monoclonal antibody. Proc Natl Acad Sci USA 76:4061, 1979

77. Reinherz EL, Weiner HL, Hauser SL: Loss of suppressor T cells in active multiple sclerosis. New Engl J Med 303:125-129, 1980

78. Rose LM, Ginsberg AH, Rothstein TL, Ledbetter JA, Clark EA: Selective loss of a subset of T helper cells in active multiple sclerosis. Proc Natl Acad Sci USA 82:7389-7393, 1985

79. Rosenkoetter M, Reder AT, Oger JJF, Antel JP: T cell regulation of polyclonally induced immunoglobulin secretion in humans. J Immunol 132:1779-1783, 1984

80.   Rotteveel TM and Lucas CJ: T cells in the CSF of MS patients. Analysis of oligoclonality with the use of a T-cell receptor cDNA probe. In: Workshop on cellular and humoral components of cerebrospinal fluid in multiple sclerosis, Hengelhoef, Limburg, Belgium, 1986, p. 42.

81.   Santoli D, Tweardy DJ, Fereiro D: A suppressor lymphokine produced by human T leukemia cells. Partial characterization and spectrum of activity against normal and malignant hemopoetic cells. J Exp Med 163:18-40, 1986

82.   Schnaper HW, Pierce CW and Aune TM: Identification and initial characterization of Concanavalin A- and interferon-induced human suppressor factors: Evidence for a human equivalent of murine soluble immune response suppressor (SIRS). J Immunol 132:2429-2435, 1984

83.   Skias DD, Kim DK, Reder AT, Antel JP, Lancki DW and Fitch FW: Susceptibility of astrocytes to class I MHC antigen-specific cytotoxicity. J Immunol 138:3254-3528, 1987

84.   Skoskiewicz MJ, Colvin RB, Schneeberger EE, and Russell PS: Widespread and selective induction of major histocompatibility complex-determined antigens in vivo by Y interferon. J Exp Med 162:1645-1664, 1985

85.   Sorensen CM and Pierce CW: Identification and characterization of a suppressor T cell hybridoma specifically inducible by $_L$-glutamic acid ^{60}-$_L$-alanine^{30}-$_L$-thyrosine^{10} (GAT). J Immunol 137:1455-1461, 1986

86.   Sun D and Wekerle H: Ia-restricted encephalitogenic T lymphocytes mediating EAE lyse autoantigen-presenting astrocytes. Nature 320:70-73, 1986

87.   Sun D and Wekerle H: Glia cells as immunoregulatory elements: Observations on the up- and down-regulating activities of astrocyte clones. J Neuroimmunol 16:167, 1987

88.   Thompson AJ, Brazil J, Whelan CA and Feighery C: Possible in vivo modulation of Leu 2a expression on suppressor T cells in active multiple sclerosis. Immunol Letters 10:189-191, 1985

89.   Titus JA, Sharrow SO and Segal DM: Analysis of C (IgC) receptors on human peripheral blood lymphocytes by dual fluorescence flow microfluorometry, Part 2 (Quantitation of receptors on cells that express the OKM1, OKT3, KT4 and OKT8 antigens), J Immunol 130:1151-1158, 1983

90.   Traugott U, Raine CS: Experimental allergic encephalomyelitis in inbred pure guinea pigs: correlation or decrease in early T-cells with clinical signs in suppressed and unsuppressed animals. Cell Immunol 34:146-155, 1977

91.   Traugott U and Lebon P: Demonstration of Alpha, Beta and Gamma interferon in active chronic multiple sclerosis lesions. J Neuroimmunol 16:172, 1987

92.   Turck CW, Kapp JA and Webb DR: Structural analyses of a monoclonal heterodimeric suppressor factor specific for $_L$-glutamic acid^{60}-$_L$-alanine^{30}-$_L$-tyrosine^{10}. J Immunol 137:1904-1909, 1986

93.   Vervliet G, Carton H, Meulepas E and Billiau A: Interferon production by cultured peripheral leukocytes of MS patients. Clin Exp Immunol 58:116-120, 1984

94.   Waldor MK, Sriram S, Hardy R: Reversal of experimental allergic encephalomyelitis with monoclonal antibody to a T-cell subset marker. Science 227:415-417, 1985

95.   Walsh MJ and Tourtelotte WM: Temporal invariance and clonal uniformity of brain and cerebrospinal IgG, IgA and IgM in multiple sclerosis. J Exp Med 163:41-53, 1986

96.   Willenborg DO and Prowse SJ: Immunoglobulin-deficient rats fail to develop experimental allergic encephalomyelitis. J Neuroimmunol 5:99-109, 1983

97.   Wong GHW, Bartless I, Clark-Lewis I, McKimm-Breschkin JL and Schrader JW: Interferon- induces the expression of H-2 and Ia antigens

on brain cells. J Neuroimmunol 2:255, 1985

98. Yamada HP, Martin PJ, Bean MA, Braun MP, Beatty PG, Sadamoto K and Hansen JA: Monoclonal antibody 9.3 and anti-CD11 antibodies define reciprocal subsets of lymphocytes. Eur J Immunol 15:1164, 1985

99. Yip YK, Pang RHL, Urban C and Vilcek J: Partial purification and characterization of human gamma (immune) interferon. Proc Natl Acad Sci USA 78:1601, 1981

100. Zamvil S, Nelson P, Trotter J, Mitchell D, Knobler R, Fritz R and Steinman L: T-cell clones specific for myelin basic protein induce chronic relapsing paralysis and demyelination. Nature 317:355-358, 1985

15

IN VIVO AND IN VITRO ABNORMAL IgG REGULATION IN MULTIPLE SCLEROSIS

Joel Oger

Division of Neurology
University of British Columbia
Vancouver, Canada

INTRODUCTION

Since it was described by Jean Mari Charcot (5) Multiple Sclerosis (MS) has remained a disease of unknown etiology. Whatever theory is put forward to explain its etiology, will have to explain: its racial predilection for Caucasians of North European ascent, its sexual predominance for females over males with a ratio of 3/2, its affecting young rather than mature adults and the variability of its evolution where all degrees of severity are represented from the benign sub-clinical forms (as can be recognized in "clinically unaffected" twins) to the acute forms which can be rapidly fatal.

None of the theories presently available to explain its etiology fulfills these criteria. It is nevertheless, clear that a case can be made for such etiological factors as: the genetic background of the individual, an infectious agent - either specific and uncommon, or aspecific and omnipresent, and some degree of dysregulation of the immune system. The immune theory of the pathogenesis of MS is attractive if one considers that changes in the immune system occur with age, that women are more susceptible to auto-immune disorders such as Lupus Erythematosus, that susceptibility to MS is linked to the major histocompatibility complex (M.H.C) and that the MHC itself not only is linked to the genetic background but also conditions the immune response, conditioning thus most responses to human pathogens. The fact that the immune system is involved in MS is shown by the inflammatory aspect that plaques have when they are active (49) and by the presence of immunoglobulin G found in excess in the plaques (65). In this regard it is worth mentioning that changes in the cerebrospinal fluid (CSF) seem to be a remote but not unfaithful picture of changes occurring in the brain parenchyma.

In the first part of this chapter we will review the body of data generated from studies of Immunoglobulins in the CSF and in the brain of MS patients with emphasis on the generation of the non-sense antibody theory, in a second part we will summarize our experience of in vitro activation of IgG secretion by lymphocytes isolated from blood and how we envisage that the mere dysregulation of IgG secretion may be sufficient to explain most of what we will have presented in the first part.

Abnormalities of Immunoglobulins in CSF Serum and Brain of MS Patients

It is Kabat (22) who originally demonstrated an increased level of IgG in the CSF of 50% of MS patients. What is characteristic of MS is an increased IgG level due to intrathecal production of immunoglobulins. This should be distinguished clinically from the increase in immunoglobulin that follows eruption of proteins in the CSF as a consequence of damage to the blood-brain barrier (BBB). In general it is the comparison of the levels of IgG with the level of albumin in the CSF which will permit to differentiate the increase of IgG due to intrathecal IgG secretion from that due to a ruptured BBB. When the BBB is leaking, albumin increases in the CSF. As the ratio of IgG to albumin is higher in serum than in CSF (.25 versus .12), when leakage of proteins from serum to CSF occurs, the albumin will increase and the IgG/albumin ratio will also increase. On a practical basis the mere increase of IgG in the CSF in the presence of normal albumin level is diagnostic of increased intrathecal IgG synthesis. If the CSF albumin is increased, indicating a damaged BBB, one has thus to use the IgG over albumin ratio or indexes. Tourtelotte (67) has proposed an empirical formula which takes into account the level of IgG and albumin measured both in CSF and serum. This formula has been most useful in permitting to follow up the effect of experimental therapeutics as well as helping in the diagnosis (64). Indexes have been reported by E. Schuller (58) and by H. Link (63).

It has been held that local production of IgG does not fluctuate in any meaningful way with disease activity but this may have to be reconsidered in the face of the improved ability to recognize disease activity since magnetic resonance imaging (MRI) has been available (Paty, this book). Reduction of intrathecal IgG production following high dose steroids has been reported (64), short term cyclophosphamide (20) and long term Azathioprine (43). A possible correlation between IgG levels in the CSF and HLA types has been suggested by us (42, 30) and Link's group (61). We found that higher level of IgG were found in DR2+ than in DR3+ patients.

Evidence that the IgGs present in the CSF are indeed produced locally has been deducted from studying the indexes mentioned above. Direct proof of this mechanism, however, has been brought by Tourtellotte and Parker (65) who compared albumin and IgG levels found in brains of MS patients with levels found controls' brain. They found that in MS brains, plaques contained twice as much IgG than white matter from controls. Confirmation has been obtained by the scandinavian groups (27, 69).

In MS, when CSF is electrophoresed, the IgG region appears non-homogeneous because of the presence of discrete bands. The original description of the phenomenon, later called oligoclonal banding, was made by Lowenthal in 1964, was confirmed and extended by Delmotte. Since then, large series have confirmed that oligoclonal bands are better seen using isoelectric focusing than agar gel electrophoresis (33). It is possible that oligoclonal bands have prognostic value in MS (36) but their diagnostic value is more striking.

Using isoelectric focusing, these oligoclonal bands in the CSF of MS have been recognized in 90 to 99% of MS patients (72, 66). Gonsette in 1984 (14) described results of the electrophoresis of CSF in over 2000 samples of CSF and Ebers (1984) over 1000 MS patients (9). Paty and ourselves have recently compared the value of CSF electrophoresis verses MRI as a diagnostic tool for MS and both seem to have the same accuracy (48).

The presence of oligoclonal bands in the CSF is not diagnostic of MS and it can be demonstrated in a variety of inflammatory neurologic conditions where specific chronic antigenic stimulation occurs. Oligoclonal bands are regularly seen in subacute sclerosis panencephalitis, neurosyphilis, rubella panencephalitis, herpes virus encephalitis; occasional bands can also be found in some patients with cerebral vascular disease as well as patients with glial tumors of the central nervous system if sensitive enough techniques are used. We have shown that if small number of peripheral blood lymphocytes are stimulated in vitro to produce IgG, those IgG also form oligoclonal bands when they are submitted to isoelectrofocusing (45).

Immunoglobulins G are secreted by B lymphocytes. Immuno-histochemical techniques used by Simpson (59) and by Tavolato (62) confirmed the early results obtained by Tourtelotte and Parker (65) where, indeed, IgG were more abundant in plaques. There was, however, some controversy regarding the nature of the binding and it was not clear whether this binding was a mere absorbtion or a specific binding through an antigen-antibody reaction. Dubois-Dalcq (8) in 1975, Mussini in 1977 (37) and Esiri in 1980 (11) showed without doubt that cells containing IgG were present within the plaques as well as around them and provided visual confirmation that brain and CSF IgG were in fact secreted by B lymphocytes infiltrating the brain.

Two sets of experiments should be reported at this point as they are relevant to the interpretation of the oligoclonal banding phenomenon. The first is a series of experiments involving rabbits inoculated with basic protein, in which undoubtedly oligoclonal bands which were not antigen specific appeared following the disease (75). The second was generated by two groups and include reports by the Chicago group (32) and the Los Angeles group (34). Both groups eluted the immunoglobulins extracted from different plaques in the brain of MS patients and submitted them to isoelectro-focusing. The results show that, on the background of similar patterns from plaque to plaque, there appear to be specific bands for each of the plaques studied. Mattson, Roos and Arnason stressed that this indicated that different clones of lymphocytes had migrated at different times in different plaques. This notion was confirmed by Olsson in 1983 (47). On their side Mehta and Tourtelotte stressed the similarity of the oligoclonal bands pattern from plaque to plaque as well as their temporal stability. They interpreted this as the result of clonal selection by the antigen itself and the specificity of the individual's genetic code. Both interpretations are not incompatible and, in fact, both have contributed to this writer's patho-physiological hypothesis.

Other evidences of dysregulation of immunoglobulin production in MS include the predominance of the IgG 1 subclass which has been found in the CSF of both MS patients and controls (68, 70). Goust and Salier (16, 56) have analyzed the concentration of different IgG allotypes showed an increase of Glm_1 homozygotes and a restriction of the expression of the IgG 3 allotype $G3m_{11}$ in MS.

Much was expected of the studies of anti-idiotypic serum. The idiotype is the antigenic determinant of the IgG molecule which is associated with the antigen binding site. Raising antibodies specifically directed to the idiotype has been fairly powerful in Myasthenia Gravis to analyze the antibodies to the Acetylcholine Receptor (3). By raising such anti-idiotypic antisera against the immunoglobulins of the CSF of MS patients, many had expected to find cross reactivity between several patients and this could have indicated similarity in the idiotype (and thus a similarity in the antigen). It could also help find good correlation between the course of the disease and the level of specific abiological finding idiotype. Such investigations have been done by Ebers 1982 (10), by Nagelkerken (38), and by Gerhard (13) but have not provided evidence for an antigenic specific of the immunoglobulins in MS. Immunoglobulin G are made of two types of light chain; the kappa (k) and lambda (1) normally in a 3 to 2 ratio. In 10 CSF (out of 39 studied), Link and Laurenzi (28) have found only bands of the k type; in 29 others where there was both k and l bands 9 times was the number of k and l chain equal. This has since been confirmed and extended by Rudick et al in 1985 (55). Alterations of this k/l ratio could be construed as a sign that a dysregulation of the DNA synthesis occurs in MS and indicate a pseudo-neoplastic state of the B cells which would have become autonomous as was originally suggested by R. Roos (51). Similar observations have been made in animals with auto-immune disorders where many end up with lymphomas. It could also be secondary to permanent stimulation of lymphocyte.

Finally, recent reports by Coyle et al have appeared that abnormal concentration of IgG and oligoclonal banding can be found in tears of MS patients (7). This would suggest that the disease is not limited to the central nervous system compartment but may be a much larger abnormality of B cells. This would concur with the conclusions of our studies of IgG secretion using peripheral blood lymphocytes.

Antibody Specificity of the Immunoglobulin Found in CSF

When the CSF of patients with subacute sclerosis panencephalitis is submitted to electrophoresis, oligoclonal bands also appear. SSPE is a persistent measles virus infection and the oligoclonal bands have been found to contain specific antibodies for the measles virus. This has originally been shown by Vandvik using absorption studies (68) and by studies of the specificity of immunoglobulins eluded from the brain. Unfortunately, the analogical reasoning that the bands would permit to recognize a specific antigen in MS did not hold to time (333), and extensive studies have in fact shown that in MS the anti-viral antibody response is directed against multiple neurotropic viruses (39, 57, 71, 6). Antibodies which have been often found increased in serum and CSF include measles, varicella zoster, rubella, HSV1, CMV. Increased Epstein-Barr virus was found by Bray (4), and mycoplasma pneumonia by Maida (31). Interestingly, in Czechoslovakia where an orbivirus is more frequent than the measles virus, Libikova (26) found that antibodies against it increased more frequently than antibodies against the measles virus. To concur with this set of evidences various groups have been trying to absorb the oligoclonal bands from the CSF of MS patients using multiple different viruses, but none has yet succeeded.

These experiments have been the basis for the so-called nonsense antibody theory which holds that the increased antibody levels found in the CSF and the serum of MS patients is not secondary to the presence of a pathogen agent but represents an heteroclitic reaction of memory cells which are dragged into secreting immunoglobulins by a phenomenon which is not antigen-specific. We think that such an aspecific phenomenon could also explain the appearance of anti-brain antibodies including antibodies to Myelin Basic protein (74), to oligodendrocytes (1, 60), to galactocerebroside and GM1 gangliosides (54) and to other auto antigens as reviewed by O'Gorman and Oger (40).

In Vitro Studies IgG Secretion in MS Following Lectin Stimulation

Peripheral blood mononuclear cells (PBMNC) cultured in vitro in the presence of pokeweed mitogen (PWM) are stimulated to divide and differentiate into immunoglobulin secreting cells and release Ig in the supernatant that can be measured. This assay has become a model of in vivo immune response and is an invaluable tool in as much as it assesses globally most of the mechanisms involved in B cell activation. The amount of IgG found in the supernatant is a function of 1) the ability of monocytes to process PWM, but also their ability to suppress B cell function, 2) the influences of T regulatory cells (T Helper/T Suppressor balance), and 3) the intrinsic ability of B cells to respond to T signals and to secrete IgG. Studies of B cell function in vitro in MS have been reported by Levitt (25), Kelly (24), Goust (15), Henricksson (19) and Hauser (18). Most of these have shown some degree of increased B cell function variably attributed to a T cell or a B cell function defect. Most of these studies have reported small groups of patients with limited clinical subgrouping.

We think this is a major problem as we have found differences in B cell stimulation by PWM in vitro between clinical situations. When MS has become chronic progressive, IgG secretion is high. When relapsing remitting patients have a clinical attack, their IgG secretion is temporarily reduced at first but increases again as time passes. MRI has permitted us to clearly demonstrate that reduced IgG secretion follows the maximum extension of the lesions. We will review our results in these 3 clinical situations.

Chronic Progressive MS

In 1982 we extended the findings of Levitt and Goust to 23 untreated MS patients (41). We already noticed that these 5 "active " MS patients had the highest response to PWM. We expanded these studies in 1986 (44) to a group of 27 consecutive chronic progressive MS (CPMS) patients. Extremely strict selection criteria were used: to be included in the study, MS patients had to fulfill 4 criteria: 1) clinically definite MS (Poser's criteria), 2) definite MS by MRI (at least 3 white matter lesions plus one in a periventricular location as in...), 3) loss of at least one point on the Kurtzke scale over the preceding 6 months (Kurtzke) and, 4) age under 50 years. Twenty-seven CPMS patients were selected on these criteria and the results of the immune function tests were compared to those obtained in a group of 21 healthy controls of similar age. The techniques used included PWM stimulated IgG secretion in vitro (IgG Sec), Concanavalin A stimulated suppressor cell function (Con A S) and monoclonal antibody enumeration of PBMNC.

Multiple sclerosis patients with chronic disorder were shown to have increased IgG Sec and reduced Con A S (see Table 1). In the same MS patients, lymphocyte subpopulations were also enumerated. The percentage of Leu 2a + cells was significantly lower than in the healthy controls: 15.5% $\pm$ 1.3% for C-MS vs 19.6% $\pm$ 1.7% for healthy controls, the percentage of OKT8 + cells was only slightly reduced. There were no changes in the percentage of cells expressing the surface markers Leu 1, Leu 3, Leu 7, Leu 11. So far, we could not enumerate the Suppressor-Inducer subpopulation, which has recently been shown to be decreased by Rose (53) and by Hafler (17). We have since confirmed (submitted) the observation of Morimoto et al that IgG Sec levels - at least in controls - correlate with the number of T helper/inducers (T4+ 2H4+).

It is not rare to notice that some MS patients with chronic disease seem to become clinically stable after a prolonged evolution. Such a clinical pattern has already been recognized in London Ontario (12). Some of their patients with MS proven at autopsy had no inflammatory

Table 1 Con A S and IgG Sec in a Group of Progressive MS with Active
        Disease

	Chronic MS		Controls		
	n	Results mean ± SEM	n	Results mean ± SEM	p
IgG secretion at 7 days (ng/ml)	27	2608 ± 278	21	1306 ± 310	<.01*
Con A induced suppression (%)	21	8.8 ± 6.1	19	27.2 ± 3.8	<.02*

* statistically different by student t test.

Table 2   PWM Stimulated IgG Secretion (10 days) in Controls and
          2 groups of MS with chronic disease

	N	Mean ± SEM		Median
Healthy Control	25	3694 ± 708		2152
CP Active *	13	6472 ± 1296	(p<.025)*	5774
CP Stable (burnt out)**	18	4368 ± 836	(n.s.)	4155
All CP MS	3	5250 ± 740		4950

Table 3   Con A Suppression (%) at 3 Time Points in the Evolution
          of Large MRI Lesions in 6 Patients

	APP.	MAX.	DECR.
Pts. with MRI Lesions	31.2 ± 6*	-49 ± 22**	26.2 ± 6.6*
Pts. with Stable MRI	45.4 ± 13	24.3 ± 12.1	49 ± 10.1

* not different from stable MRI
** p<.05 from stable MRI

infiltrate in or around the plaques and had not had oligoclonal bands in their CSF de vivo. In table 2 we present results of IgG Sec in CPMS broken down in 2 groups: in the first group each patient had lost 1 point on the Kurtzke scale over the preceding year (Burnt out or CP-Stable). The CP-Active group was statistically different from controls while the CP-Stable was not (46, 98, 46 - see table 2).

We conclude that MNC from progressive MS patients secrete large amounts of IgG in response to PWM and have reduced Suppressor function when compared to healthy controls. A possibility exists that the former is a mere consequence of the latter. The patients who become stable as the disease burns itself out cease to exhibit this increase of in vitro IgG Secretion. It is possible that this could parallel the absence of the OCB in CSF as it was reported by the group in London (Ontario).

Relapsing Remitting MS

Clinical attacks are the hallmark of multiple sclerosis and if immune factors are to play a role in the pathophysiology of the disease one would expect to find fluctuations of immune functions with disease activity. This relationship, however, can be difficult to identify since, in MS, a reduced performance can come not only from attacks, but also from other factors such as increased exercise, increased body temperature or even depression.

Lesions that appear and disappear without clinical correlate sometimes recognized on double dose delayed contrast enhanced CT scan as large area of contrast leakage. We have shown more recently that MRI can permit one to follow their appearance and disappearance over a matter of a few months (see Paty this volume). We have thus followed prospectively with MRI and clinical examinations of the head a series of 15 relapsing remitting patients (monthly or twice monthly over a period of 6 months). Clinical attacks have been rare (n=4) but 10 patients showed distinct changes in their MRI. Six of them developed large (approximately 1 cm in diameter), round and isolated lesions, most often away from the ventricle wall and that could be seen on cuts encompassing 1 cm or more of thickness. The striking characteristic of these lesions was their timing: they appeared and disappeared over a matter of 3-6 weeks. There was no clinically correlated symptom even if most patients who developed such lesions would, at some close time point, develop some clinical signs (Willoughby et al in preparation). As these 15 patients were followed up immunologically, it became evident that fluctuations of immune function occurred with time and that the largest changes accompanied the appearance and disappearance of these lesions (46).

We have analyzed the results of immune function studies with respect to their time relation with the appearance and disappearance of these large lesions (see Table 3 and Table 4). Data from the six patients who had a single lesion appearing and disappearing were pooled and analyzed at the 3 time points corresponding to a) first appearance (App.), b) maximum size (Max.), and c) first measurement of a decreasing lesion (Decr.). Values obtained at these 3 time points were then compared to values which had been generated at a matched date from patients whose MRI showed no changes (see Table 3 and Table 4).

IgG secretion in vitro varied in both MRI stable MS and in those who developed large lesions. However, there was an impressive reduction in the amount of IgG secreted when large MRI lesions appeared. Depending on

Table 4   IgG Secretion in Culture (ng/ml) Induced by PWM at
7 and 10 Days and by SAC

	APP.	MAX.	%[1]	DECR.	%[1]
MS with Large MRI Lesions					
PWM 7	978±288 (5)	373±99 (5)	62	429±120 (5)	56
PWM 10	2476±828 (4)	1146±719 (4)	54	929±483 (4)[3]	63
SAC 10	5835±2301 (4)	3290±1390 (4)	44	2096±1644 (4)	64
MS with stable MRI					
PWM 7	1455±645 (5)	2379±1380 (5)	NR[2]	897±291 (5)	38
PWM 10	6333±3175 (4)	7109±3204 (4)	NR	3796±818 (4)	40
SAC 10	6368±2151 (4)	7036±2404 (4)	NR	7567±2236 (4)	NR

[1] % reduction when compared to levels of IgG secretion found at the time
of appearance of the lesion.
[2] NR:no reduction.
( ) In bracket number of data point available for calculation.

[3] Different from MS with stable MRI, t=3.02, p<.01.

Table 5   PWM Stimulated IgG Secretion in RR-MS During Remission.
Patients were Separated in 2 Groups According to the
Distance From the Last Relapse

	N	Mean ± SEM[1]
Controls	31	4108 ± 648
Recent Relapse (< 3 weeks)	6	2691 ± 956
Older Relapse (3-10 weeks)	7	8742 ± 1262[2]

[1] PWM induced IgG Sec. at 10 days
[2] p<.01 from controls and from recent relapses

Table 6   IgG Secretion in Stable MS Patients with RR Disease.
Patients are Sub-Grouped According to the Results of
Further Follow-Up

		Level of Response	
Group		HR(>1000 ng/ml)	LR(<1000 ng/ml)
Group 1 (Clinical Attack)	7	6	1
Group 2 (No Clinical Attack)	13	6	7

the mitogen used, there was a 44 to 62% reduction in the level of IgG secreted when results at time "Max." were compared to results obtained at time "App.". In the group with stable MRI there was no reduction in any of the assays at time "Max.", whereas in the group with large lesions all the assays were reduced by at least 40%.

We have defined high response in the PWM stimulation assay on the 7th day as being greater than 1000 ng. A high responder (HR) secretes more than 1000 ng, a low responder (LR) secretes less than 1000 ng. Controls are relatively stable over time and generally remain HR or LR. Among the MRI stable patients 2/5 were HR and did not change to LR. To the contrary, 4/5 of the patients showing large MRI lesions were HR at some point during the 6 month study but 0/5 during the period of time when the large lesion was cycling.

A large reduction in Con A inducible suppression also occurred during the progression of the MRI lesion. The maximum reduction of Con A S occurred at the time of maximal extent of the lesion. Con A suppression in the chronological controls (MRI stable MS) also showed some variation but not of the magnitude seen in the patients with large lesions (Table 4).

We have also measured IgG Sec in another group of patients at various time points after the onset of a clinical attack. We observed (Table 5) that, in some patients who had most recently experienced a clinical attack (ie. <3 weeks), IgG Sec was significantly lower than that in patients who had experienced a clinical attack over a longer period of time (ie. >3 weeks). These results based on clinical follow-up parallel the results generated in MRI assisted serial studies (46) and further confirm that attacks are followed by depressed in vitro IgG secretion.

It is of interest that the level of secreted IgG varied directly with the time elapsed since the onset of the clinical relapse (correlation coefficient r=.775). This is an indication that during attacks IgG Sec is low, but as time passes it progressively starts to increase again.

So far the usefulness of measuring IgG Sec in terms of the prognosis of RR-MS is not clear. We looked prospectively at IgG Sec in a third group of clinically stable patients and analyzed their results 6 months later, subgrouping them according to the appearance of a further attack during this follow-up period (Table 6).

Six out of 7 patients (86%) who later went on to develop further attacks were initially HR versus only 6/13 (46%) patients who remained clinically stable. These results suggest that RR-MS patients who are HR in the PWM assay are more likely to develop a clinical relapse than patients who are LR.

CONCLUSIONS

The level of reactivity of B lymphocytes isolated from the peripheral blood of MS patients changes with disease activity. When MS is progressive it is heightened - when attacks occur IgG secretion decreases from being high before the attack to being low at the maximum of the extent of the lesion.

A possible interpretation is that increased IgG secretion is secondary to reduced suppressor cell function as reported by others (2, 21) and we confirmed that both functions varied inversely in chronic progressive MS (44). It is, however, striking that our serial studies with MRI revealed increased IgG secretion <u>and</u> suppressor cell function occurring at the maximum of the extent of the lesion as seen by MRI (46). We hypothesize that a substance, secreted in the large lesions recognized by MRI, leaks in the circulation and is a specific inhibitor of immune function - we also hypothesize that when this substance disappears the immune functions increase, again on a very progressive fashion and that, past a certain threshold this increased reactivity of blood lymphocytes results in a new lesion to appear in the brain - this could explain the self sustained activity of the disease. A genetically inherited tendency to "over-react" to down regulation could then be the basis for a loose inheritance.

ACKNOWLEDGEMENTS

I thank T. Aziz, M. O'Gorman, K. Eisen, and R. Farquhar, for their assistance. The original work reported here was supported by the Medical Research Council of Canada, the Multiple Sclerosis Society of Canada and the Jacob W. Cohen Fund for MS research. I also thank A. Shaffer for preparation of the manuscript.

REFERENCES

1.   Abramsky O, Lisak RP, Silberberg DH, Pleasure DE: Antibodies to oligodendroglia in patients with multiple sclerosis. New Engl J Med 297:1207, 1977
2.   Antel JP, Arnason BGW, Medof ME: Suppressor cell function in MS: Correlation with clinical disease activity. Ann Neurol 5:338, 1979
3.   Blair D, Mihovilovic M, Agius M, Fairclough R, Richman D: Human X human hybridomas from patients with myasthenia gravis; possible tools for idiotypic therapy for myasthenia. Ann NY Acad Sci 505:155, 1987
4.   Bray PF, Bloomer LC, Salmon VC, Bagley MH, Larsen PD: Epstein-Barr virus infection and antibody synthesis in patients with multiple sclerosis. Arch Neurol 40:406, 1983
5.   Charcot JM: Gaz Hop Paris 41:554, 1868
6.   Clanet M, Puel J, Brette F, Abbal M, Blancher A, Rascol A: Antibody to viral antigens in multiple sclerosis demonstrated by ELISA method. IN: Gonsette RE, Delmotte P, Eds. Immunological and clinical aspects of multiple sclerosis. MTP Press Boston, pp 15-27, 1984
7.   Coyle PK, Sibony P, Johnson C: Oligoclonal IgG in tears. Neurology 37:853, 1987
8.   Dubois-Dalcq M, Schumacher G, Worthington EK: Immunoperoxidase studies on multiple sclerosis brain. Neurology 25:496, 1975
9.   Ebers GC: Cerebrospinal fluid electrophoresis in multiple sclerosis. IN: CM Poser, Eds. The Diagnosis of Multiple Sclerosis. ed. New York: Thieme, pp 179-184, 1984
10.   Ebers GC: A study of CSF idiotypes in multiple sclerosis. Scand J Immunol 16:151, 1982
11.   Esiri M: Ig containing cells in MS plaques. Lancer 1:478, 1977
12.   Farrell MA, Kaufmann JCE, Noseworthy JH, Armstrong HA, Ebers GC: Oligoclonal bands in MS clinical pathologic correlation. Neurology 35:212, 1985
13.   Gerhard W, Taylor A, Wroblewska Z, Sandberg-Wollheim M, Koprowski H: Analysis of a predominant immunoglobulin population in the cerebrospinal fluid of a multiple sclerosis patient by means of an anti-idiotypic hybridoma antibody. Proc Natl Acad Sc USA 78:3225, 1981
14.   Gonsette RE, De Smet Y, Delmotte P: Isoelectric focusing of the

CSF gammaglobulins in 2594 patients. IN: Gonsette RD, Delmotte Eds. Immunological and clinical aspects of multiple sclerosis. Boston: MTP Press, pp 409-421, 1984

15. Goust JM, Hogan EL, Arnaud P: Abnormal regulation of IgG production in multiple sclerosis. Neurology 32:228, 1982

16. Goust JM, Salier J-H, Glynn P, Cusner ML: Random distribution of IgG allotypes producing B-cell clones in plaque evaluates from multiple slcerosis brains. Neurology 32:148, 1982

17. Morimoto C, Hafler D, Weiner HL, Letvin NL, Hagen M, Daly J, Schlossman SF: Selective loss of the suppressor-inducer subset of T cells in progressive multiple sclerosis. New Engl J Med 316:67, 1987

18. Hauser SL, Ault KA, Johnson D, Hoban C, Weiner HL: Increased IgG secretion by unstimulated mononuclear cells in active multiple sclerosis and functional assessment of the T8 subset. Clin Immunopath 37:312, 1985

19. Henriksson A, Kam-Hansen S, Link H: IgM, IgA and IgG producing cells in cerebrospinal fluid and peripheral blood in multiple sclerosis. Clin Exp Immunol 62:176, 1985

20. Hommes OR, Prick JJG, Lamerss KJB: Treatment of the chronic progressive form of multiple sclerosis with a combination of cyclophosphamide and prednisone. Clin Neurosurg 8:59, 1975

21. Huddlestone JR, Oldstone MBA: T suppressor (TG) lymphocytes fluctuate in parallel with changes in the clinical course of patients with multiple sclerosis. J Immunol 123:1615, 1979

22. Kabat EA, Moore DH, Landow H: An electrophoretic study of the protein components in cerebrospinal fluid and their relationship to the serum proteins. J. Clin Invest 21:571, 1942

23. Keightley R, Cooper M, Lawton A: The T cell dependence of B cell differentiation induced by Pokeweed Mitogen. J. of Imm 117:1538, 1976

24. Kelley RE, Ellison GW, Myers LW, Goymerac V, Tarrick SB, Kelly CC: Abnormal regulation of in vitro IgG secretion in multiple sclerosis. Ann Neurol 9:267, 1982

25. Levitt D, Giffin NK, Egan ML: Mitogen-induced plasma cell differentiation in patients with multiple sclerosis. J. Immun 124:2117, 1980

26. Libikova H, Heinz F, Ujtazyova D: Orbiviruses of the kemerovo complex and neurological diseases. Med Microbiol Immun (Berl) 166:255, 1978

27. Link H: Oligoclonal IgG in MS brain. J. Neurosci 16:103, 1972

28. Link H, Laurenzi MA: Immunoglobulin class and light chain type of oligoclonal bands in CSF in multiple sclerosis determined by agarose gel electrophoresis and immunofixation. Ann Neurol 6:107, 1979

29. Lowenthal A: Agar gel electrophoresis in neurology. Amsterdam: Elsevier/North Holland, 1964

30. Madigand M, Oger J, Fauchet R, Sabouraud O, Genetet B: HLA profiles in multiple sclerosis suggest two forms of disease and the existence of protective haplotypes. J. Neurol Sci 53:519, 1982

31. Maida E: Immunological reactions against mycoplasma pneumoniae in multiple sclerosis. In: Pedersen E, Clausen J, Oades L (eds). Actual Problems in Multiple Sclerosis Research. Copenhagen: FADL's Forlag, pp 281-285, 1983

32. Mattson DH, Roos RP, Arnason BGW: Isoelectrofocusing of IgG eluded from multiple sclerosis and subacute sclerosing panencephalitis brains. Nature (London) 187:335, 1980

33. Mattson DH, Roos RP, Arnason BGW: Comparison of agar gel electrophoresis and isoelectric focusing in MS and SSPE brains. Ann Neurol 9:34, 1981

34. Mehta PD, Miller JA, Tourtellotte WW: Oligoclonal IgG bands in plaques from multiple sclerosis brains. Neurology 32:372, 1982

35. Morimoto C, Letvin NL, Rudd CE, Hagan M,Takeuchi T, Schlossman SF: The role of the 2H4 molecule in the generation of suppressor function in Con A activated T cells. J. Immunol 137:3247, 1986

36. Moulin D, Paty DW, Ebers G: The predictive value of cerebrospinal fluid electrophoresis in possible multiple sclerosis. Brain 106:809, 1983

37. Mussine MM, Haw JJ, Escourolle R: Immunofluorescence studies of intra-cytoplasmic Ig binding cells in the CNS. Acta Neuropathol 40:227, 1977

38. Nagelkerken LM, Aalberase RC, VanWalbeek HKV, Out TA: Preparation of antisera directed against the idiotype(s) of immunoglobulin G from the cerebrospinal fluid of patients with multiple sclerosis. J Immunol 125:384, 1980

39. Norrby E, Link H, Olsson JE: Comparison of antibodies against different viruses in cerebrospinal fluid and serum samples from patients with multiple sclerosis. Infect Immun 10:688, 1974

40. O'Gorman M, Oger J: Cell-mediated immune function in multiple sclerosis. Pathol Immunopathol Rev: in press

41. Oger J, Antel, J, Kuo HH, Arnason BGW: Effect of Imuran therapy on in vitro immune function of MS patients. Ann Neurol 11:177, 1982

42. Oger J, Charles G, Bansards Y, Sabouraud O, Raingeard P, Kerbaol M: Relation between intrathecal secretion of IgG and HLAB phenotype in multiple sclerosis. In: Dausset J. (ed). HLA and Disease. INSERM ed. Paris: INSERM 73, 1976

43. Oger J, Cloarec L, Chatel M, Sabouraud O: Modification du taux des gamma-globulines du LCR dans la sclerose en plaques traitee par les immunosuppresseurs au long cours. Nelle Pr Med (Paris) 2:443, 1973

44. Oger J, Kastrukoff L, O'Gorman M, Paty DW: Progressive multiple sclerosis: abnormal immune functions in vitro and aberrant correlation with enumeration of lymphocytes subpopulation. J. Neuroimmun 12:37, 1986

45. Oger J, Mattson DH,Roos R: Isoelectric focusing of IgG secreted by peripheral blood lymphocytes in multiple sclerosis and controls. Neurology 21:144, 1981

46. Oger J, Willoughby E, Paty DW: Serial studies of relapsing multiple sclerosis: reduced IgG secretion in vitro and reduced suppressor cell function correlate with disease activity as recognized by MRI. Ann Neurol 22:152, 1987

47. Olsson T, Link H, Kostulas V, Henriksson KG: Direct tissue isoelectric focusing of nervous system and muscle sections for detection of IgG patterns. Acta Neurol Scand 67:202, 1983

48. Paty DW, Oger J, Kastrukoff L, Hashimoto S, Hooge JP, Eisen AA, Eisen KA, Purvis SJ, Low MD, Brandejs V, Robertson WD, Li DKB: MRI in multiple sclerosis; a prospective study with comparison of clinical evaluation, evoked potentials, oligoclonal bands and computerized tomography. Neurology (in press)

49. Poser CM, Paty DW, Scheinberg L, McDonald WI, Davis FA, Ebers GC, Johnson KP, Sibley WA, Silberberg DH, Tourtellotte WW: New diagnostic criteria for multiple sclerosis: guidelines for research protocols. Ann Neurol 13:227, 1983

50. Prineas JW, Wright RG: Macrophages, lymphocytes and plasma cells in the perivascular compartment in multiple sclerosis. Lab Invest 38:409, 1978

51. Roos RP: B-cell abnormalities in multiple sclerosis. Arch Neurol 42:73, 1985

52. Rose AS, Ellison GN, Myers LW, Tourtellotte WW: Criteria for the clinical diagnosis of MS. Neurology 26(Suppl 2):20, 1976

53. Rose LM, Ginsberg AH, Rothstein TL, Ledbetter JA, Clark EA: Selective loss of a dubset of T-Helper cells in active multiple sclerosis. Proc Natl Acad Sci USA 82:7389, 1985

54. Rostami A, Eccleston PA, Silberberg DH, Manning MC, Burns JB, Lisak RP: Absence of antibodies to galactocerebroside in the sera and cerebrospinal fluid of human demyelinating disorders. Neurology 33:130, 1983

55. Rudick RA, Peter DR, Bidlack JM, Knutson DW: Multiple sclerosis: free light chains in the cerebrospinal fluid. Neurology 35:1443, 1985

56. Salier JP, Goust JM, Pandey JP: Preferential synthesis of the Gm(1) allotype of IgG1 in the central nervous system of multiple sclerosis patients. Science 213:1440, 1981

57. Salmi A, Panelius M, Vainopaa R: Antibodies against different viral antigens in the CSF of patients with multiple sclerosis and other neurological diseases. Acta Neurol Scand 36:261, 1979

58. Schuller E, Sagar HJ: Local synthesis of CSF IgA neuroimmunological classification. J. Neurol Sci 51:361, 1981

59. Simpson JF, Tourtellotte WW,Kokmen E, Parker JA, Itabashi HH: Fluorescent protein tracing in multiple sclerosis brain tissue. Arch Neurol 20:373, 1969

60. Steck AJ, Link H: Antibodies against oligodendrocytes in serum and CSF in multiple sclerosis and other neurological diseases - ^{125}I-protein A studies. Acta Neurol Scand 69:81, 1984

61. Stendhal L, Link H, Mooler E: Relation between genetic markers and oligoclonal IgG in CSF of optic neuritis. J. Neurol Sci 27:93, 1976

62. Tavolato BF: Immunoglobulin G distribution in multiple sclerosis brain - an immunofluorescence study. J. Neurol Sci 24:1, 1975

63. Tibbling G, Link H, Ohman S: Principles of albumin and IgG analyses in neurological disorders: establishment of reference values. Scand J. Clin Lab Invest 37:385, 1977

64. Tourtellotte WW, Murphy K, Brandejs D: Schemes to eradicate the MS/CNS immune reaction. Neurology 6:59, 1976

65. Tourtellotte WW, Parker JA: Multiple sclerosis brain IgG and albumin. Nature (London) 214:683, 1967

66. Tourtellotte WW, Staugaitis SM, Walsh MJ, Shapshak PL, Baumhefner RW, Potvin AR, Syndulko K: The basis of intro-blood-brain-barrier IgG synthesis. Ann Neurol 17:21, 1985

67. Tourtellotte WW: On cerebrospinal fluid immunoglobulin-G (IgG) quotients in multiple sclerosis and other diseases - review and a new formula to estimate the amount of IgG synthesized per day by the central nervous system. J. Neurol Sci 10:279, 1970

68. Vandvik B,Norrby E,Nordal HK: Oligoclonal measles virus-specific IgG antibodies isolated from cerebrospinal fluids, brain extracts, and sera from patients with subacute sclerosing panencephalitis and multiple sclerosis. Scand J. Immunol 5:979, 1976

69. Vandvik B, Reske-Nielsen E: Immunochemical and immunohistochemical studies of brain tissue in subacute panencephalitis and multiple sclerosis. Acta Neurol Scan 51:413, 1972

70. Vartdal F, Vandvik B: Multiple sclerosis - subclasses of intrathecally synthesized IgG and measles and varicella zoster virus IgG antibodies. Clin Exp Immunol 54:641, 1983

71. Vartdal F,Vandvik B, Norrby E: Viral and bacterial antibody responses in MS. Ann Neurol 8:248, 1980

72. Walsh MJ, Staugaitis S, Shapshak P, Tourtellotte WW: Ultrasensitive detection of IgG in unconcentrated cerebrospinal fluid utilizing silver nitrate staining after isoelectric focusing and immunofixation. Ann Neurol 12:114, 1982

73. Walsh MJ, Tourtellotte WW: IgG, IgM, IgA temporal clonal stability in multiple sclerosis. Neurology 33:123, 1983

74. Warren KG, Catz I, McPherson TA: CSF myelin basic protein levels in acute optic neuritis and multiple sclerosis. Can J. Neurol Sci 10:235, 1983

75. Whitacre CC, Mattson DH, Day ED, Peterson DJ, Paterson PY, Roos RP, Arnason BG: Oligoclonal IgG in rabbits with experimental allergic encephalomyelitis: Non-reactivity of the bands with sensitizing neural antigens. Neurochem Res 10:1209, 1982

16

STUDIES OF MYELIN BREAKDOWN IN VITRO

D. Johnson, R. Toms, and H. Weiner

Center for Neurologic Diseases, Brigham and Women's Hospital
Harvard Medical School
Boston, MA

INTRODUCTION

The present chapter describes that aspect of our studies that is concerned with the mechanism of myelin sheath damage in inflammatory demyelinating diseases such as multiple sclerosis (MS) and Guillain-Barre syndrome (GBS). Our work has been largely conducted in vitro, and most of it has been concerned with biochemical changes occurring in myelin under a variety of pathologically relevant challenges. Although not intended as a detailed review of the biochemistry of demyelination, the results we have obtained will be discussed with respect to related studies, the problems in interpreting these findings, and ways in which our results may relate to the pathogenesis of human disease. Inflammatory diseases can affect both central nervous system (CNS) and peripheral nervous system (PNS) myelin, and although most of the experiments described here involve the CNS, thee mechanism of PNS sheath breakdown may well have features in common.

Myelin Structure

The myelin sheath is a greatly elaborated plasma membrane which is wrapped around the axon in a spiral fashion to form a multilamellar structure consisting of closely apposed layers of plasma membranes. The myelin membranes originate from, and remain part of, the Schwann cell in the PNS, and the oligodendrocyte in the CNS. While the Schwann cell myelinates a single PNS axon, the oligodendrocyte can myelinate as many as forty CNS axons. The compact myelin forms a lipid-rich (greater than 70% of dry weight) sheath around the axon. Some myelin components, such as myelin basic protein (BP), the myelin-associated glycoprotein (MAG), and galactocerebroside (GC), are common to both the CNS and PNS. In higher vertebrates, the major structural proteins, proteolipid protein (PLP), and P_0 are unique to the myelin of CNS and PNS respectively. An additional basic protein, P_2, is found predominantly in PNS myelin. More detailed descriptions of myelin structure are available (1-3).

Multiple Sclerosis

Although there are several disorders that can be classified as inflammatory demyelinating diseases, our primary interest lies in MS.

MS is a complex disease, with varying severities and time courses, and has a similarly complex pathology. There have been many pathological studies of MS tissue, and a detailed review is outside the scope of this article. Extensive accounts of the pathology of MS are available elsewhere (4-6). The main pathological feature of MS is the presence of plaques, or areas of CNS demyelination. These plaques are generally centered around blood vessels, suggesting that the factors initiating demyelination enter the CNS from the bloodstream. In active plaques there is perivascular infiltration comprised predominantly of macrophages, lymphocytes, plasma cells and large mononuclear cells. Macrophages appear to play a major role in myelin damage, as macrophage processes can be found invading myelin sheaths, appearing to strip myelin from the axon. They also appear to play a role in clearance of debris, and macrophages containing myelin fragments are frequently observed. Axons, neuronal cell bodies, and other constituents of the CNS appear to be largely spared in MS.

MS is of unknown etiology, although it is generally thought that autoimmune processes are involved. The apparently specific attack of the immune system on the CNS myelin sheath in MS is consistent with a specific myelin antigen acting as a target for the immune system or of a selective vulnerability of the sheath to a local insult, followed by scavenging by macrophages and activation of the immune system. It is still not clear whether the obvious signs of immune activity in the plaque region reflect a direct attack on the myelin sheath, or represent a secondary event following prior myelin damage caused by some other agent. An insult to the oligodendrocyte, such as viral infection, could cause destabilization of associated myelin sheaths. Although detailed pathological studies of MS tissue suggest intimate contact between macrophages and damaged myelin is more common, abnormal sheaths are also observed in the vicinity of macrophages that are apparently not in direct contact with them. This may mean that the myelin is damaged by macrophage secretion products, or that the macrophage is attracted to an already damaged myelin sheath. It is not known whether macrophages only attack damaged sheaths, or whether they are capable of initiating the damage themselves.

Experimental Considerations

Although demyelination is typically defined in pathological terms we have been interested in the sequence of biochemical events that underlie the loss of the myelin sheath, as an understanding of potential molecular mechanisms involved may help to understand the pathogenesis of demyelinating diseases. There have been many previous studies examining mechanisms of demyelination and a wide range of substrates have been employed. These range from whole animals, through tissue culture systems, to biochemical preparations such as isolated myelin, or purified proteins. Many of these models have been recently reviewed (7). It is obvious that the more complex a system is, the harder it is to derive unequivocal results, and the more simple it is the harder it becomes to extrapolate the results to the in vivo situation. Although this is a problem with many sorts of experiments, it seems to be particularly relevant to demyelination, where one is faced with the problem of distinguishing artefactual damage from that which may occur in vivo, where intact myelin represents a unique target for immune-mediated breakdown. Thus experiments using isolated preparations of myelin are fraught with complications of interpretation. Many biochemical experiments have used isolated myelin, and examined the effects of various challenges on protein and lipid composition, and membrane breakdown. The myelin must necessarily undergo a certain amount of physical trauma during isolation and thus any vulnerability demonstrated during such experiments must be carefully interpreted. PNS myelin, for instance, is surrounded in vivo by a layer of Schwann cell cytoplasm, a Schwann cell plasma membrane, and a basal lamina, and it is difficult to experimentally maintain the same sorts of topographical considerations in isolated preparations.

Despite the caveats of working with an isolated myelin preparation, it does enable us to study a chemically defined substrate, presumably including material similar to that attacked in demyelinating diseases. The majority of our studies have employed a preparation of myelinated axons from guinea pig CNS, and we have examined changes in myelin biochemistry under a variety of conditions. We have focused on the loss of BP from these preparations, using it as ann indicator of demyelination. BP is of interest for several reasons; it is a major component of CNS myelin, where it is thought to play a role in maintaining compaction. It is an extrinsic membrane protein located on the cytoplasmic side of the compacted myelin membrane. BP may be lost more easily than proteins such as PLP which are integral membrane proteins. BP is located throughout compact myelin, so that a significant loss would presumably reflect considerable disruption of the multilamellar structure. Extensive loss of BP has been demonstrated in the areas of MS plaques using both immunocytochemical and biochemical techniques (8,9).

As macrophages are so prominent in MS lesions, we were particularly interested in how they might interact with myelin, and how macrophage enzymes may attack the myelin sheath. In order to further investigate the possible events involved in myelin breakdown in vivo we have employed a preparation of myelinated axons, usually prepared from guinea pig brainstem and suspended in isotonic saline, as a "target". Myelinated axons were used because they can be easily prepared at relatively high purity, avoiding the osmotic shock of conventional myelin preparations. Physical treatment of the myelin sheath during isolation can affect its susceptibility to exogenous proteases (10-12). The protein composition of the myelinated axon preparation is similar to that of more highly purified myelin, and although the myelinated axon does contain axolemmal constituents, the predominant component is myelin, and the two major proteins are PLP and BP (Figure 2).

Effects of Ca^{2+} on Myelinated Axons

We have examined the effects on myelinated axons of incubation with cultured macrophage supernatants, and with agents such as complement and peroxide which may be released by macrophages adjacent to myelin sheathes. Macrophages normally play a major role in host defense, and secrete many agents, such as proteases, reactive oxidative intermediates and cytotoxic agents, when stimulated (13,14). They have a range of activation states, and for the majority of our studies we have used mouse peritoneal macrophages elicited by thioglycollate, an agent that induces a population of inflammatory macrophages with high levels of protease secretion. Conditioned media obtained from cultured thioglycollate-elicited macrophages contained considerable proteolytic activity when assayed by incubating with isolated preparations of bovine BP. This activity was maintained for at least four days in culture (Figure 1). Myelin proteins in this and subsequent experiments were separated by gel electrophoresis, stained with Coomassie blue and quantitated by densitometry. In our initial experiments with lyophilized bovine myelin isolated by the widely used procedure of Norton and Poduslo (15), the myelin was incubated with conditioned media from thioglycollate-elicited macrophages. We observed a significant loss of BP compared to unincubated control samples; however, when control samples were incubated

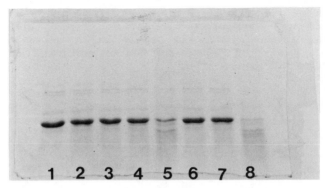

Figure 1   Degradation of Isolated BP by Macrophage Supernatants.
Purified bovine BP was incubated with dialyzed and concentrated
conditioned supernatants collected from thioglycollate-elicited
peritoneal murine macrophages following the first or second period of two
days in culture.   Incubations were for 16 hours at 37 degrees C, in 25 mM
Tris (pH 7.5), and 6 mM $CaCl_2$, with increasing volumes of supernatant.
Lanes show: 1) unincubated control; 2) incubated in buffer alone;
3-5) incubated with 1 1, 5 1 and 25 1 respectively of the first 48
hour supernatant; 6-8) were incubated with similar volumes from the
second period of 48 hours in culture.

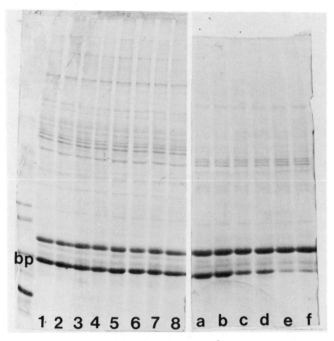

Figure 2   Comparison of the Effects of Ca^{2+} on myelin and myelinated
axons. Myelin (15) and myelinated axons (19) were prepared from guinea
pig CNS. Frozen and thawed myelin and freshly prepared myelinated axons
were incubated in 0.15 M saline, 20 mM HEPES (pH 7.4) and in the presence
or absence of 2.5 mM $CaCl_2$ or $MgCl_2$. Incubations were for 2 hours at
37 degrees C. Incubations were terminated by addition of buffered saline
and centrifugation. SDS gels are of the pelleted material. Lanes (1-8)
= myelinated axons. Lanes (a-f) = myelin.1 Lanes are: (1-2, c-d)
without Ca^{2+}; (3-4, e-f) plus Ca^{2+}; (5-6) plus Mg^{2+}; (7-8) plus Ca^{2+} and
MG^{2+} and (a-b), unincubated controls.

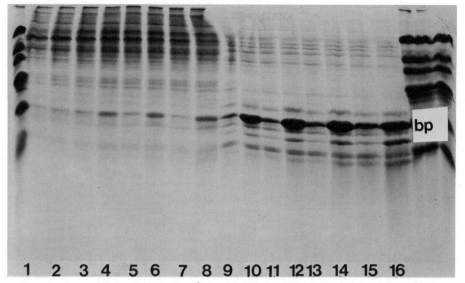

Figure 3   Comparison of Ca^{2+}-Stimulated Release of Peptides from Myelin and Myelinated Axons.   Incubations were carried out as for Figure 2, but for varying lengths of time. Gel shows non-pelleted Coomassie Blue-staining material obtained after centrifugation of incubations. From left to right, lanes are molecular weight standards; supernatants from myelinated axons incubated - and + Ca^{2+} for 0, 30, 60 and 180 min (lanes 1-8) and myelin incubated under identical conditions (lanes 9-16).

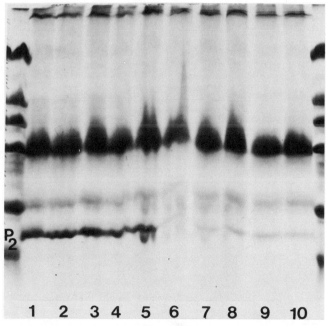

Figure 4   Release of P_2 from Bovine Myelin.   Lyophilized bovine PNS myelin was resuspended at 4 mg protein/ml and incubated for 3 hours at 37 degrees C under the following conditions: (1-5) 20 mM HEPES, pH 7.4, and Control (1, 2), 2.5 mM Ca^{2+}, (3, 5) and A23187 (4, 5).   (6-10) 20 mM HEPES, 0.15 M NaCl, and Control (6, 7) 2.5 mM Ca^{2+} (8,10) and A23187 (9, 10).   Samples were centrifuged, and pellets were electrophoresed.

in unconditioned media there was also a reduction in BP. Further examination of this effect led to the conclusion that the active factor was calcium, a normal constituent of our culture media. That the Ca^{2+}-induced loss of BP may be due to activation of some endogenous Ca^{2+}-dependent process is supported by reports that BP is susceptible to the action of a myelin-associated Ca^{2+}-activated neutral protease (11,16-18).

What prevents the Ca^{2+}-induced loss of BP from normal myelin in vivo? Perhaps structural constraints prevent Ca^{2+}/salt access to the myelin. In order to examine this possibility, we compared the effects of Ca^{2+} on two different myelin preparations. One of these was isolated using the standard Norton and Poduslo procedure (15) and the other was a preparation of myelinated axons prepared by the method of De Vries and co-workers (19). These preparations differ primarily in the absence of an osmotic shock step in the De Vries technique, so that the morphology of the sheath is presumably somewhat closer to the in vivo state than that of the shocked myelin. Absence of osmotic shock means that the preparation comprises portions of intact myelinated axons rather than isolated myelin. In addition, the myelinated axons we used were freshly prepared while the preparation of purified myelin had been frozen. While myelinated axons and purified myelin have similar protein compositions (Figure 2), the loss of BP was considerably greater than from myelinated axons. The purified myelin lost a significant amount of BP when incubated with saline alone, but in the presence of Ca^{2+} there was an almost total loss of BP. In the case of myelinated axons, BP loss was negligible in the absence of Ca^{2+}, and was only detectable (and then at lesser levels than in isolated myelin) when Ca^{2+} was present. These experiments support earlier descriptions of how structural disruption of the myelin sheath affects the ease with which BP is lost (10-12). This is relevant to experimental systems, as well as to studies in pathology, where local inflammation may cause sufficient disruption for these mechanisms to be operative.

In order to compare material released from the myelinated axons with that remaining in the pellet, our incubations were terminated by washing out the Ca^{2+}, and pelleting the myelinated axons by centrifugation. Loss of BP from the insoluble pellet could be explained by proteolytic degradation (with generation of soluble fragments) as well as by dissociation of the intact BP from the myelin. This latter explanation is made more likely by the recent studies of Glynn et al (20,21) who suggested that a likely sequence of events is a dissociation of intact BP from the myelin membrane, which is promoted by Ca^{2+}, and subsequent proteolysis by a myelin-associated protease. In their studies the myelin substrate was shocked and frozen and the release of intact BP was followed by the appearance of a distinct pattern of proteolytic breakdown fragments following longer incubations. We observed a similar pattern with the same substrate, but when we analyzed the pattern of release from our standard fresh myelinated axon preparations there was a much slower release of intact BP. The pattern of peptides apparently generated from BP was also different (Figure 3).

Most of our incubations used isotonic saline, in order to resemble in vivo conditions. There have been reports of the dissociation of BP from myelin membranes in the presence of isotonic saline (20,22), perhaps because of ionic interference with the BP-lipid interactions that maintain the BP in the membrane. However, one of these studies (22) was in Krebs-Ringer bicarbonate, which contains physiological Ca^{2+} in addition to isotonic saline. Although subsequent studies of Glynn (20) and Smith (23) have shown that appropriate concentrations of sodium chloride alone are sufficient to dissociate some of the BP, these studies were performed on shocked, sometimes frozen, myelin, which would be

expected to have significantly greater susceptibility to BP loss than
native myelin. In our incubations of myelinated axons, isotonic saline
had little effect on BP content.

Effects of Complement

It was of interest to examine ways in which increased calcium access
to compact myelin could occur pathologically, and whether this could
promote loss of BP. One possibility could involve complement fixation.
It has been shown that CNS myelin can fix complement via the classical
pathway (24,25), leading to assembly of the pore-forming membrane attack
complex (MAC), and that assembly of the terminal complex is needed for
antibody-mediated demyelination of myelinated CNS culture systems (26).
If the MAG were to be assembled on the myelin this could lead to an
increased influx of Ca^{2+} down a concentration gradient and stimulate the
Ca^{2+}-dependent loss of BP. Using human or rabbit serum as a source of
complement we were able to demonstrate a marked increase in the loss of
BP, an effect that could be abolished by heating at 56 degrees C for 30
minutes, conditions known to inactivate complement. As serum contains
many proteases which could themselves degrade BP, it was possible that
this increase was due to exogenous proteolysis by heat-inactivable
proteases. In order to exclude this possibility, we used sera that were
deficient in a specific complement component; human serum immunologically
depleted of C5, and rabbit serum genetically lacking C6. When myelinated
axons were incubated with these depleted sera, the increased loss of BP
was not seen, but when the sera were specifically reconstituted with the
appropriate purified complement component, the increase in BP loss was
restored (Table 1). This appears to demonstrate that the presence of
terminal complement components, and presumably the formation of MAC, was
necessary for the increased BP loss to occur. Cammer et al found that
complement was able to potentiate the action of plasmin on freshly
isolated CNS myelin, and that an intact complement pathway was necessary
(12). We do not know how this relates to our observations, but it is of
interest that reference was made in their study to the ability of human
serum alone to degrade fresh myelin. When we conducted a time course of
the complement effect, it was seen that the majority of the changes
occurred in the first 15 minutes, (Table 2) and that subsequently, loss
of BP occurred at the same rate in both the presence add absence of
complement (results not shown). This suggests that either the complement
was inactivated, or that there was a fraction of the myelinated axons
containing a finite pool of BP that was susceptible to complement-
promoted removal. Once this had been released, the remaining BP would
then be lost by similar, slower mechanisms that we see in myelinated
axons not exposed to complement. One possible explanation would be that
all the complement binds to the outermost lamellae of the myelinated axon
fragments, leading to loss of BP from the myelin, and at the same time
absorbing the complement. Electron micrographs show that our myelinated
axon fragments are an average of 4-6 membrane pairs so that perhaps 25%
of the BP would be lost, which is in agreement with our experimental
data.

We also attempted to examine the form in which BP is released
following complement treatment, and to see how this compares with BP-
derived peptides released from myelinated axons incubated with Ca^{2+}
alone. This has been difficult to analyze because the preparations of
complement we use are essentially whole serum, so that even if BP were
released intact, it would be exposed to serum proteases. We have
attempted to examine the effect of increased Ca^{2+} in myelin by using the
calcium ionophores, A23187 and ionomycin, to mimic the effect of
complement, but they appear to have less of an effect on BP release. In
preliminary experiments, we observed a slight increase in BP loss when

Table 1   Effects of Complement on BP Loss from Myelinated Axons

Condition	% Loss of BP
Incubated	
$\quad$ - Ca^{2+}	9
$\quad$ + Ca^{2+}	21
Human Serum	44
$\quad$ - C5	11
$\quad$ + C5	33
$\quad$ + HI	10
Rabbit Serum	61
$\quad$ - C6	29
$\quad$ + C6	69
$\quad$ + HI	10

Myelinated axons were incubated as for Figure 2. Incubations contained 2.5 mM Ca^{2+} unless stated, and additional components as listed. Serum concentration was 25%, and was either: 1) normal; 2) lacking individual components (-C5, -C6); 3) serum from #2 reconstituted with the appropriate component (+C5, +C6). HI = Heat inactivated. Results are the mean of duplicate determinations.

Table 2   Effect of Serum on Ca^{2+}-Dependent BP Loss from Myelinated Axons

	Amount of BP	
	Unincubated	Incubated
Myelinated Axons + 0	1.09 ± 0.14	1.04 ± 0.12
$\quad$ + Ca	1.07 ± 0.14	1.08 ± 0.05
$\quad$ + Ca/Human Serum	1.08 ± 0.10	0.80 ± 0.07
$\quad$ + Ca/Rabbit Serum	1.00 ± 0.09	0.65 ± 0.09

Myelinated axons were incubated as described for Figure 2 for 15 minutes at 37 degrees C. Incubations were centrifuged and aliquots of the pellets were electrophoresed. Units are relative amounts of BP quantitated by densitometry of SDS gels, and are the mean of quadruplicate determinations.

Table 3   Effects of Ionophores on BP in Myelinated Axons

	% Loss of BP
$-Ca^{2+}$	7.5
$+Ca^{2+}$	33.9
Ionomycin	
1 μm	32.2
5 μm	40.3
25 μm	41.9
A23187	
1 μm	34.5
5 μm	32.6
25 μm	30.6

Myelinated axons were incubated as for Figure 2, for 1 hour, in the presence or absence of Ca^{2+} and various concentrations of ionophores. Numbers represent the % loss of BP compared to unincubated controls, quantitated by densitometry, and are the mean of duplicate determinations.

Table 4   Effects of Peroxide on BP Loss from Myelinated Axons

First Incubation	Second Incubation	% Loss of BP
Control	Ca^{2+}	80
	Ca^{2+}/MO	78
H_2O_2	Ca^{2+}	0
	Ca^{2+}/MO	74

Freshly prepared myelinated axons were pre-incubated in 25 mM Tris (pH 7.5), 2.5 mM $CaCl_2$, 0.1 M NaCl, with or without 5 μm H_2O_2 and 0.34 M $CuSO_4$, for 2.5 hours at 37 degrees C. Additional H_2O_2/Cu was added once during the incubation. The peroxide was washed out, and samples incubated for an additional 2.5 hours, in the presence of 2.5 mM Ca^{2+}, with (MO), or without conditioned medium from thioglycollate-elicited macrophages. Results are expressed as % loss of BP compared to unincubated controls, and are the mean of duplicate determinations.

myelinated axons were incubated with ionomycin, although the effect smaller than that of complement (Table 3). This may be due to the large pores formed by complement compared to the actions of the ionophores, which act as ion carriers. The holes formed by MACs are large enough to permit passage of BP (27) and it is possible that they can effect disruption of the myelin both by permitting Ca^{2+} influx and by providing a means for BP to be lost from the membrane.

The amount of BP loss that occurs in the presence of complement is high (65% in 3 hours), and one would expect there to be some sort of corresponding disruption of myelin structure. In order to study this we have begun to examine ways of assessing overall damage to the myelin sheath. Electron microscopy shows that the relatively simple isolation procedure itself causes considerable structural damage and that it is difficult to detect additional damage that may be induced by complement. In collaboration with Dr. Dan Kirschner (Children's Hospital, Boston) we have been attempting to quantitate changes in myelin organization using X-ray diffraction. This technique provides a means to non-invasively measure the extent to which multilamellar myelin is preserved in our various preparations (28). The structural results can then be correlated with the biochemical changes. In our preliminary experiments, the degree of loss of ordered structure of the compact myelin detected by X-ray diffraction appears to be proportional to Ca^{2+}-induced BP loss assessed by SDS electrophoresis, and the most disruption occurs when myelinated axons are incubated with active complement. This suggests that the loss of BP is associated with actual damage to the myelin sheath. It is perhaps to be expected that significant loss of a major protein would be reflected in a disruption of the overall structure of the myelin sheath. However, we cannot tell from these studies whether the loss of BP is the initial event following complement fixation, or if there are intermediate steps involving other myelin components, such as lipids that interact with BP.

PNS Myelin

There are similarities between inflammatory breakdown of myelin in the CNS and PNS. Demyelination appears to be effected by invading macrophages, and vesiculation of the myelin sheath occurs in the vicinity of these cells (4). The mechanism of myelin breakdown is unknown and it was of interest to examine whether the potential mechanisms of myelin breakdown that we had identified in the CNS were operative in the PNS. This seemed to be a possibility for two reasons. Exposure of peripheral nerves to Ca^{2+} ionophores causes PNS myelin damage (29,30), and in addition, isolated PNS myelin can fix complement, although unlike CNS myelin, it does so via the alternative pathway (31).

It was not possible to use myelinated axons from the PNS for these experiments, due to the presence of substantial amounts of connective tissue, which made thee appropriate tissue disruption techniques impractical. In addition, although PNS myelin contains the same BP as CNS, it is present in lower amounts, and is probably not involved in membrane compaction (3,32). There is an additional basic protein, P_2, present in PNS myelin. It is a significant component in most species except rodents, which have only trace amounts. Like BP, P_2 is an extrinsic membrane protein, and a potent immunogen for experimental autoimmune disease (33). We have examined the susceptibility to Ca^{2+}-dependent protein loss in various PNS preparations, including whole nerve and isolated myelin. We found that P_2 is easily lost from isolated lyophilized bovine myelin in the presence of isotonic saline; a simple washing step is sufficient to remove the majority of the P_2 (Figure 4). This is in agreement with a previous report (34). In further experiments, we found that calcium ionophores had no additional effect, and that the P_2 was released intact. In contrast, P_2 is not lost from fresh bovine spinal roots incubated with saline. In fact, the isotonic saline seems to protect the nerve from P_2 loss compared to incubation in water, where there is presumably osmotic shock damage. Incubation of PNS preparations that are resistant to saline-induced loss of P_2 (e.g. fresh bovine spinal roots) with physiological levels of calcium, with or without ionophores, and under isotonic conditions, has little or no effect on myelin proteins. This is somewhat surprising considering the effects of such treatment on CNS myelin, and the rapid effects of calcium ionophores on PNS myelin morphology (29,30). Potential effects on lipids were not studied in the present experiments and it is possible that the effects of ionophores on whole nerve are due to activation of phospholipase (29). It will be of interest to test the ability of physiological treatments to mimic the effects that the myelin purification procedure has on the association of P_2 with the myelin membrane. We have not examined the effects of exogenous proteases on PNS myelin, although there have been several previous studies on this subject. The major protein of PNS myelin, P_0 glycoprotein, is susceptible to proteolytic cleavage by neutral proteases known to be secreted by macrophages (35,36). Plasmin can also degrade P_0, as well as BP, in bovine PNS, although P_2 seems to be resistant (36).

Macrophages and Myelin Breakdown

As described earlier, macrophages appear to be central to inflammatory myelin breakdown, and macrophages containing myelin debris are evident in demyelinating disease. As it is unlikely that they could phagocytose a whole sheath, a certain amount of extracellular "predigestion" of a small portion of the myelin sheath must occur. This may happen independently of the macrophage, perhaps via mechanisms described above, or the macrophage itself may effect some of this damage. Myelin damage in MS brains has been observed in the presence of invading cells but prior to obvious phagocytosis. Previous studies have found that BP in lyophilized myelin is susceptible to proteolytic activity in cultured macrophage supernatants. This activity can be increased by incubation with plasminogen, suggesting that macrophage plasminogen activator is generating plasmin. In unlyophilized myelin, BP is relatively resistant to the activity of macrophage supernatants however, and significant breakdown is only seen in the presence of exogenous complement or phospholipase (12). Trotter and Smith examined the effects of secreted macrophage phospholipases on myelin vesicles and found that myelin lipids were also relatively resistant to degradation (38). The myelinated axon preparations that were used in our experiments also showed some resistance to degradation by macrophages; with the exception of peroxide preincubation or heat inactivation (see below), the loss of BP from myelinated axons incubated with macrophage supernatants was not greater than that observed when incubations were in Ca^{2+}-containing media alone. This was the case even when the supernatants were from inflammatory macrophages, and concentrated several-fold.

Another potential mechanism of myelin damage is via the release of reactive oxidative intermediates (ROI). Circulating monocytes, such as may be attracted to inflammatory lesions, and tissue macrophages activated by ã-interferon, which might also be present in active lesions, generate large amounts of these damaging metabolites (13,14). Various oxidative reagents are used by macrophages to kill cells by damaging the plasma membrane through lipid peroxidation. As myelin is lipid-rich it may be susceptible to such an attack. These free radicals are very unstable, and cultured macrophage supernatants may not maintain sufficient levels to cause detectable myelin damage in experimental incubations. Isolated BP is aggregated by incubation with excess hydrogen peroxide (39), and Konat et al have recently shown that incubation of lyophilized myelin with a ROI-generating system leads to a general aggregation of myelin proteins (40). We have employed a similar cupric ion-peroxide system (41) to examine the effects of ROIs on freshly isolated myelinated axons. In our initial studies we found that myelin proteins in the presence of ROIs were readily aggregated in our incubations, although PLP seemed to be more susceptible than BP. As it seemed likely that any respiratory burst _in vivo_ would be accompanied by the release of protease(s), we studied the effects of incubation of myelinated axons with the peroxide system followed by incubation with cultured macrophage supernatants. Prior to these studies, we titered the ROI system against myelinated axons to find conditions that had a minimal effect on protein aggregation. Interestingly, this led to the use of peroxide concentrations of 5 m, considerably lower than those used in previous studies (39,40), but close to levels generated by cultured monocytes (42). These experiments led to two interesting results. First it was found that pre-incubation of myelinated axons with the ROI system caused an almost complete inhibition of the Ca^{2+}-dependent loss of BP described earlier (Table 4). This could be due to a direct effect on any enzymes involved in the process; the inhibition of proteases by peroxide has recently been reported (43). Alternatively, the ROIs could affect myelin lipids and alter the ability of BP to dissociate from the myelin membrane. The other interesting finding is that the BP in the pretreated myelinated axons becomes susceptible to macrophage proteases (Table 4). This was of interest because we have seen similar effects when heat inactivating our preparations; that is, inactivation of endogenous Ca^{2+}-dependent BP loss, and exposure of the BP to exogenous proteases. These are the only conditions that we have studied in which macrophage proteases have any affect on BP in myelinated axons, and may reflect an increase in accessibility of the BP.

Mast Cells

During the course of these studies, we became aware of the ability of certain basic peptides to degranulate mast cells (44). As BP and related peptides are known to be released from myelin sheaths undergoing demyelination (45), and as mast cells occur in both the PNS and CNS where they are often adjacent to myelin (46,47), it seemed possible that degranulation of mast cells could be stimulated by basic myelin proteins. Although nervous system mast cells have not been as well characterized as other tissues, what is known suggests that they are of the serosal or connective tissue type, and for the mast cell studies described here we used serosal mast cells isolated from rat peritoneum. Initial studies did in fact demonstrate that purified BP, and also P_2, were able to stimulate degranulation of mast cells. The degranulation was assayed by following the release of the granule enzyme, ß-hexosaminidase. To ensure that the release of this enzyme was not in fact due to cell death, release of the cytosolic enzyme, lactate dehydrogenase (LDH) was also followed. Both BP and P_2 were able to stimulate degranulation of mast cells with no effect on LDH release. On a molar basis they appeared as effect secretagogues as the standard agent, Compound 48/80, a synthetic polycation (48). Whether the ability of these myelin proteins to degranulate mast cells has any significance in the demyelinating process is not clear. It may be that the release of these proteins from a damaged myelin sheath could cause some degranulation which could then exacerbate local inflammatory reactions in ways discussed below. In addition to ß-hexosaminidase, and other hydrolytic enzymes, mast cell granules contain vasoactive amines and a large amount of a neutral protease. On appropriate stimuli, mast cells can also synthesize and release chemoattractants such as leukotrienes.

To examine the potential of mast cell proteases to degrade myelin, purified rat serosal mast cells were degranulated with Compound 48/80 and the supernatant incubated with myelinated CNS axons. There was a significant loss of BP following a 2 hour incubation of freshly prepared guinea pig CNS myelinated axons with mast cell supernatants (Table 5). When the experiment was repeated using PNS myelin from bovine spinal roots, there was a similarly large loss of a major myelin protein, in this case P_0 (Table 5). Thus in both the CNS and PNS, mast cell proteases were able to degrade a significant amount of a structural myelin protein, to an extent that would be almost certain to cause disruption of the myelin sheath if it were to occur _in vivo_. The greater myelinolytic ability of mast cell proteases compared to macrophages may represent a different substrate specificity, or may just be a reflection of the extremely high neutral protease content of mast cell granules. Our mast cell studies have led us to hypothesize that mast cells may be involved in the initial stages of inflammatory demyelinating lesion. They are often situated around blood vessels, and on degranulation, can release vasoactive amines to open the blood-brain or blood-nerve barrier, synthesize and release chemoattractants to attract inflammatory cells, and might also secrete myelinolytic enzymes to expose myelin epitopes to sensitized cells. The role of mast cells in PNS demyelination is supported by our recent finding that the use of a mast-cell stabilizing drug can cause a significant reduction in the severity of symptoms in experimental allergic neuritis (49).

Relevance of Model Studies to the Disease Process

As discussed above, there are many considerations when extrapolating these experimental findings to _in vivo_ conditions. At the very least these studies tell us about potential interactions that can occur in pathological states, and at best they can provide clear clues as to the pathogenesis of disease. It is not clear whether the changes that have been observed in this and related studies are sufficient to account for _in vivo_ myelin damage by macrophages. Although numerous studies have shown that macrophages are involved at some stage of myelin breakdown, the high lipid content and multilamellar structure of myelin may initially confer a degree of resistance to attack by macrophages. Macrophages appear not to be as proficient in lipid degradation as they are in proteolysis, and in MS brains, macrophages are often seen to contain a large amount of accumulated lipid (4-6). Pathological evidence of myelin loss in inflammatory diseases in the absence of macrophages is rare, but it is not known whether macrophages are necessary for breakdown or whether initial myelin destabilization is followed too rapidly by macrophage migration to be detected as a separate event. Although oligodendroglial sparing has been reported in MS (4-6), it would seem that as myelin is an intracellular organelle, the macrophage would first have to penetrate the plasma membrane of the oligodendrocyte before it could gain access to the myelin. If this is the case, it would seem reasonable to assume that the oligodendrocyte would somehow be affected, and may then be less able to maintain the integrity of the myelin. Perhaps the oligodendrocyte could "jettison" a damaged sheath, and somehow seal and withdraw the appropriate supporting process. Another important consideration when interpreting these results is that _in vitro_ conditions may not sufficiently reproduce the _in vivo_ situation. Macrophages make a tight connection with their target (50) so that local concentrations of secretion products may reach much higher levels than can be obtained _in vitro_. In addition, the macrophage challenge to myelin _in vivo_ may proceed over a considerable length of time. For the purposes of this concluding section, issues such as the initial event in demyelination, the attraction of inflammatory cells, and ways in which they might cross the blood-brain/nerve barrier are not addressed. We will concentrate on how the various mechanisms of myelin damage that we have described might operate in vivo, in animal models or in human disease. An example of how these kinds of studies can lead to a practical result comes from the studies of Brosnan, Cammer and colleagues who applied their observation of macrophage-mediated BP loss from isolated myelin (37), to successfully treat EAE with protease inhibitors (52). In regard to the potential significance of our observations on the ability of complement to stimulate BP loss; complement may be able to enter the CNS from serum, following lesions in the blood-brain barrier. The presence of terminal complement complexes has been reported in CSF of MS patients (52). Macrophages can synthesize many of the components of the complement pathway (53,54), and may use these to attack myelin. Other pore-forming molecules, such as the perforans of cytolytic T-cells, could lead to destabilization, through a similar mechanism to complement. What is not clear is whether this potentially humorally regulated Ca^{2+} influx/destabilization of myelin can occur in anything other than somewhat structurally damaged myelin; i.e., does this mechanism play a role in a priori destabilization of myelin, or does it only occur once myelin breakdown has been initiated by other mechanisms? Although it is known that purified preparations of both PNS and CNS myelin can fix and activate C', it has not been shown that intact myelinated axons can do so. In the PNS, the basal lamina may prevent the sheath from C fixation. In experiments on intact CNS myelin sheaths in culture, it was shown that C' had no effect in the absence of myelin-specific antibodies (26). If this were true for the in vivo state, it may be that C'-mediated BP loss such as we have observed only occurs after primary damage to the oligodendrocyte-myelin unit and exposure of C' binding residues, or after the appearance of the appropriate antibody. It is not known to what extent the BP in MS CSF (45) is derived from the type of selective BP loss we and others have observed in vitro, and how much is derived from macrophage phagocytosis of myelin and release of breakdown products. Although we observed the potentially synergistic effect of reactive oxygen intermediates and macrophage proteases in CNS myelinated axons, it is of interest that superoxide dismutase, which acts to remove damaging superoxides, can be used to treat experimental allergic neuritis, a disease with known macrophage involvement (55). Experiments investigating the susceptibility of PNS myelin oxygen metabolites and proteases are in progress.

We have also been examining the potential of in situ degranulation of mast cells to damage myelin. This has been difficult to unequivocally demonstrate at the level of the light microscope, and we are currently conducting ultrastructural studies in thee PNS of animals with EAN, and in preparations of desheathed nerve. We have also co-cultured mast cells with cultures of myelinated dorsal root ganglia, and stimulated degranulation using Compound 48/80. In this latter system, we have been unable to demonstrate mast cell-mediated demyelination but this may be due to an unfocused release of mast cell proteases or to the presence of high levels of other potential protein substrates in the culture medium. In the peripheral nerve, mast cells tend to be located between myelinated axons so that degranulation products would be more concentrated in the area of the myelin sheath. Although intact PNS myelin is surrounded by a basal lamina, it is known that mast cell proteases can degrade basal lamina and collagen (56,57). We have been able to successfully treat EAN with mast-cell stabilizing drugs, suggesting that mast cells do play a role in disease development, (49). However, mast cells have a range of properties, and it is not yet clear whether their myelinolytic potential is as important to disease progression as their secretion of vasoactive amines and chemoattractants. It would seem that they are not present in sufficient numbers in vivo to have anything other than a minor role in the overall breakdown of myelin.

Table 5   Degradation of Myelin Proteins by Mast Cell Proteases

Treatment	CNS Myelinated Axons % Loss of BP	PNS Myelin % Loss of P_0
Lysate	84 ± 1.4	85 ± 0.9
Degranulation products	58 ± 1.6	72 ± 1.7
Control incubation	4 ± 9.1	3 ± 2.6

Substrates (guinea pig CNS or bovine PNS) were incubated in 10 mM HEPES (pH 7.4), 0.15 M NaCl, and 2 mM $CaCl_2$ for 3 hours at 37 degrees C. Incubations also contained mast cell lysate, degranulation products isolated from rat serosal mast cells, or Tyrodes buffer as control. Changes in myelin proteins were quantitated using electrophoresis and densitometry. The experiments are described in more detail elsewhere (48).

SUMMARY

We have conducted a series of _in vitro_ studies on isolated preparations of myelin, to examine potential mechanisms of myelin breakdown in inflammatory demyelinating diseases. In the process of examining the effects of macrophages on myelin, we observed that incubation with physiological levels of calcium was dependent on the physical treatment of the myelin during isolation. Although the actual mechanism of the loss is not known, it may involve dissociation of intact BP from the myelin, and a subsequent degradation by a myelin-associated protease. Incubation of myelinated axons with active complement causes a significant increase in loss of BP. The effect is removed if complement is heat-inactivated, or if terminal complement components are missing. This suggests formation of membrane attack complexes on the myelin is necessary, and thus complement may act by increasing access of Ca^{2+} to compact myelin, and allowing BP to escape through membrane pores. Myelinated axons were generally resistant to the action of macrophage supernatants other than a loss of BP that could be explained by Ca^{2+} in the medium. Preincubation of myelinated axons with peroxide, however, prevented the Ca^{2+}-promoted loss of BP, while making them susceptible to macrophage-mediated BP loss. Mast cell proteases were extremely active in proteolysis of both CNS and PNS myelin. These results are discussed in the context of their possible relevance to disease processes.

ACKNOWLEDGEMENTS

We are grateful to Dr. Pierrette Seeldrayers for her part in the mast cell experiments, and to Dag Yasui for his help in the studies on PNS myelin. This work was supported by grants from the PHS/NIH to DJ.

REFERENCES

1. Morell P, Norton WT: Myelin Sci Am 242:88, 1980
2. Raine CS: Morphology of myelin and myelination. In: Myelin. Morell P (ed.), Plenum, New York, 1-50, 1984

3.   Norton WT, Cammer W: Isolation and characterization of myelin. ibid 147-196

4.   Raine CS: The neuropathology of myelin diseases. ibid 259-310

5.   Lassmann H: Comparative neuropathology of chronic experimental allergic encephalomyelitis and multiple sclerosis. Berlin: Springer-Verlag, 1983

6.   Prineas JW: The neuropathology of multiple sclerosis. In: Koetser JC (ed.) Handbook of Clinical Neurology. Demyelinating Diseases, New York, Elsevier, 213, 1985

7.   Smith ME, Benjamin JA: Model systems for perturbations of myelin metabolism. In: Myelin, Morell P (ed.) New York, Plenum, 441, 1984

8.   Itoyama Y, Sternberger NH, Webster HdeF: Immunocytochemical observations on the distribution of myelin associated glycoprotein and myelin basic protein in multiple sclerosis lesions. Ann Neurol 7:267, 1980

9.   Johnson D, Sato S, Quarles RH: Quantitation of the myelin-associated glycoprotein in human nervous tissue from control and multiple sclerosis patients. J Neurochem 46:1086, 1986

10. Marks N, Grynbaum A, Lajtha A: The breakdown of myelin-bound proteins by intra- and extracellular proteases. Neurochem Res 1:93, 1976

11. Banik NL, McAlhaney WW, Hogan EL: Calcium-stimulated proteolysis in myelin: evidence for a Ca^{2+}-activated neutral proteinase associated with purified myelin for rat CNS. J Neurochem 45:581, 1935

12. Cammer W, Brosnan CF, Basile C: Complement potentiates the degradation of myelin proteins by plasmin: implications for a mechanism of inflammatory demyelination. Brain Res 364:91, 1986

13. Nathan C: Secretory products of macrophages. J Clin Invest 79:319, 1987

14. Adams DO, Hamilton TA: The cell biology of macrophage activation. Ann Rev Immunol 2:283, 1984

15. Norton WT, Poduslo SE: Myelination in the rat brain: method of myelin isolation. J Neurochem 21:749, 1973

16. Sato S, Miyatake T: Degradation of myelin basic protein by calcium-activated neutral protease (CANP)-like enzyme in myelin and inhibition by E-64 analogue. Biomed Res 3:461, 1982

17. Kerlero de Rosbo N, Carnegie PR, Bernard CCA: Quantitative electroimmunoblotting study of the calcium-activated neutral protease in human myelin. J Neurochem 47:1007, 1986

18. Berlet HH: Calcium-dependent neutral protease activity of myelin from bovine spinal cord: evidence for soluble cleavage products of myelin proteins. Neurosci Lett 73:266, 1987

19. De Vries GH: Isolation of axolemma-enriched fractions from mammalian CNS. In: Marks N, Rodnight R (eds.) Research Methods in Neurochemistry, New York, Plenum, 3, 1981

20. Glynn P, Chantry A, Groome N, Cuzner ML: Basic protein dissociating from myelin membranes at physiological ionic strength and pH is cleaved into three major fragments. J Neurochem 48:752, 1987

21. Chantry A, Earl C, Groome N, Glynn P: Metalloendoprotease cleavage of 18.2 and 14.1 kilodalton basic proteins dissociating from rodent myelin membranes generates 10.0- and 5.9-kilodalton fragments. J Neurochem 50:688, 1988

22. Smith ME: Studies on the mechanism of demyelination: myelin autolysis in normal and edematous CNS tissue. J Neurochem 28:341, 1977

23. Smith R, Braun PE: Release of proteins from the surface of bovine central nervous system myelin by salts and phospholipases. J Neurochem 28:341, 1977

24. Cyong JC, Witkin SS, Rieger B, Barbarese E, Good RA, Day NK: Antibody-independent complement activation by myelin via the classical pathway. J Exp Med 155:587, 1982

25. Vanguri P, Koski CL, Silverman, Shin ML: Complement activation by isolated myelin: activation of the classical pathway in the absence of myelin specific antibodies. Proc Natl Acad Sci 79:3290, 1982

26. Liu WT, Vanguri P, Shin ML: Studies on demyelination in vitro: the requirement of the membrane attack components of the complement system. J Immunol 131:778, 1983

27. Ramm LE, Mayer MM: Life-span and size of the trans-membrane channel formed large doses of complement. J Immunol 124:2281, 1980

28. Kirschner DA, Ganser AL, Caspar DLD: Diffraction studies of molecular organization and membrane interactions in myelin. In: Morell P (ed.) Myelin. New York, Plenum, 51, 1984

29. Smith KJ, Hall SM: Peripheral demyelination and remyelination initiated by the calcium-selective ionophore ionomycin: in vivo observations. J Neurol Sci 83:37, 1988

30. Smith KJ, Hall SM, Schauf CL: Vesicular demyelination induced by raised intracellular calcium. J Neurol Sci 71:19, 1985

31. Koski CL, Padmavathy V, Shin ML: Activation of the alternative pathway of complement by human peripheral nerve myelin. J Immunol 134:1810, 1985

32. Kirschner DA, Ganser AL: Compact myelin exists in the absence of basic protein in the shiverer mutant mouse. Nature 283:207, 1979

33. Kadlubowski M, Hughes RAC: Identification of the neuritogen for EAN. Nature 277, 1979

34. Uyemara K, Kato-Yamanaka T, Kitamura K: Distribution and optical activity of the basic protein in bovine peripheral nerve myelin. J Neurochem 29:61, 1977

35. Smith ME: Effects of proteolytic enzymes on PNS myelin. Trans Am Soc Neurochem 11:213, 1980

36. Cammer W, Brosnan CF, Bloom BR, Norton WT: Degradation of the P_0, P_1, and P_r proteins in peripheral nervous system myelin by plasmin: Implications regarding the role of macrophages in demyelinating diseases. J Neurochem 36:1506, 1981

37. Cammer W, Bloom BR, Norton WT, Gordon S: Degradation of basic protein in myelin by neutral proteases secreted by stimulated macrophages: a possible mechanism of inflammatory demyelination. Proc Natl Acad Sci 75:1554, 1978

38. Trotter JL, Smith ME: The role of phospholipases from inflammatory macrophages in demyelination. Neurochem Res 11:349, 1985

39. Cammer W, Bieler LZ, Norton WT: Proteolytic and peroxidatic reactions of commercial horseradish peroxidase with myelin basic protein. Biochem J 169:567, 1978

40. Konat GW, Wiggins RC: Effect of reactive oxygen species on myelin membrane proteins. J Neurochem 45:1113, 1985

41. Chan PC, Peller OG, Kesner L: Copper (II)-catalyzed lipid peroxidation in liposomes and erythrocyte membranes. Lipids 17:331, 1982

42. Gately CL, Wahl SM, Oppenheim JJ: Characterization of hydrogen peroxide-potentiating factor, a lymphokine that increases the capacity of human monocytes and monocyte-like cell lines to produce hydrogen peroxide. J Immunol 131:2853, 1983

43. Dawson G, Dawson SA, Siakotos AN: The biochemical defect in Batten's disease. Trans Am Soc Neurochem 19:174, 1988

44. Dufton MJ, Cherry RJ, Coleman JW, Stanworth DR: The capacity of basic peptides to trigger exocytosis from mast cells correlates with their ability to immobilize band 3 proteins in erythrocyte membranes. Biochem J 223:67, 1984

45. Cohen SR, Herndon RM, McKhann GM: Radioimmunoassay of myelin basic protein in spinal fluid: an index of active demyelination. N Engl J Med 295:1455, 1976

46. Ibrahim MZM: The mast cells of the central nervous system., 1. Morphology, distribution and histochemistry. J Neurol Sci 21:431, 1974

47. Olsson Y: Mast cells in the nervous system. In: Bourne GH, Danielli JF, Jeon KW (eds.) International review of cytology, New York, Academic Press, 27, 1968

48. Johnson D, Seeldrayers PA, Weiner HL: The role of mast cells in demyelination. 1. Myelin proteins are degraded by mast cell proteases and myelin basic protein and P_2 can stimulate mast cell degranulation. Brain Res 444:195, 1988

49. Seeldrayers PA, Johnson D, Weiner HL: Treatment of experimental allergic neuritis with a mast cell stabilizing drug. Neurology 38:238, 1988

50. Wright SD, Silverstein SC: Phagocytosing macrophages exclude proteins from the zones of contact with opsonized targets. Nature 309:359, 1984

51. Brosnan CF, Cammer W, Norton WT, Bloom BR: Proteinase inhibitors suppress the development of experimental allergic encephalomyelitis. Nature 285:235, 1980

52. Morgan BP, Campbell AK, Compston DAS: Terminal component of complement (C9) in cerebrospinal fluid of patients with multiple sclerosis. Lancet i, 251, 1984

53. Sanders ME, Koski CL, Robbins D: Activated terminal complement in cerebrospinal fluid in Guillain-Barre syndrome and multiple sclerosis. J Immunol 136:4456, 1986

54. Ferluga J, Schorlemmer HU, Baptista LC, Allison AC: Production of complement cleavage product, C3a, by activated macrophages and its tumorolytic effects. Clin Exp Immunol 31:587, 1978

55. Newell SL, Shreffler DC, Atkinson JP: Biosynthesis of C4 by mouse peritoneal macrophages. J Immunol 129:653, 1982

56. Hartung HP, Schafer B, Heininger K, Toyka K: Suppression of experimental autoimmune neuritis by the oxygen radical scavengers superoxide dismutase and catalase. Ann Neurol 23:453, 1988

57. Seppa H, Vaanen K, Korhonen K: Effect of mast cell chymase of rat skin on intracellular matrix: a histochemical study. Acta Histochem 64:65, 1979

58. Sage H, Woodbury RG, Borstein P: Structural studies on human type IV collagen. J Biol Chem 254:9893, 1979

17

AUTOANTIBODIES TO MYELIN BASIC PROTEIN

IN MULTIPLE SCLEROSIS

K.G. Warren and I. Catz

Multiple Sclerosis Care and Research Clinic
Division of Neurology, University of Alberta
Edmonton, Canada

INTRODUCTION

Multiple sclerosis (MS) was initially described in the nineteenth century in Europe and subsequently in North America. The diary and letters of Sir Augustus Frederick d'Este (1794-1848), the grandson of George III of England and a cousin of Queen Victoria, illustrate the earliest personal description of this disease. Over a period of 28 years he experienced recurring symptoms including blurring of vision, numbness of limbs, loss of balance, bladder and bowel disturbances, staggering gait and finally paralysis, all highly characteristic of the clinical course of MS (21). The Parisian physician Jean Martin Charcot (1825-1893) and his colleague Vulpian are generally given the credit for bringing this unique disease to the attention of the medical community and separating it form the multitude of undifferentiated paralytic disorders of that era (16). Attempts to understand the etiology-pathogenesis of this disease must consider the following facts:

(a) MS as a clinical-pathological entity has been known for close to 200 years and therefore its precipitating factor(s) must have been present for an equal period of time.

(b) MS is an inflammatory-demyelinating disease with associated gliosis and relative neuronal-axonal sparing, which is restricted to the central nervous system (CNS) of human beings.

(c) MS is characterized by clinical-pathological variability. This disease may be entirely benign causing only the mildest of neurological symptoms, it may follow a more classical course causing progressive disability over years to decades, or it may be extremely fulminant causing premature death within weeks to months from the onset. MS may be characterized by relapses and remissions of neurological dysfunction, stepwise progression of neurological disability, or insidious progression. From the clinical point of view this illness may strike only one neuroanatomical locus, such as the optic nerves or spinal cord, or it may affect multiple loci simultaneously or sequentially.

(d) MS has epidemiological features which have been clarified during the twentieth century. It is a disease of young adults with a mean age of 30±20 years. It affects females more than males with a ratio of approximately 3 to 2. MS is more common in temperate climates than in tropical areas. It is far more common in caucasians than in oriental races, suggesting that the former race is genetically predisposed and the latter is genetically resistant to this illness.

The best overall idea to date regarding the etiology-pathogenesis of MS is that it is a virus or a virus-like precipitated autoimmune mediated demyelinating disease of the CNS. Consequently, during the last several years, we have been attempting to determine whether there are antibodies directed against myelin basic protein (MBP), a potential autoantigen of the CNS.

NEUROPATHOLOGICAL CHARACTERISTICS

The first pathological descriptions of MS are generally attributed to Jean Cruveilhier (13) and to Sir Robert Carswell (6). The underlying inflammation associated with MS was well recognized by the time James W. Dawson presented his findings on "the histology of disseminated sclerosis" to the Pathological Society of Great Britain and Ireland in 1915 (15) when "a paucity of immunoglobulin producing plasma cells relative to lymphocytes" was noted.

During recent years there have been extensive research efforts to further understand the inflammatory characteristics of the CNS demyelination in MS patients. Tourtellotte has found the following inflammatory cell types statistically more frequent in the cerebrospinal fluid (CSF) of MS patients than in normal subjects; large lymphocytes (2.8 times), small lymphocytes (3.3 times), monocytes (2.5 times), plasma cells (8 times) and histocytes/macrophages (12 times). If individual MS plaques within a single patient are histologically examined, a large degree of variation of the inflammatory cell makeup may be observed. An enormous number of inflammatory cells may be observed in the Virshow-Robin space of a centrally located vein, while another plaque showing degenerating myelin may contain no inflammatory cells whatsoever. If these inflammatory cells are producing antibodies or lymphokines involved in the pathogenesis of demyelination, such factors may circulate by way of the CSF to distant sites within the CNS to produce plaques of tissue destruction quite remote from the cells they originated from. Immunoglobulins and complement accumulate in areas of active demyelination (28). Inflammatory cells including T and B lymphocytes, plasma cells and macrophages (33, 34, 37),immunoglobulin containing cells (19, 29, 29) as well as bound IgG may occur in plaques or plaque borders in MS patients, but autoantibodies to MBP or other myelin antigens within central nervous tissue have not been reported to date.

HUMAN CENTRAL NERVOUS SYSTEM MYELIN

Myelin of the CNS is a lipoprotein complex produced by the oligodendrocyte plasma membrane wrapping consecutively around neuronal axons. Chemically, myelin consists of 70% lipid and 30% protein. Major proteins within myelin include Folch-Lees proteolipid protein, high molecular weight Wolfgram protein, some myelin associated enzymes, glycoproteins and MBP. Because of its ability to induce experimental allergic encephalomyelitis (EAE), MBP has been extensively studied. This protein is species specific. Human MBP contains 170 amino acids with a determined sequence (5) and a molecular weight of 18 300 daltons (3). It is characterized by a high content of basic amino acids with only one tryptophan residue in position 116 near the center of the molecule. There is a relatively high proportion of proline with a rare triproline sequence in positions 99-101 which together with another proline nearby (in position 96) forms a hairpin-like turn in the molecule (4). The arginine residue in position 114, near this tripoline sequence may appear either unmethylated or mono or di-methylated or even phosphorylated, offering this molecule some degree of micro-heterogeneity. Because of their occurrence in normal myelin, the biological importance of these molecular features vis-a-vis MS is probably insignificant. If MS is an autoimmune disease directed against MBP, then this mechanism would be directed against MBP in situ within the myelin sheath. Once MBP is released into the CSF, its tertiary structure is likely to be altered, the molecule may be digested by enzymes and consequently some of its epitopes may be altered or even lost allowing MBP or fragments of it to coexist in free forms with unbound antibodies to other epitopes.

In the 1880's Louis Pasteur recognized that encephalomyelitis could be induced in rabbits by inoculation with homogenates of spinal cord containing rabies virus. Subsequently in 1933, EAE, the experimental model of MS, was discovered by Rivers and coworkers (35). Myelin basic protein contains encephalitogenic determinants which produce both EAE and experimental optic neuritis (EON) in animals (46, 57, 59). Since the relapsing form of EAE has many clinical and pathological features in common with MS, interest in MBP has therefore intensified (58). Antibodies to MBP have been observed in primates with EAE (2). Circulating T-cells sensitized to MBP (11) as well as immune complexes containing MBP as their antigenic component (12, 14) have also been described in MS patients.

DETECTION OF MYELIN BASIC PROTEIN AND ANTI-MYELIN BASIC PROTEIN ANTIBODIES IN CEREBROSPINAL FLUID

CSF Hydrolysis

In order to detect both free (F) and bound (B) fractions of MBP and anti-MBP, CSF was subjected to acid hydrolysis in order to dissociate possible preformed complexes containing these proteins. Results obtained prior to acid hydrolysis represent free (unbound) forms of MBP and anti-MBP, while values obtained subsequent to acid hydrolysis represent total (free and bound) quantities. Bound fractions can be calculated by subtracting free from corresponding total values. CSF acid hydrolysis was performed as follows: 300-500 ul of CSF were acidified to pH = 2.9-3.2 with 2N acetic acid and incubated for 1 hour at room temperature, when theoretically preformed complexes were dissociated. This mixture was then neutralized to pH = 6.8-7.1 with 100-300 ul 2N borate buffered saline (BBS, pH = 11.2) and immediately used in the MBP and anti-MBP assay; assuming that at least some of the complexes remained dissociated, values obtained in this mixture were considered to be total fractions.

Myelin Basic Protein Assay

MBP was prepared from normal human brain (17). This antigen was used as standard, iodinating material and antigenic factor to produce suitable antibodies for radioimmunoassay (RIA). Anti-MBP antibodies were produced in young white New Zealand rabbits (54); iodination was performed by chloramine T method (23) and the specific activity of the final product was 40-60 uc/ug.

The MBP assay was performed as follows: 200 ul of MBP standard in 0.2 ml Tris buffer (T$_3$) or the same volume of control or unknown CSF were incubated for 2 hours at 37 degrees C with an appropriate amount of rabbit anti-MBP in the presence of 1% carrier serum and 100 ul of T3 concentrated tenfold. Following a second 2 hour incubation at 37 degrees C with MBP I^{125}, a pretitered amount of a precipitating antibody (goat anti-human IgG) was used to separate the components. Each assay contained 3 internal CSF controls containing known amounts of 5, 10 and 20 ug MBP/l. Intra-assay variability calculated as coefficient of variation (cv) using these controls was 6.1, 4.1 and 2.5 respectively, for 25 sets of duplicates; similarly calculated inter-assay variability (cv) of 1 set of duplicates over 25 different assays was 9.4, 5.2 and 3.3 respectively. Nonspecific binding was 2-5%, indicating that the radioligand was not damaged by incubation at 37 degrees C.

Anti-MBP Radioimmunoassay

Anti-MBP was determined by a solid phase RIA. Final IgG concentration was adjusted to 0.010 g/l in all samples before they were assayed. Immulon microtiter plates were coated with 1 ug/well of human MBP (17); Staphylococcus A protein iodinated with I^{125} (23) was used as radiolabel.

100-200 ul aliquots of diluted CSF (0.010 g IgG/l), before or after acid hydrolysis, were incubated for 2 hours at room temperature in MBP coated wells; after 5 washes, goat anti-human Ig was added to enhance the immune response, and incubation was continued for 1 hour at room temperature; after another 5 washes, the radiolabel was added (20-30,000 cpm/well) and the plates were incubated for 2 more hours at room temperature, and finally after 5 more washes the plates were individually counted and results were expressed as percent bound of total radioactivity. Serum from a hyperimmunized rabbit with human MBP served as positive control; negative controls were also used in each assay. Intra-assay variability (cv) for 1 set of duplicates over 10 runs was 6.2; blanks, performed to determine nonspecific binding to uncoated wells, were performed with each sample and subtracted from the matched counts in MBP coated wells. When samples with initially high IgG and anti-MBP values were serially diluted, anti-MBP paralleled IgG concentrations and recovery of anti-MBP varied between 81-105%. When increasing amounts of CSF from a sample with initially high anti-MBP were added to another CSF with undetectable antibody titer, anti-MBP recovery varied between 86-98%. Absorption of CSF with MBP prior to anti-MBP assay resulted in complete elimination of initially high antibody titers. Dilution factors were applied for calculations of MBP and anti-MBP after acid hydrolysis.

Lack of a Relationship Between Cerebrospinal Fluid Myelin Basic Protein and Intrathecal IgG Parameters in Multiple Sclerosis

Over 90% of patients with MS synthesized IgG within the blood-brain barrier (BBB) (41, 49). Kabat and colleagues (25) observed that 85% of MS patients had an abnormal CSF IgG/total protein ratio. Subsequently, 59-90% of MS patients have been found to have an abnormal CSF IgG/albumin ratio (22,24,42,43). Link and Tibbling (27) found that 80% of patients with clinically definite MS have an elevated IgG index. Daily rate of intrathecal IgG synthesis has been estimated to be increased in 92% of patients with clinically definite MS (44).

We measured levels of free CSF-MBP, CSF/serum albumin ratio, absolute level of CSF IgG and 3 estimates of intrathecal IgG synthesis in 4 subgroups of clinically definite MS patients. The numbers and percentage

of patients with elevated test values in each of the clinical subgroups
are shown in Table 1, and the mean and range of these parameters are
illustrated in Figures 1-4. No correlation between levels of free CSF-
MBP and the absolute CSF IgG, CSF IgG/albumin ratio (CSF IgG/alb), IgG
index or daily rate of CSF IgG synthesis (IgG synth) was found in these 4
clinical subgroups of MS patients. In 21 patients in remission, CSF MBP
and CSF/serum albumin ratio were normal, while IgG levels were elevated
in the majority of them (Fig 1, Table 1). CSF MBP and IgG values were
significantly elevated in the majority of 22 patients with
polysymptomatic exacerbations and 22 patients with monosymptomatic
attacks (Figs 2 and 3, Table 1) and all mean values were higher in the
group with polysymptomatic exacerbations. Forty-one patients with
chronically progressing disease had modest elevations of free CSF MBP and
CSF/serum albumin (Fig 4, Table 1). Their mean CSF IgG level was similar
to that of patients with polysymptomatic exacerbations, while CSF
IgG/alb, IgG index and IgG synth were highest in this clinical subgroup.
In our laboratory, longitudinal case studies confirmed that unbound MBP
titers in CSF do no correlate with parameters of intrathecal IgG
synthesis. During the recovery phase from acute relapses when CSF MBP
levels are returning towards normal, IgG parameters may be either
increasing or decreasing in individual patients. This lack of
correlation between CSF MBP titers and total intrathecal IgG synthesis
suggests that there are multiple immune processes and disease activity is
likely to correlate with specific subunits of the total intrathecally
synthesized IgG.

Table 1   Number (percentage) of multiple sclerosis patients
exhibiting elevated values of CSF unbound myelin basic protein; CSF/serum
albumin and four parameters indicative of intrathecal IgG synthesis.
Reprinted with permission of the editor of Annals of Neurology

		R	PE	ME	CP
	TOTAL NUMBER	21	22	22	41
	CSF-MBP	1 (5)*	21 (96)	16 (73)	19 (46)
NUMBER ELEVATED	CSF/S alb	6 (29)	11 (50)	9 (41)	16 (39)
	CSF IgG	12 (57)	20 (91)	18 (82)	30 (73)
	CSF IgG/alb	14 (67)	29 (91)	18 (82)	35 (85)
	IgG Index	17 (81)	21 (96)	20 (91)	36 (88)
	IgG Synth	10 (48)	22 (100)	20 (91)	35 (85)

Myelin Basic Protein as an Indicator of Multiple Sclerosis Disease Activity

Although not specific for MS, elevated levels of CSF unbound MBP are
a valuable indicator of disease activity. Cohen et al (9,10) found that
free (unbound) CSF MBP is elevated in MS patients with monosymptomatic or
polysymptomatic attacks.

Similarly, Whitaker (55,56) detected P_1 fragment of MBP (residue 43-
48) in most patients with exacerbations of less than 2 weeks duration.
We analyzed CSF for unbound MBP levels in a large cross sectional study
of 325 MS patients divided into 78 with polysymptomatic exacerbations
(PE), 43 with monosymptomatic exacerbations (ME), 15 with initial attacks

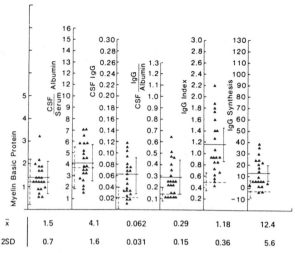

Figure 1   Cerebrospinal fluid (CSF) myelin basic protein and matched
values for the CSF/serum albumin ratio, and four CSF IgG parameters in 21
MS patients in clinical remission.   (Vertical bars = mean ± 2SD of the MS
group; vertical dashed bars = mean± 2SD of non-neurological disease
group).

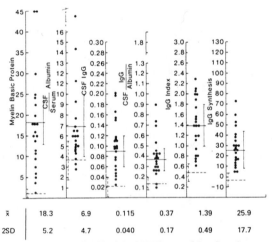

Figure 2   Cerebrospinal fluid (CSF) myelin basic protein and matched
values for the CSF/serum albumin ratio and four CSF IgG parameters in 22
MS patients with polysymptomatic exacerbations. (Symbols as in Figure 1).

causing internuclear ophthalmoplegia (INOP) and 68 with optic neuritis
(ON) as their initial attack, 86 with chronically progressing disease
(CP) and 35 patients in clinical remission (R), and 193 controls
including 100 patients with non-neurological disease (NND), 53 with
various neurological diseases (ND) and 40 with recent head trauma (HT)
(Fig 5).   Normal free CSF MBP levels in patients with NND was 1.0 ± 1.0
ug/l (mean ± 3SD).   Patients with various ND exclusive of MS may have
elevated MBP levels (3.4 ± 4.2).   Highest levels in this group occurred
in 2 patients with postinfectious encephalomyelitis and in 1 with vitamin
B12 deficiency with neurological complications.   All but 1 patient with
HT had elevated levels (33 ± 30).   All 78 patients with PE had elevated
free CSF MBP levels (17.4 ± 8.9); 40 of 43 patients with ME also had

elevated values (4.9 ± 2.7). Similarly 12 of 15 patients with INOP and 63 of 68 with ON had elevated unbound CSF MBP (5.1 ± 1.8 and 5.2 ± 2.9). Only 47 of 87 patients with CP had unbound MBP levels above normal (3.1 ± 1.8), while most patients in clinical remission (28 of 35) had normal CSF MBP levels. Each group of patients with active MS had significantly higher CSF MBP than MS patients in remission or the group of controls with NND.

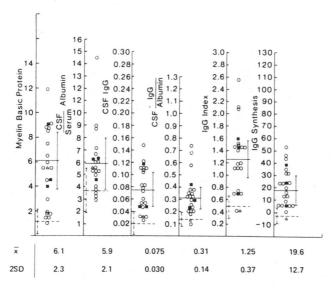

Figure 3   Cerebrospinal fluid (CSF) myelin basic protein and matched values for the CSF/serum albumin ratio and four IgG parameters in 22 MS patients with monosymptomatic exacerbations (open circles), optic neuritis (black squares) and acute internuclear ophthalmoplegia (open triangles). (Symbols as in Figure 1).

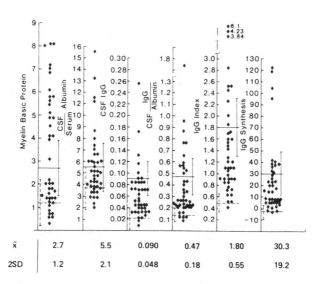

Figure 4   Cerebrospinal fluid (CSF) myelin basic protein and matched values for CSF/serum albumin ratio and four IgG parameters in 41 patients with progressing MS. (Symbols as in Figure 1).

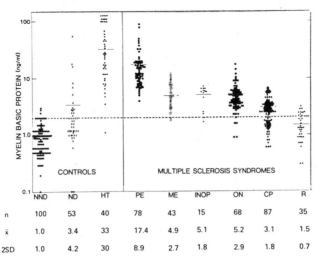

	NND	ND	HT	PE	ME	INOP	ON	CP	R
n	100	53	40	78	43	15	68	87	35
x̄	1.0	3.4	33	17.4	4.9	5.1	5.2	3.1	1.5
2SD	1.0	4.2	30	8.9	2.7	1.8	2.9	1.8	0.7

Figure 5   Cerebrospinal fluid free myelin basic protein levels in 6 groups of MS patients and 3 groups of controls. (NND = non-neurological disease; ND = neurological disease; HT = head trauma; PE = polysymptomatic exacerbations; ME = monosymptomatic exacerbations; INOP = internuclear ophthalmoplegia; ON = optic neuritis; CP = progressing MS; R = remission; horizontal bar = mean of group; dashed line = 2SD's above mean of NND group; n = number of patients per group; x = mean).

Diagnostic Value of Cerebrospinal Fluid Anti-Myelin Basic Protein in Multiple Sclerosis

CSF anti-MBP titers are elevated in MS patients with active disease; this is relatively specific for MS, but anti-MBP may also be found in patients with subacute sclerosis panencephalitis (SSPE) and some patients with postinfectious encephalomyelitis. However, CSF anti-MBP was undetectable in a large group of control patients. In order to determine the diagnostic value of CSF anti-MBP we studied 146 MS patients and 112 controls. All MS patients had clinically definite disease (39). This group consisted of 33 patients in clinical remission, 20 with clinically stable disease, 47 with progressing MS and 46 experiencing exacerbations. All patients in remission were asymptomatic at the time of the study, with no residual deficits. Patients judged to be stable had residual deficits but were not clinically deteriorating on a year-to-year basis, while those with progressing MS were becoming clinically worse on a monthly or yearly basis. All patients with exacerbations had experienced acute development of one or more neurological symptoms with associated signs, which persisted for a minimum of 24-48 hours and they were studied within 2 weeks from the onset of the attack. The control group of 112 patients consisted of 44 with psychoneurosis, 32 who were having myelograms for degenerative disc disease and 36 with miscellaneous neurological disorders exclusive of MS. In 44 patients with psychoneurosis, the total anti-MBP was 1.5 ± 1.1 with corresponding free (F) and bound (B) values of 0.4 ± 0.3 and 1.2 ± 1.0 respectively, and a F/B ratio of 0.333 ± 0.21 (Table 2). These levels were considered normal, with none of these patients having elevated values. Similar results were observed in 32 patients with degenerative disc disease; their total anti-MBP level was 1.6 ± 0.6 with free and bound values of 0.4 ± 0.3 and 1.1 ± 0.4 respectively and a F/B ratio of 0.327 ± 0.25. Again, there was no exception with elevated anti-MBP. Despite elevated levels of total intrathecal IgG synthesis, none of 33 MS patients in

Table 2 CSF anti-MBP in four clinical groups of multiple sclerosis patients and three groups of controls. Free and total anti-MBP were obtained before and after acid hydrolysis respectively. Bound anti-MBP levels were calculated by subtracting free from the matched total value. Free/bound antibody ratios are illustrated for each clinical subgroup. All results expressed as mean ±2SD. Student's t test versus patients with psychoneurosis. *p<0.05 **p<0.01 ***p<0.001 ****p<0.0001

	n	CSF ANTI-MBP (% BOUND RADIOACTIVITY)							
		TOTAL		FREE		BOUND		FREE/BOUND RATIO	
			#† (%)		#† (%)		#† (%)		#† (%)
CONTROL GROUPS:									
PSYCHONEUROSIS	44	1.5 ± 1.1	0/44 (0)	0.4 ± 0.3	0/44 (0)	1.2 ± 1.0	0/44 (0)	0.333 ± 0.21	0/44 (0)
DEGENERATIVE DISC DISEASE	32	1.6 ± 0.6	0/32 (0)	0.4 ± 0.3	0/32 (0)	1.1 ± 0.4	0/32 (0)	0.327 ± 0.25	0/32 (0)
NEUROLOGICAL DISEASE	36	3.8 ± 0.6	3/36 (8)	0.5 ± 0.3	0/36 (0)	3.2 ± 0.7	3/36 (8)	0.241 ± 0.28	0/36 (0)
MS CLINICAL GROUPS:									
REMISSION	33	1.7 ± 0.9	0/33 (0)	0.5 ± 0.3	0/33 (0)	1.3 ± 1.0	0/33 (0)	0.380 ± 0.24	0/33 (0)
STABLE	20	8.5 ± 3.4*	20/20 (100)	1.0 ± 0.7	5/20 (25)	7.7 ± 3.3*	20/20 (100)	0.348 ± 0.10	0/20 (0)
PROGRESSIVE	47	25.0 ± 4.7***	47/47 (100)	7.6 ± 3.8*	29/47 (62)	17.0 ± 5.6**	47/47 (100)	0.490 ± 0.24	8/47 (17)
EXACERBATIONS	46	21.9 ± 5.0***	46/46 (100)	17.8 ± 7.5**	46/46 (100)	4.0 ± 2.3*	16/46 (35)	8.200 ± 4.6	46/46 (100)

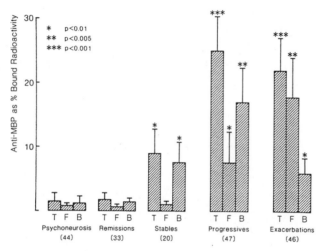

Figure 6 Total (T), free (F) and bound (B) anti-myelin basic protein (anti-MBP) in a group of patients with psychoneurosis and 4 groups of MS patients. Numbers of patients in parentheses.

remission had detectable CSF anti-MBP in either free or bound form. All 20 patients with clinically stable disease had elevated levels of total CSF anti-MBP (8.5 ± 3.4); 5 of these 20 patients had increased free (1.0 ± 0.7) and all 20 had increased bound (7.7 ± 3.3) anti-MBP levels (Table 3, Fig 6) and their F/B ratio remained normal (Fig 7). Highest total anti-MBP levels were detected in 47 patients with progressing MS. All of these patients had elevated total and bound values, only 62% of them (29 of 47) having elevated free values (Table 2, Fig 6) and their F/B ratio (Fig 7) was at the upper limits of normal (0.490 ± 0.24). The characteristic feature of this group was higher levels of bound anti-MBP (17.0 ± 5.6) than of free antibody (7.6 ± 3.8) (Fig 6). The group of 46 MS patients with exacerbations had significantly different results: in contrast with patients with progressing disease, all these 46 patients had elevated free anti-MBP (17.8 ± 7.5) while only 16 of 46 had elevated bound values (4.0 ± 2.3) (Fig 6). Consequently, the F/B anti-MBP ratio was dramatically elevated in this group (8.2 ± 4.6); all 46 patients had F/B ratios above 1.0 (Fig 7). Similar to these patients, previously normal individuals experiencing first attacks of acute idiopathic optic neuritis also show increased levels of CSF anti-MBP with elevated F/B ratios. It is important that their CSF is obtained in the acute phase when CSF MBP is also elevated (Table 3).

In conclusion, all patients with clinically active MS have elevated total CSF anti-MBP in either free or bound form. Patients whose disease is in remission have undetectable anti-MBP levels and patients with clinically stable disease with residual disability may have detectable antibody titers. Chronically progressing MS is usually associated with high levels of antibody in bound rather than in free form, resulting in a low or normal F/B ratio. In contrast, MS exacerbations are characterized by relatively high levels of free CSF anti-MBP resulting in a high F/B antibody ratio. It can be hypothesized that acute exacerbations of MS are associated with a "sudden pulse" of intrathecal anti-MBP synthesis within one or more sites of the CNS, whereas insidiously progressing disease is associated with "steady state" anti-MBP synthesis.

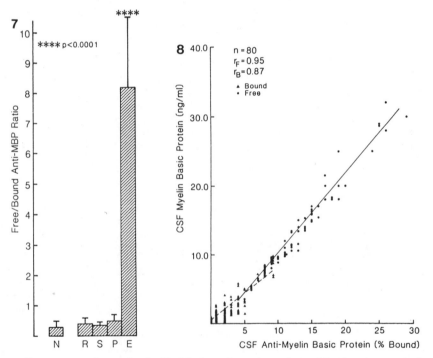

Figure 7   Cerebrospinal fluid free/bound anti-myelin basic protein (anti-MBP) ratio in a normal control (N) group and a clinical group of MS patients. (R = complete clinical remission; S = stable phase of disability; P = progressing neurological deterioration; E = exacerbation).

Figure 8   Correlation of myelin basic protein (MBP) and anti-MBP in MS exacerbations. (r_F = correlation coefficient between free MBP and free anti-MBP; r_B = correlation coefficient between bound MBP and bound anti-MBP).

Table 3   CSF myelin basic protein (MBP) and anti-MBP in a group of patients with acute idiopathic optic neuritis as well as normal and multiple sclerosis controls.   (n=number of patients per group; #=number of patients with elevated values per group).

	n	CSF MBP Free	#	CSF MBP Bound	#	CSF Anti-MBP Free	#	CSF Anti-MBP Bound	#	CSF Anti-MBP Free/Bound	#
• Controls	76	0.6±0.3	7/76	nil	N/A	0.4±0.2	0/76	1.2±0.4	0/76	0.3±0.2	0/76
MS Clinical Groups:											
• Remission	40	1.5±0.3	5/40	nil	N/A	0.5±0.2	0/40	1.5±0.2	0/40	0.4±0.2	0/40
• Progressing	47	4.9±3.2	35/47	14.0±6.1	47/47	7.6±3.8	29/47	17.0±5.6	47/47	0.5±0.3	8/47
• Relapsing	46	15.8±9.2	45/46	3.0±2.7	12/46	17.8±7.5	46/46	4.0±0.3	16/46	8.2±4.6	46/46
• Optic neuritis	20	9.0±5.3	13/20	1.5±0.9	5/20	8.8±4.8	17/20	1.7±0.9	9/20	5.8±2.4	17/20

Cerebrospinal Fluid Anti-Myelin Basic Protein Titers Correlate with Disease Activity

CSF anti-MBP titers correlate with multiple sclerosis disease activity. In order to demonstrate this, CSF anti-MBP titers were compared with CSF MBP, a valuable nonspecific indicator of disease activity. The correlation coefficient between F anti-MBP and F MBP (r_F) in the CSF of 80 MS patients with exacerbations was 0.95 (Fig 8). Without exception, relatively low levels of F anti-MBP were associated with similarly low levels of F MBP, and conversely, relatively high levels of F anti-MBP occurred with high levels of F MBP. Although bound fractions of MBP and anti-MBP were present in much lower levels in this group, there was also a high correlation ($r_B = 0.87$) between them (Fig 8). The correlation coefficient between B anti-MBP and B MBP (r) levels in 100 patients with progressing MS was 0.93 (Fig 9). Again, low levels of bound antibody were associated with similarly low levels of B MBP while high levels of B anti-MBP occurred with high levels of B MBP. In

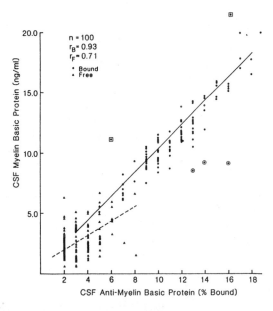

Figure 9   Correlation of myelin basic protein (MBP) and anti-MBP in progressing MS. (Symbols as in Figure 8).

these patients, although F anti-MBP and F MBP were present in relatively low levels, there was a moderately high correlation ($r_F = 0.71$) between them. These high correlation coefficients between CSF anti-MBP and CSF MBP in MS patients with active disease, and the fact that anti-MBP is undetectable in MS patients in remission are suggestive that these antibodies may be involved in the pathogenesis of MS.

   Longitudinal case studies also illustrated a high correlation between CSF MBP and anti-MBP. Figure 10 illustrates the levels of free and bound MBP and anti-MBP found in the CSF of a patient with the hyperacute form of MS (type A of McAlpine) (30). During a period of 35 weeks, this patient had 16 CSF analyses that showed persistently high levels of free MBP and anti-MBP and lower levels of bound fractions. The case of a typical patient with progressing MS (type D, McAlpine) (30) is illustrated in Figure 11. In contrast to the patient just described, this one shows persistently high values for bound anti-MBP and MBP and correspondingly low free levels. These results were consistent in 14 CSF analyses over a period of 80 weeks. A patient with relapsing-progressing MS (type C, McAlpine) (30) is illustrated in Figure 12. Twenty-two CSF analyses over a period of 118 weeks were studied in this chronically hospitalized MS patient. The initial lumbar puncture performed during a clinical relapse illustrates high levels of free MBP and anti-MBP with correspondingly low bound values. During subsequent months, after his return to a chronic care hospital, he had 6 additional CSF analyses (weeks 21-46) showing high levels of bound and correspondingly lower levels of free antibody and MBP which were associated with the progressing phase of his illness. Between weeks 52 and 68, this patient had 5 more CSF analyses because of a sudden increase of neurological symptoms. A reversal of the MBP and anti-MBP pattern occurred at this time; free levels of MBP and antibody became higher than did the corresponding bound fractions. Between weeks 70 and 118, when the patients disease again entered a progressing phase, the remaining 10 CSF analyses showed high levels of bound anti-MBP and MBP with correspondingly lower free values. Neurological exacerbations were

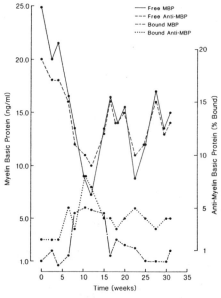

Figure 10   Longitudinal case study of hyperacute MS (type A of McAlpine).

248

associated with high F/B ratios, while conversely, progressing MS was associated with low F/B ratios. The changing pattern of free and bound MBP and anti-MBP in this patient remains biologically perplexing. For example with a sudden pulse of intrathecal anti-MBP synthesis a rise in the free anti-MBP level would be expected, but a drop in the corresponding bound fraction was not anticipated.

In conclusion, CSF analyses are an important diagnostic tool in the assessment of patients with confirmed and suspected MS. Not only is there evidence of increased intrathecal IgG synthesis as indicated by elevated IgG index or daily CNS IgG synthesis and oligoclonal banding, but in addition autoantibodies to the myelin protein, MBP, can be detected in patients with active disease. The effect of putative MS therapies on these autoantibody titers can now be determined.

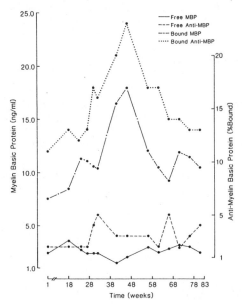

Figure 11   Longitudinal case study of progressing MS (type D of McAlpine).

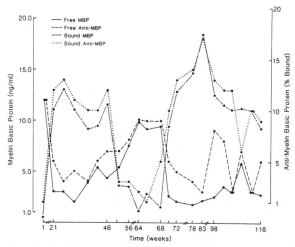

Figure 12   Longitudinal case study of relapsing-progressing MS (type C of McAlpine).

The Effect of Synthetic Corticosteroids on Titers of Cerebrospinal Fluid Anti-Myelin Basic Protein

We have studied the effects of methylprednisolone (MP) therapy on CSF anti-MBP in MS patients with acute relapses and progressing disease. MS patients with acute relapses were treated with high-to-mega dosages of MP for 10 days (Table 4). Changes in CSF results after treatment were compared to a nontreated control group. The untreated controls showed no change in either free, bound or F/B anti-MBP levels. Highly significant reductions ($p<0.05$ to $p<0.001$) were observed in free and F/B anti-MBP levels when these patients were treated with 160 mg/day or higher (Table 4, Fig 13, 14); mega doses were not required for this effect. In contrast to free anti-MBP in these patients the bound fractions showed a significant elevation. The combination of decreasing free and rising bound fractions contributed to a significant reduction of F/B ratios in patients treated with high-to-mega doses of MP (Fig 14). Longitudinal case studies of MS patients with exacerbations confirm results obtained in cross-sectional studies. Figure 16 illustrates two exacerbations of MS in a 28 year old female. In 1983 she had a prolonged severe

Table 4   Effects of methylprednisolone (MP) on CSF parameters (mean ± 2SD) in MS patients with exacerbations. (n=number of patients per group; Pre = data prior to commencing treatment (day 0); Post = data after 10 days of treatment). Student's t test p value: *$p<0.05$ ** $p<0.01$ *** $p<0.001$.

MP / day	No treatment		80 mg		160 mg		250 mg		1000 mg		2000 mg	
n	10		12		12		7		3		10	
	Pre	Post	Pre	Post	Pre	Post	Pre	Post	Pre	Post	Pre	Post
IgG Index	1.40± 0.5	1.46± 0.6	1.32± 0.5	1.20±0.4	1.21± 0.19	0.82±0.16**	1.31±0.3	0.86±0.1**	2.59± 0.3	1.13±0.5***	1.20±0.2	0.82±0.1**
IgG Synth	31.3 ±10.0	34.7 ±14.9	19.0 ±10.0	14.2 ±7.1*	25.4 ± 5.6	15.9 ±2.8**	29.8 ±4.6	19.7 ±2.8**	101.7 ±10.7	45.6 ±9.2***	21.1 ±2.6	5.8 ±2.2***
MBP Free	11.8 ± 7.8	11.5 ± 7.8	14.2 ± 5.5	10.5 ±4.8*	15.9 ± 2.8	5.9 ±1.6***	11.6 ±2.8	4.7 ±1.3***	17.1 ± 5.5	5.0 ±1.6***	15.2 ±2.9	4.6 ±1.6***
MBP Bound	3.6 ± 2.2	3.0 ± 2.0	1.9 ± 1.2	2.3 ±1.2	1.7 ± 1.0	5.5 ±4.2	2.6 ±1.2	7.8 ±3.6	0.9 ± 0.5	6.3 ±2.7	1.0 ±0.4	1.4 ±0.8
Anti-MBP Free	11.0 ± 6.0	11.0 ± 6.0	14.0 ± 4.0	10.0 ±3.0*	16.0 ± 2.0	5.0 ±2.0***	11.0 ±2.4	4.0 ±1.2***	16.0 ± 5.1	6.0 ±1.9***	14.0 ±2.1	5.0 ±1.8***
Anti-MBP Bound	2.0 ± 3.0	3.0 ± 2.0	2.0 ± 1.1	3.0 ±1.4	3.0 ± 1.5	6.0 ±4.0	5.0 ±1.5	8.0 ±3.1	1.0 ± 0.5	6.0 ±4.0	0.9 ±0.3	3.0 ±1.2
Anti-MBP F/B	7.7 ± 3.3	6.8 ± 3.2	7.8 ± 4.4	4.3 ±2.2*	6.0 ± 2.0	2.6 ±1.6**	3.9 ±1.2	0.91±0.74***	14.5 ± 5.0	5.9 ±1.8***	15.4 ±6.4	2.3 ±1.4***

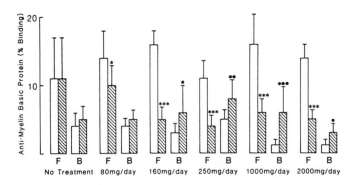

Figure 13   MS exacerbations: Free (F) and bound (B) CSF anti-MBP levels pre (day 0) and post (day 10) treatment for each dosage group. (White area = pre-treatment values; student's t test p values: *,· $p<0.05$; **,·· $p<0.01$; ***,··· $p<0.0001$).

250

exacerbation during which she was given multiple types of steroid therapy. Initially, she was treated with 2 courses of steroid therapy. Initially, she was treated with 2 courses of ACTH (A) (60 units/day for 10 days) and only modest reductions in F anti-MBP titers occurred. Similarly, the administration of 50 mg prednisone every 2 days for 12 days also produced insignificant reduction of free antibody titers (B). Three months after the patient was into this severe exacerbation she was treated with a megadose of 2000 mg MP/day (C) and even then, after 10 days there was only a modest anti-MBP reduction. Subsequently, the patient went into remission, but 6 months later she had another attack at the onset of this relapse, she was again treated with 2000 mg MP/day for 10 days, which this time was associated with a precipitous fall in free anti-MBP (C). Similar to clinical observations, this study suggests that corticosteroid therapy should be administered as soon as possible after the onset of an acute relapse in order to obtain maximal beneficial effects. Figure 17 illustrates a study of a 29 year old male MS patient with progressing disease and superimposed relapses. He had 19 CSF analyses over a period of 2 years. During periods of acute relapse, megadose of MP (A) administered early was associated with a significant fall of free anti-MBP and a rise of bound anti-MBP, while relapses treated with 80 mg MP/day (D), even though administered at the onset were not necessarily associated with reduction in free anti-MBP titers. This case also illustrates that administration of MP in either high (B) or mega (C) doses during the progressing phase of MS is not associated with reduction of anti-MBP titers.

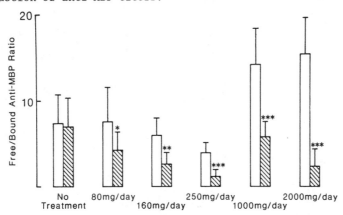

Figure 14   Free/bound anti-MBP ratio pre (day 0) and post (day 10) treatment in MS exacerbations.   (Symbols as in Figure 13).

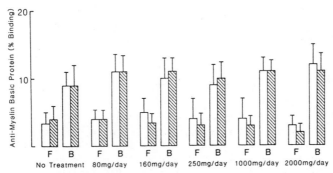

Figure 15 Progressing MS: Free (F) and bound (B) CSF anti-MBP levels pre (day 0) and post (day 10) treatment for each dosage group. (White are = pre-treatment values; hatched area = post-treatment values).

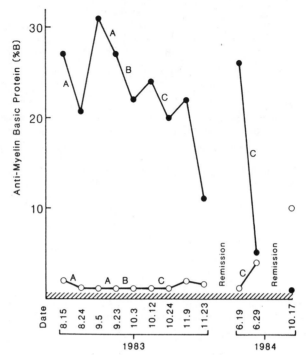

Figure 16   Longitudinal case study (12 CSF analyses over 14 months) of
anti-MBP levels of a patients with 2 MS exacerbations. Free anti-MBP
levels of a patient with 2 MS exacerbations.   Free anti-MBP always
exceeds the bound fraction. See text for details.   (o = free anti-MBP; o
= bound anti-MBP.   Course of treatment: A = ACTH: 60 units/day; B =
Prednisone: 50 mg/2 days; C = Methylprednisolone: 2000 mg/day).

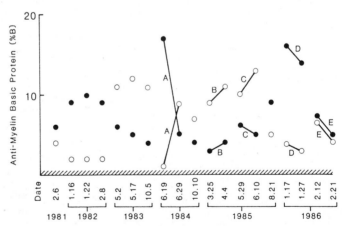

Figure 17   Longitudinal case study (19 CSF analyses over 5 years) of a
patient with progressing-relapsing MS.   Free anti-MBP levels exceed bound
fractions during relapses, and bound levels exceed free levels during the
progressing phase.   See text for details.   (o = free anti-MBP; o = bound
anti-MBP.   Course of treatment: A = Methylprednisolone (MB):   2000
mg/day; B = MP: 1000 mg/day; C = MP: 160 mg/day; D = MP: 80 mg/day; E =
MP: 250 mg/day).

The effects of various high-to-mega doses of MP on anti-MBP titers in MS patients with progressing disease are illustrated in Table 5 and Figure 15. Both free and bound anti-MBP titers were not significantly affected by any dosage of MP. This data illustrates that titers of free anti-MBP, the molecule potentially capable of causing demyelination, are more resistant to change under the influence of methylprednisolone in this group of patients. Figure 18 illustrates a 27 year old male hospital confined MS patient who had 23 CSF analyses over 3-1/2 years. This patient received low doses of 50 mg PR every 2 days with insignificant changes occurring in titers of both free and bound anti-MBP.

Table 5   Effects of methylprednisolone (MP) on CSF parameters (mean ± 2SD) of MS patients with progressing disease. (n=number of patients per group; Pre = data prior to commencing treatment (day 0); Post = data after 10 days of treatment). Student's t test p value: *p<0.05 **p<0.01 ***p<0.001.

MP / day	No treatment		80 mg		160 mg		250 mg		1000 mg		2000 mg	
n	8		16		19		11		9		12	
	Pre	Post	Pre	Post	Pre	Post	Pre	Post	Pre	Post	Pre	Post
IgG Index	1.10± 0.4	1.06± 0.3	1.0 ±0.41	0.89±0.43	3.65± 1.9	1.18±0.72*	2.32±0.9	1.07±0.7*	1.49±0.4	0.72±0.1***	1.24±0.2	0.77±0.2**
IgG Synth	12.9 ±10.0	13.9 ±12.0	11.9 ±5.9	9.3 ±4.3	25.1 ±15.9	16.4 ±9.2*	29.1 ±6.8	12.9 ±5.9**	29.3 ±6.3	6.4 ±3.2***	23.6 ±5.1	9.1 ±2.9***
MBP — Free	3.1 ± 1.6	3.2 ± 1.2	4.1 ±0.9	3.6 ±1.3	3.7 ± 2.2	2.4 ±1.6	2.8 ±1.6	1.8 ±0.5	3.7 ±0.9	2.3 ±0.9*	2.6 ±1.3	2.0 ±0.6
MBP — Bound	9.2 ± 1.9	8.6 ± 2.8	11.2 ±2.9	11.0 ±3.2	10.3 ± 2.1	10.5 ±2.1	9.7 ±3.8	10.0 ±3.8	11.4 ±2.0	10.0 ±1.9	11.8 ±2.1	11.7 ±2.2
Anti-MBP — Free	3.0 ± 1.9	4.0 ± 1.4	4.0 ±1.3	4.0 ±1.3	5.0 ± 1.9	3.0 ±1.8	4.0 ±0.9	3.0 ±0.8	4.0 ±2.3	3.0 ±1.1	3.0 ±0.9	2.0 ±0.8
Anti-MBP — Bound	9.0 ± 2.0	9.0 ± 3.1	11.0 ±2.4	11.0 ±2.2	10.0 ± 2.9	11.0 ±2.0	9.0 ±2.7	10.0 ±2.8	11.0 ±2.1	11.0 ±1.5	12.0 ±2.7	11.0 ±2.7
Anti-MBP — F/B	0.39± 0.3	0.48± 0.28	0.36±0.11	0.35±0.14	0.5 ± 0.2	0.3 ±0.2	0.51±0.17	0.29±0.13	0.40±0.1	0.23±0.07	0.41±0.05	0.20±0.06

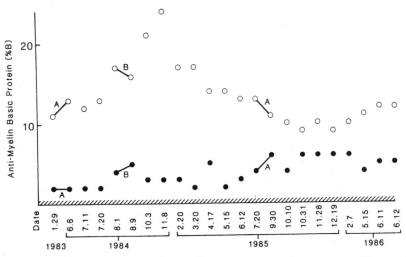

Figure 18   Longitudinal case study (23 CSF analyses over 30 months) of anti-MBP levels in a patient with progressing MS. Bound anti-MBP levels exceed free fractions. (o = free anti-MBP; o = bound anti-MBP. Course of treatment: A = Prednisone : 50 mg/2 days; B = Methylprednisolone: 2000 mg/day).

SUMMARY

    Although MS is characterized by increased total intrathecal IgG
synthesis, to date there has been no specific antibody associated with
its pathogenesis. Initially, in 1948, increased intra-BBB IgG synthesis
was observed by Kabat et al (25) and this was subsequently confirmed by
other laboratories (27,43,44,49,51). Increased intrathecal IgG synthesis
is commonly associated with chronic inflammatory diseases of the CNS such
as MS, subacute sclerosing panencephalitis (SSPE), or neurosyphilis but
does not occur in noninflammatory chronic diseases such as amyotrophic
lateral sclerosis, Parkinson's disease or Alzheimer's disease for
example. Theoretically, in MS, intra-BBB IgG synthesis should have a
specific purpose, but information to date shows that specific CSF
antibodies represent only 0.1% of total intrathecal IgG (45). It can
safely be concluded that the specificity of the majority of IgG has not
been identified. The kinetic relationship between anti-MBP and total
intrathecal IgG has not been determined to date, but a positive
correlation should not be anticipated since we have observed that as the
titer of anti-MBP varies in one direction that of intrathecal IgG may
vary in the same or opposite direction.

    Increased titers of antibodies to many viral and brain targets have
been identified in the CSF IgG of MS patients. These are thought to be
nonspecific and may be the result of polyclonal activation of B cells
which are recruited to the CNS by the event or events that cause MS (45).
Following the observation of increased titers of measles antibodies (1),
antibodies to other viruses have been identified. For example, in the
study of Norrby et al (31) 15% of the cases had antibodies to mumps, 19%
to rubella and 11% to herpes simplex. Titers to other viruses have also
been found to be elevated at times. About 50% of MS patients have a high
titer to one virus, about 15% to 2 viruses and about 5% to 3 or more
viruses (26,38,50). Unfortunately, viral antibodies have not been
reliably detected in high titer in the majority of MS patients and they
may be related to agents that commonly infect and/or invade the CNS and
have no relevance for the pathogenesis of MS (47).

    CSF antibodies to non-viral antigens have also been detected in MS.
Antibodies to myelin-associated glycoprotein (MAG) have been identified
(48). There is no evidence of antibodies to galactocerebroside (36)
while antibodies to ganglioside were found in 9 of 34 MS patients (18).
Antibodies against oligodendrocytes have been identified in MS CSF but
their titers are not significantly increased relative to controls (40).
None of these specific antibodies appear to be strongly associated with
disease activity.

    Autoantibodies directed against myelin basic protein may or may not
be involved in the pathogenesis of MS and consequently treatment attempts
to lower or eliminate them from the intrathecal compartment may or may
not be of benefit to patients. Although there have been negative
reports, antibodies to MBP are detectable in MS CSF (32) in greater titer
(per milligram IgG) than in the serum and they seem to be synthesized
within the intrathecal compartment (7). Consequently any treatment
modality developed to lower these autoantibodies will probably have to
cross the BBB. MS exacerbations are characterized by a high (above
unity) free to bound anti-MBP ratio, while progressing MS is
characterized by a low (below unity) free to bound anti-MBP ratio (52).
This fact together with the common clinical observation that relapsing MS
occurs in the early years while the progressing form develops many years
later suggests that there are multiple mechanisms involved in MS
pathogenesis. For example, MS exacerbations seem to be characterized by

a sudden pulse of anti-MBP with its subsequent elimination by unknown mechanisms allowing the patient to go into remission, whereas progressing MS is characterized by the chronic persistence of these autoantibodies in low titers. Separate treatment modalities designed to eliminate autoantibodies to MBP from these different phases of MS may be required. It has also been determined that titers of anti-MBP correlate with those of MBP in patients with active disease (53) suggesting that these autoantibodies are indeed involved in the pathogenesis of MS. However, it is also possible that their appearance in patients with active MS may simply be due to polyclonal B cell activation restricted to the active phases of the disease. Antibodies to MBP may also be detected in patients with SSPE (32,52). This, however, does not negate the fact that they may also be involved in the pathogenesis of MS since their appearance in patients with SSPE may be a consequence of the measles virus infection resulting in release of MBP into the systemic compartment, while in MS their development may be due to altered immunoregulation of normally covert autoantibodies. Furthermore, the MBP epitope(s) against which these antibodies are directed may be distinctly different in these two diseases. Chou et al (8) utilizing a solid phase radioimmunoassay measured antibodies to intact MBP as well as MBP fragments and found comparable levels of binding in the CSF of MS patients and controls. This is in distinct contrast to our large control population in which these autoantibodies have not been detected. It has to be borne in mind that the development of autoantibodies in MS patients may not be restricted to MBP since autoantibodies to MAG have also been detected (48). It is also possible that multiple autoantibodies develop against different epitopes on the human MBP molecule. It seems clear that the exact pathogenesis of demyelination in MS patients should be determined prior to attempts to develop a highly successful treatment modality.

ACKNOWLEDGEMENTS

Assistance with patient care was provided by P. Shaw, J. Christopherson, J. Rodgers and Dr. D.J. Carroll. Support for this research was provided by P. and L. May and the Friends of MS Patient Care & Research Clinic, C. and R. Giese, E. Laforge and the Tegler Foundation, G. Gerth and friends, Barrhead, and L. Atkins.

REFERENCES

1. Adams JM, Imagawa DT: Measles antibodies in multiple sclerosis. Proc Soc Exp Bio Med III:562-566, 1962
2. Alvord EC, Shaw CM, Hruby S, Sites L, Slimp JC: The onset of experimental allergic encephalomyelitis as defined by clinical, electrophysiological and immunochemical studies. In Alvord CE, Kies MW, Suckling AJ (eds): EAE a useful model for MS. Alan R Liss Inc, New York, 1984, pp 461-466
3. Braun PE, Brostoff SW: Proteins of myelin. In Morell P (ed): Myelin, Plenum Press, New York, 1977, pp 201-230
4. Brostoff SW, Eylar EH: The proposed amino acid sequence of the P_1 protein of rabbit sciatic nerve myelin. Arch Biochem Biophys 153:590-598, 1972
5. Carnegie R: Amino acid sequence of the encephalitogenic protein of human myelin. Biochem J 123:57-67, 1971
6. Carswell R: Pathological anatomy: Illustrations on the elementary forms of the disease. Longman, Orwe, Brown, Green and Longman, London, 1838

7. Catz I, Warren KG: Intrathecal synthesis of autoantibodies to myelin basic protein in multiple sclerosis. Can J Neurol Sci 13:21-24, 1986

8. Chou CHJ, Tourtellotte WW, Kibler RF: Failure to detect antibodies to myelin basic protein or its peptic fragments in CSF of patients with MS. Neurology 33:24-28, 1983

9. Cohen SR, Herndon RM, McKhann GM: Radioimmunoassay of myelin basic protein in spinal fluid. N Engl J Med 295:1455-1457, 1976

10. Cohen SR, Brooks BR, Herndon RM, McKhann GM: A diagnostic index of active demyelination: Myelin basic protein in cerebrospinal fluid. Ann Neurol 8:25-31, 1980

11. Colby SP, Sheremat W, Bain B, Eylar EH: Cellular hypersensitivity in attacks of MS. 1. A comparative study of migration inhibitory factor production and lymphoblastic transformation in response to MBP in multiple sclerosis. Neurology 27:132-139, 1977

12. Coyle PK: CSF immune complexes in multiple sclerosis. Neurology 35:429-432, 1985

13. Cruveilhier J: Anatomie pathologique du corps humain. Bailliere, Paris, 1842

14. Dasgupta MK, Catz I, Warren KG: Myelin basic protein: A component of circulating immune complexes in multiple sclerosis. Can J Neurol Sci 10:239-243, 1983

15. Dawson JW: The histology of disseminated sclerosis. Published as part of: Transactions of the Royal Society of Edinburgh, 1916

16. Dejone RN: Multiple sclerosis: History, definition and general considerations. In Vinken PJ, Bruyn GW (eds): Handbook of Clinical Neurology, North Holland, Amsterdam, 1970, vol 9, pp 45-62

17. Diebler GE, Martenson RE, Kies MW: Large scale preparation of myelin basic protein from central nervous tissue of several mammalian species. Prep Biochem 2:139-165, 1972

18. Endo T, Scott DD, Stewart SS, Kundu SK, Marcus DM: Antibodies to glycosphingolipids in patients with MS and SLE. J Immun 23:1793-1797, 1984

19. Esiri MM: Immunoglobulin containing cells in multiple sclerosis plaques, Lancet II:478-481, 1977

20. Esiri MM: Multiple sclerosis: A quantitative and qualitative study of immunoglobulin-containing cells in the central nervous system. Neuropath Appl Neurobiol 6:9-21, 1980

21. Firth D: The case of Augustus d'Este, Cambridge University Press, London, 1948

22. Hershey LA, Trotter JL: The use and abuse of the CSF IgG profile in the adult: A practical evaluation. Ann Neurol 8:426-434, 1980

23. Hunter WM, Greenwood FC: Preparation of I^{131} labelled human growth hormone of high specific activity. Nature 195:495-496,1980

24. Johnson KP: CSF and blood assays of diagnostic usefulness in MS. Neurology 30:106-109, 1980

25. Kabat EA, Glusman M, Knaub V: Quantitative estimation of the albumin and gamma globulin in normal and pathologic cerebrospinal fluid by immunochemical methods. Am J Med 4:653-662, 1948

26. Leibowitz S, Hughes RAC: Multiple sclerosis. In Immunology of the Nervous System. Arnold E (ed) London 1983, pp 185-214

27. Link H, Tibbling G: Principles of albumin and IgG synthesis in neurological disorders. II. Evaluation of IgG synthesis within the central nervous system in multiple sclerosis. Scand J Clin Lab Invest 37:397-401, 1977

28. Lumsden CE: The immunogenesis of the multiple sclerosis plaque. Brain Res 28:365-370, 1971

29. Mehta PD, Frisch S, Thormar H, Tourtellotte WW, Wisniewski HM: Bound antibody in multiple sclerosis brains. J Neurol Sci 49:91-98, 1981

30. McAlpine D, Lumsden CE, Acheson ED: Multiple sclerosis: Cause

and prognosis. In Multiple sclerosis: A reappraisal. Churchill Livingstone, London, 1972, pp 214-220

31. Norrby E, Link H, Olsson JE: Measles virus antibodies in multiple sclerosis. Arch Neurol 30:285-292, 1974

32. Panitch HS, Hooper CS, Johnson KP: CSF antibody of myelin basic protein: Measurement in patients with MS and SSPE. Arch Neurol 37:206-209, 1980

33. Prineas JW, Connell F: The fine structure of chronically active multiple sclerosis plaques. Neurology 28:68-75, 1978

34. Prineas JW, Wright RG: Macrophages, lymphocytes and plasma cells in the perivascular compartment in chronic multiple sclerosis. Lab Invest 38:409-421, 1978

35. Rivers TM, Sprunt DH, Berry GP: Observations on attempts to produce acute disseminated encephalomyelitis in monkeys. J Exp Med 58:39-53, 1933

36. Rostami A, Eccleston PA, Silberberg DH, Manning MC, Burns JB, Lipak RP: Absence of antibodies to galactocerebroside in the sera and CSF of human demyelinating disorders. Neurology 33:130, 1983

37. Ryland H, Mork S, Matre R: In situ characterization of mononuclear cell infiltrates in lesions of multiple sclerosis. Neuropath Appl Neurobiol 8:403-411, 1982

38. Salmi AA: Viral antibodies in multiple sclerosis. In Problems in Multiple Sclerosis Research, Pedersen E, Clausen J and Oades L (eds) FADL's Forlag, Copenhagen, 1983, pp 368-370

39. Schumacher GA, Beebe G, Kibler RE: Problems of experimental trials of therapy in MS. Ann NY Acad Sci 122:552-568, 1965

40. Steck AJ, Link H: Antibodies against oligodendrocytes in serum and CSF in MS and other neurological diseases. I^{125} protein A studies. Acta Neurol Scand 69:81-89, 1984

41. Tibbling G, Link H, Ohlman S: Principles of albumin and IgG analyses in neurological disorders. I. Establishment of reference values. Scand J Clin Lab Invest 37:385-390, 1977

42. Tourtellotte WW: Multiple sclerosis cerebrospinal fluid. In Vinken PJ, Bruyn GW (eds): Handbook of Clinical Neurology, North Holland, Amsterdam, 1970, vol 9, pp 324-383

43. Tourtellotte WW: On cerebrospinal fluid IgG quotients in multiple sclerosis and other diseases. J Neurol Sci 10:279-304, 1970

44. Tourtellotte WW, Ma B: Multiple sclerosis: The blood-brain barrier and the measurement of de novo central nervous system IgG synthesis. Part II. Neurology (NY) 28:76-83, 1978

45. Tourtellotte WW: The CSF in MS. In Koetsier JC (ed): Handbook of Clinical Neurology: Demyelinating diseases, North Holland, Amsterdam, 1985, vol 3, pp 79-130

46. Traugott W, Stone SH, Raine CS: Chronic relapsing experimental allergic encephalomyelitis. J Neurol Sci 41:17-29, 1979

47. Vartdal F, Vandvik B, Norrby E: Viral and bacterial antibody responses in MS. Ann Neurol 8:255, 1980

48. Wajgt A, Gorny M: CSF antibodies to myelin basic protein and to myelin-associated glycoprotein in multiple sclerosis - evidence of the intrathecal production of antibodies. Acta Neurol Scand 68:337-343, 1983

49. Walsh JM, Tourtellotte WW: The cerebrospinal fluid in multiple sclerosis. In Hallpike JF, Adams CWM, Tourtellotte WW (eds): Multiple sclerosis. Williams & Williams, Baltimore, 1982, pp 275-358.

50. Walsh MJ, Tourtellotte WW: IgG, IgM and IgA temporal clonal stability in multiple sclerosis. Neurology 33 (suppl 2):123, 1983

51. Warren KG, Catz I: An investigation of the relationship between cerebrospinal fluid myelin basic protein levels and IgG measurements in multiple sclerosis patients. Ann Neurol 17:475-480, 1985

52. Warren KG, Catz I: Diagnostic value of cerebrospinal fluid anti-myelin basic protein in multiple sclerosis patients. Ann Neurol 20:20-25, 1987

53. Warren KG, Catz I: A correlation between CSF myelin basic protein and anti-myelin basic protein in multiple sclerosis patients. Ann Neurol 21:183-189, 1987

54. Whitaker JN: The antigenicity of myelin basic protein: Production of antibodies to encephalitogenic protein with DNA-encephalitogenic protein complexes. J Im Neurol 144:823-900, 1975

55. Whitaker JN: Myelin encephalitogenic protein fragments in the cerebrospinal fluid of persons with multiple sclerosis. Neurology 27:911-920, 1977

56. Whitaker JN, Lisak RP, Bashir RM et al: Immunoreactive myelin basic protein in the cerebrospinal fluid in neurological disorders. Ann Neurol 7:58-64, 1978

57. Wisniewski HM, Bloom BR: Experimental allergic optic neuritis in the rabbit: A new model to study primary demyelinating diseases. J Neurol Sci 24:257-263, 1975

58. Wisniewski HM: Morphogenesis of the demyelinating process. In Davidson AN, Humphrey JH, Liversedge LA, McDonald WI, Porterfield JS (eds): Multiple Sclerosis Research, HMSO, London, 1975, pp 132-141

59. Wisniewski HM, Keith AB: Chronic relapsing experimental allergic encephalomyelitis: An experimental model of multiple sclerosis. Ann Neurol 1:144-148, 1977

18

MAGNETIC RESONANCE IMAGING IN DEMYELINATION

D.W. Paty

Division of Neurology
University of British Columbia
Vancouver, Canada   .

INTRODUCTION

The ability of magnetic resonance imaging (MRI) to delineate the localized areas of abnormality in multiple sclerosis (MS) has created a considerable amount of excitement in the MS community. The most immediate impact of MRI in the MS field is going to be on the accuracy of diagnosis. In addition, the ability to very precisely outline the areas of abnormality and to follow those areas over time is going to make a considerable impact on the follow up of the disease, the understanding of the evolution of the pathological process, and the adjudication of clinical trials. The clinical course in MS evolves over many years. The clinical course is also so complex that the neuropathologist has usually been unable to correlate the history of the patient to the pathology found at autopsy. It is in the area of clinical pathological correlation that MRI is going to have its major long term impact. Pathological correlation studies (1) are validating the sensitivity and specificity of the technique, and it is clear that MRI will eventually demonstrate the evolution of the pathological process. For example, systematic MRI studies are going to help us to answer the question of whether the clinical nature of MS, as a phasic disorder, reflects the actual evolution of pathology. At this point, there is reason to believe that the pathological process is one that continues uninterrupted, despite the only intermittent clinical symptoms in the majority of patients.

The most specific and characteristic pathological change in MS is demyelination. Specific demyelination occurs when there is loss of myelin with preservation of axons. In addition to demyelination, we also know that, in acute lesions, though pathological experience is limited, inflammation is quite intense (2). This inflammation is marked by the presence of immunoglobulins, lymphocytes, and plasma cells. Macrophages are active in the process of damage to myelin. The presence of immunoglobulins is one of the characteristics of MS resulting in a pattern of proteins in the cerebrospinal fluid (CSF) of oligoclonal banding (OB) that is very useful in diagnosis. CSF OB is a necessary component to satisfy the diagnostic criteria for laboratory supported definite MS (LSDMS) (3). Though little is understood about the process, the very first event in the development of the acute MS lesions may be breakdown in the blood brain barrier (BBB) (4). In spite of the fact that the enhanced CT scan demonstrates BBB disruption, the CSF albumin is usually not elevated. This lack of CSF albumin elevation indicates that the breakdown in the BBB is a localized phenomenon. Probably because of its exquisite sensitivity to changes in water content, MRI has the capability of being able to detect the acute areas of abnormality in MS. It may also be able to distinguish, in the future, between lesions that are primarily inflammatory as opposed to those that are primarily demyelinating and/or gliotic in nature.

Gliosis develops in the later stages of lesion development in MS. The gliotic scar is produced by astrocytic proliferation which begins in the subacute phase of the lesion and then progresses to form the chronic contracted scar characteristic of the end stage lesion. It is not known exactly how much the process of gliosis contributes to the neurological deficit, but Fog (5) has proposed that the chronic evolving clinical deficit in MS may be at least partly be due to gliosis. If chronic progressive disability in MS is due to the development of gliosis and not to immunological factors, the concept of treatment with immunosuppressives in chronic progressive MS (CPMS) will require revision. Newer strategies of therapy for CPMS should perhaps include use of inhibitors of astrocytic proliferation as well as the use of immunosuppressives and immunomodulators. We have compared the evolution of the process, as revealed by MRI, in relapsing and remitting and CPMS patients (see below).

Brain Scanning. Positron Emission Tomography and CT Scanning in MS

The first application of imaging to the study of MS was the use of radionuclide brain scanning. This technique has generally been very insensitive to MS lesions. In spite of that, there have been a number of reports of positive radionuclide brain scans in MS (6). Some of these positive scans were thought to indicate neoplastic disease. We now know that a positive brain scan in MS is only a marker for gross disruption in the BBB. Even though the ability to show MS lesions on the brain scan has been limited, the fact that MS lesions could be seen at all produced a considerable amount of interest in that perhaps the lesion of MS might be finally amenable to quantitative detection and follow up. The first abnormal radionuclide brain scan in MS was reported in 1954 (7).

Positron emission tomography (PET) has also been found to be abnormal in some patients with MS (8). A reduction in grey matter oxygen utilization (fluorodeoxyglucose) has been thought to be due to the undercutting of grey matter by large demyelinated lesions. Some reports have suggested that reduction in oxygen utilization and cerebral blood flow are linked (9). However, as in brain scanning, the sensitivity of PET in MS is low. In spite of this, the use of sensitive markers for BBB disruption using PET may be useful in the future because, as noted above, disruption in the BBB may be a primary lesion in the MS pathology.

A major advance in imaging lesions in MS occurred with the advent of computerized tomography (CT). Early studies suggested that 36% of patients had low density lesions in their periventricular white matter (10). Earliest studies did not report contrast enhancement, but subsequent studies, particularly using high volume delayed contrast (HVD) have shown that multiple enhancing lesions are not at all unusual (11). It is now accepted that multiple enhancing lesions in the white matter characteristic of MS can be used to satisfy the diagnostic criteria for dissemination of lesions in space. The enhancing CT lesions can resolve quickly and seem to be very susceptible to the effects of corticosteroids. This finding suggests that corticosteroids can help to heal a disrupted BBB, particularly when the steroids are administered intravenously (12).

In addition to focal low density and enhancing lesions, many patients with longstanding MS show ventricular enlargement and/or cortical atrophy on CT. The ventricular enlargement has been confirmed in several autopsy cases. There have also been numerous reports of biopsy and autopsy confirmation of CT lesions. Comparative studies using unenhanced, conventionally enhanced, and HVD enhanced CT scans have show that in 72% of MS cases more information could be derived from HVD CT than from ordinary single dose enhanced scans. In addition, the enhancing lesions seen on HVD CT are probably more likely to occur in acute relapsing patients than in chronic and/or stable ones. Unfortunately, CT is not sensitive enough to be helpful in the majority of patients for either diagnosis or follow up (13). Approximately 72% of patients in acute relapse may show multiple enhancing lesions, but the frequency of abnormality drops to 30 or 40% in patients during remission. CT has also shown that the pattern of lesions in MS can be quite varied. Ring enhancement, mass formation with edema, and multilobed lesions suggesting abscess have not been unusual findings.

Multifocal enhancing lesions seen on CT are not, of course, specific for MS. Metastatic neoplasm, infarcts, abscess, primary brain tumors, and other inflammatory neurological diseases can all produce multifocal enhancement. Therefore, the multiple lesions seen on CT can be used only as an anatomical marker for dissemination in space. The pathological process that is going on at the site of the enhancing lesions has not yet been identified.

When used in the follow up of patients, both CT and brain scanning can be used to study some of the characteristics of the disease process. Both of these techniques can identify disruption in the BBB. CT can also identify edema.

MRI In MS

MRI uses a strong magnetic field and radio frequencies to produce an image (14). The hydrogen atoms in the body will align themselves and spin like tops along a magnetic field, acting as tiny polar magnets. Radio frequencies, tuned to the resonance frequency of those atoms can cause them to spin at a different angle. The duration of the signal that emanates from the relaxation of that angle can be measured. Computer techniques can then detect not only the location of the source of the signal, but the quantity of atoms in question and some of the characteristics of the tissue surrounding the atoms. MRI measures mostly the water content and the molecular state of that water in tissues. Images can be constructed in many different ways. For MS studies the most commonly used image (spin echo) is one in which the lesions of MS stand out as high signal intensity areas (white) against a darker background. The darker background is made up of three contrasting areas; the grey matter being lighter in signal, the white matter being darker in signal, and the CSF showing as either black or white according to the radio frequency excitation parameters used.

One exciting aspect of MRI is that the image slices can be produced in multiple planes (see figure 1), therefore allowing the construction of a three-dimensional model of not only brian substance but of the distribution of MS lesions. Subsequently, by adding careful serial scans, a fourth dimension, that is of time, can be added.

Diagnosis (see Table 1)

The first report of the use of MRI in MS was by Young et al in 1981 (15). Eight patients with clinically definite MS (CDMS) were studied.

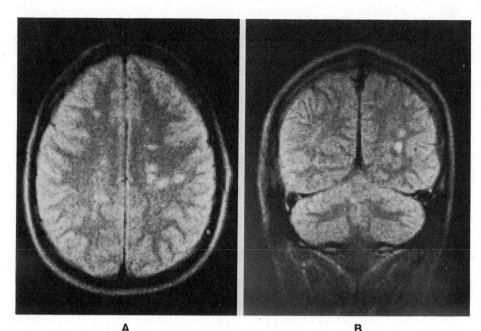

| A | B |

Figure 1A and B.   These two views show typical MRI MS lesions.   "A" shows several (white spots) lesions at a level above the ventricle in an axial view.   "B" shows two of the same lesions seen in a coronal view at the level of the parietal lobe.

Table 1 MRI IN DIAGNOSIS*

Conditions Studied	# Cases Studied	# Cases Positive	%Positive	Comments
CDMS	689	590	86%	Standards of what is considered positive on MRI vary considerably.
Suspected MS	533	317	59%	
ON	106	72	68%	
CPM	82	50	61%	

* Summary from the literature, for full reference list see reference 38.

Table 2 DIAGNOSIS OF MS COMPARING CT TO MRI*

	CT			MRI		
Condition Studied	# Pts. Studied	# Pts. Positive	% Pos.	# Pts. Studied	# Pts. Positive	% Pos
CDMS	313	162	52%	298	235	79%
Suspected MS	272	63	23%	272	149	55%

* Summary from the literature, for full reference list see reference 38.

All patients were found to have multiple focal areas of abnormality on inversion recovery (IR) scans. Additional patients were reported by the same laboratory in 1982 (16).

It was evident from those early studies that MS lesions could not only be detected, but the sensitivity of MRI for detection of lesions was as high as 10 times greater than that of CT. This differential sensitivity was greatest in the brainstem. Subsequent reports have supported the concept that MRI is more sensitive than is CT in the detection of MS lesions (17). In addition, experience has shown that the spin echo (SE) technique is more sensitive than is IR (18). At this point, MRI cannot distinguish between acute and chronic lesions but with increasing experience this differentiation may become possible. In contrast, enhanced CT shows something that current MRI techniques cannot, that is, BBB disruption. Perhaps a combination of MRI and CT will identify not only most of the MS lesions but also distinguish BBB disruption as well. The use of paramagnetic contrast agents such as gadolinium will also provide an MRI marker for BBB disruption.

One useful way of classifying the MR abnormalities seen in MS is that of Robertson and Li (18), which is as follows:

1. MRI Strongly Suggestive of MS (SSMS)
    a) Four lesions present
    b) Three lesions present, one periventricular in location

2. MRI Suggestive of MS
    a) Three lesions present
    b) Two lesions present, one periventricular in location

3. MRI Possible MS
    a) Two lesions present
    b) One lesion present, periventricular in location

4. One Lesion Only

Typical MS lesions are of high intensity on SE, measuring greater than 3 mm in diameter, and are prominently white matter in location. Caution must be exercised in the use of this classification. Multiple periventricular white matter lesions are characteristic of MS, but only in the proper clinical context (19). At this point one cannot differentiate with absolute certainty between MS lesions, multiple infarcts and such non specific patterns as subcortical arteriosclerotic encephalopathy (Bingswanger's disease). These non-specific lesions increase in frequency above the age of 50 to 7-10% of the normal population. Recent studies have shown that large (>6mm) lesions, brainstem lesions, and lesions located near the body of bilateral ventricles are more specific for MS. It is the pattern of lesions that is highly suggestive of MS. It must be remembered that tissue characterization is not yet possible on MRI. Therefore, critical clinical judgement must be used in the interpretation of the MRI appearance.

Our group at UBC has just completed a prospective study comparing HVD, CT, MRI, visual and somatosensory evoked potentials, and CSF analysis for OB in the evaluation of patients with suspected MS (20). MRI was the most sensitive technique for detecting asymptomatic lesions and demonstrating dissemination in space. In 200 cases of suspected MS; 40% had "MRI SSMS" scans. Sixty seven percent of the "MRI SSMS" patients also had OB. Evoked potentials were less helpful than MRI in detecting asymptomatic dissemination in space. HVD CT scan was abnormal in only

38% of the 98 patients who had "MRI SSMS" scans (see Table 2).

When looking at specific clinical groups, MRI and CSF analysis for OB were most helpful in chronic progressive myelopathy (CPM) and optic neuritis (ON). MRI was "definite" in 60% of 52 patients with CPM, and 66% of 38 patients with ON.

Follow up of these 200 suspected MS patients at one year has shown that 19 out of 200 (10%) had converted to CDMS by having a second clinical episode during the year. Of these 19 patients, 18 (95%) were predicted as having MS by having "MRI definite" scans during the study. Only long term clinical follow up will be able to establish the ability of any of these tests to accurately predict the diagnosis of CDMS. MRI scans have been done in identical twins that are non-concordant for MS (21,22). If asymptomatic disseminated MRI lesions means underlying MS in this context, the concordance rate in identical twins can possible be doubled by using MRI.

MRI Assessment of Disease Activity

It is with the use of serial MRI scans that more precise information is accumulating concerning the nature of the lesions. Johnson et al (23) and Li et al (24) both in 1984 reported that MRI lesions could diminish in size over time and that new lesions could appear asymptomatically. A series of systematic serial studies has now been undertaken at the University of British Columbia (25,26,27). The first serial study was of seven mostly relapsing CDMS patients who had monthly MRI scans, neurological evaluations, and immunological tests over a period of six months (25). There were five clinical relapses in three patients. Four of the patients were clinically stable throughout the study. It was found that very careful repositioning of patients was necessary in order to get reproducible and quantitative MRI data.

There were many new asymptomatic MRI lesions seen to develop in five of these patients. In the patients with clinical relapses, there was no correlation between the location of the new MRI lesions and the clinical findings. Two of the new MRI lesions were quite large ones. One large new cerebral MRI lesion occurred just prior to a spinal cord clinical relapse. Immunological studies done in parallel with the scans showed that major changes in functional immunological studies (Con A generated suppressor cell activity, natural killer (NK) cell activity, and immunoglobulin secretion in vitro) paralleled the appearance of major new MRI lesions (28,29). There were no immunological correlations with clinical activity. There were no consistent correlations between lymphocyte phenotypic markers and either clinical or MRI evidence for disease activity.

A second serial study (26) involved eight mildly affected relapsing and remitting (R&R) CDMS patients scanned every two weeks over six months. There were 12 new or enlarging MRI lesions seen in five patients, all of them asymptomatic. There were two clinical relapses both in the same patient. A third serial study has just been completed (27). This study was in a group of eight patients with chronic progressive (CP) MS. The same sort of MRI activity as seen in both the relapsing and CP patients, but the CP patients had twice the rate of appearance of new lesions as was seen in the R&R patients. There was no clinical correlation with new MRI lesions in any of the three studies.

In summary, 24 patients have been studied systematically with serial MRI, clinical, and immunological observations (see Table 3). There were seven clinical relapses. In contrast there were 41 new MRI lesions and

Table 3 SERIAL MRI STUDIES*

Summary of Findings in 24 Patients

1. New MRI lesions appear at a much faster rate than do clinical relapses.
   - la) The rate of development of new and increasing size lesions in 16 relapsing patients was 5/patient/year.
   - lb) The rate of development of new or increasing size lesions in 8 CP patients was 22/patient/year.
   - lc) The clinical relapse rate was 0.6 relapses/patient/year.

2. Many new MRI lesions reduce in size or disappear over time (40%).

3. The time scale of the appearance of new lesions is gradual enlargement over 4-6 weeks and then decline in size over the next 6-8 weeks.

4. Immune function changes seem to parallel the evolution of large new MRI lesions, not clinical events or small MRI lesions.

* Summary from UBC studies; references 25,26,27,28,29.

Table 4   SUMMARY OF FINDINGS QUANTITATIVE MRI STUDIES
WITH CLINICAL CORRELATION*

1) The "burden of disease" as estimated by MRI does not correlate well with clinical estimates of severity.

   R values range from 0.022 to 0.558.

2) There are measurable changes in serial studies that show that the extent of the MS process gets greater over time (25% in 2 years).

3) Comparison of MRI lesions in "clinically benign" patients with matched moderately severely disabled CP patients shows that in 20% of the matched pairs the "burden of disease" is greatest in the benign cases.

4) The major MRI difference between benign and disabled patients is the high degree of confluence of lesions in the CP patients and the high degree of "clinical expression" of MRI detected brainstem lesions in the CP patients.

* Summary from UBC studies; references 30,31,32.

Table 5   BIOPSY AND AUTOPSY CORRELATION STUDIES: CT AND MRI*

Type of Lesion	# Cases
CT Low Density	3
CT Enhancing	9
CT:Cord Enhancement	1
MRI	8

* Summary from the literature, for full reference list see reference 38.

43 significantly enlarging or reappearing MRI lesions (previously disappearing lesions that reappeared). Chronic progressive patients had about twice as many "active" lesions as did the relapsing patients, but they also had more new lesions as well. These studies show that the frequency of new MRI lesions is five to 10 times the frequency of clinical relapses. Based upon this striking phenomenon, we propose that for accurate assessment of disease activity in future studies, serial MRI scans will be necessary. Even though there is no information available to identify the actual pathological process that is going on in these "active" and new lesions, they must represent a very fundamental part of the acute MS process. Therefore, serial MRI studies will provide an objective method of detecting new areas of abnormality (whatever they may represent pathologically). This technique should decrease the amount of time that must be spent in order to detect significant changes over time. Even though expensive, this method of assessment should increase the objectivity, and decrease the amount of time and/or the numbers of patients that are necessary in order to reach statistical significance in clinical trials.

These findings also show that disease activity in relapsing MS as detected by MRI is a much more continuously active process than had previously been thought. The fact that new lesions are coming and going asymptomatically shows that there is underlying smoldering activity that cannot be detected clinically. Exactly what the pathology of these new lesions is not apparent. They probably represent acute areas of breakdown in the BBB with inflammation and edema. Perhaps demyelination and remyelination also play a role (29).

Many of these newly detected lesions (40%) diminished in size or disappeared over time. The usual tempo of evolution of the new lesions was gradual enlargement over two to four weeks followed by a gradual decline in size over the next six to ten weeks. In addition, there were 38 instances in which there was MRI evidence for some lesions to be enlarging while others were decreasing in size simultaneously.

Quantitation of Size of Lesions and Clinical Correlations (see Table 4)

As noted above, it was felt early in our experience with MRI in MS that perhaps this technique could be used to measure the "disease burden" in individual patients. Such techniques have now been developed (30) using computer assisted methods of quantitation. These methods allow the computer to assist in the calculation and storage of the location, size, and intensity of lesions as well as quantitation of the total extent of disease.

Our experience has shown poor correlation between extent of disease as measured by MRI and the severity of disease as determined by clinical scoring methods (30). Reproducibility studies of the quantitation system noted above have shown that a skilled and experienced technician can outline the MS lesion with a reproducibility error of about 6%. Different observers, even experienced ones, have different ways of outlining lesions. Therefore, the error between observers can be as high as 17-20%. For this reason it is important that a single observer be used for any particular study in order to minimize systematic quantitation errors.

One very critical requirement in serial quantitation studies is that of patient repositioning. Very careful repositioning must be undertaken in order to minimize artifacts and errors. The current procedure at UBC is to reposition the patient based on two angles of external landmarks (the cantho meatal line and the nasion-tragal line). A midline sagittal slice (pilot) is then obtained and the head position is checked by an internal angle (the angle between the superior surface of the cerebellum and the anterior sphenoid sinus). If this angle differs by more than two degrees from previous scans, the patients is repositioned and the pilot scan is repeated. Using the available system software, the slices for the follow up scan are then programmed so that the middle slice of a simultaneous 12 slice series is tangential to the top of the cerebellum. This method usually results in a follow up series of slices that match the original slices to within the 2-3 mm.

These quantitative methods are now being used in clinical trials. Palmer et al (31) have found that there are significant increases to be measured by this method over two years. There was an average increase in "MS burden" in placebo treated patients of 25% over two years. In addition, the quantitative method has been used to examine and compare scans of patients with different manifestations of MS. Koopmans et al (32) has compared a matched group of benign (n=32) and progressive (n=32) MS patients. He has shown that in these pairs (matched for age, sex, and duration of disease), that 80% of the pairs showed a heavier "MS burden" in the progressive patient than in the benign patient. However, the total extent of disease overlapped considerably between the two groups. It is of interest that in 20% of cases there was a heavier "disease burden" in the benign patient than in the matched CP control. This latter finding shouldn't surprise anyone experienced in MS pathology. A single lesion in the spinal cord can cause severe motor disability, whereas large lesions in cerebral hemispheres can be totally asymptomatic. This lack of clinical correlation is one of the reasons for thinking that quantitative MRI may lend itself to more objectivity and accuracy in determining the "disease burden" and disease activity over time than do our standard clinical measurements.

The problems posed by the findings in the serial studies for the quantitation method are very perplexing. If the acute lesions are coming and going over time the measure of "disease burden" will be subject to considerable variability. This variability will, therefore, produce difficulties in statistical evaluations. However, these variability problems are no greater than the ones posed by the spontaneously fluctuating nature of the clinical features of MS.

Pathological Correlation Studies (see Table 5)

As mentioned above, there have been a number of biopsy and autopsy confirmations of CT abnormalities in MS. MRI pathological correlation studies have shown (1,33) that MRI accurately measures the extent of demyelination (Figure 1). Lesions as small as 3 mm in diameter can be detected. Unfortunately, there is nothing at this point that can help to distinguish acute from chronic lesions.

Preliminary data from a survey of demyelinated lesions in eight cases (1) has suggested that longer T_1 and T_2 values can be seen in the more heavily gliotic lesions. A protocol that involves immediate post mortem MRI scanning and then scanning of the fixed brain has shown that the extent of disease as detected by these two methods is very similar to that seen on pathological examination. Fixing the brain in formalin does not seem to interfere with the ability to visualize demyelinating lesions using SE sequences. IR sequences, however, do not demonstrate the lesions in fixed brain as well as in the unfixed state. In acute inflammatory experimental lesions, fixation destroys the ability to visualize the images by MRI (34,35). In the last few years there have been several attempts at producing experimental inflammatory demyelination in order to understand the contribution of the various pathological changes to the MR signal. Pathological correlation studies in human MS can help understand chronic demyelination, but acute inflammatory lesions, being very unusual in end stage MS, must be investigated by experimental methods.

Demyelination causes an increase in T_1 and T_2 relaxation times, and as noted above, pathological correlation studies have suggested that the more heavily gliotic lesion has longer relaxation times than does the ordinary demyelinating lesion. The MRI differentiation between white and grey matter is probably related to the relative water and lipid contents of the two tissues. White matter is high in lipid and low in water content. White matter also has shorter T_1 and T_2 values than does normal grey matter. Grey matter, being higher in water content and lower in lipid than white matter, also has longer T_1 and T_2 values.

The acute MS lesion probably involves disruption of the BBB and local edema. In addition, pathological studies of acute experimental demyelination have shown that there is an increased amount of immunoglobulin present. Our serial MRI studies would suggest that the primary and fundamental lesion in MS is an acute edematous and inflammatory one, probably resulting from local disruption of the BBB.

When demyelination occurs, there is disruption of the heavily compacted lipid-rich myelin membrane with degradation into free lipid and long chain fatty acids. The eventual development of gliosis is probably associated with increasing free water content, but there also must be an increase in bound water. This means that the pathological evolution of MS lesions involves a sequence of events such as; breakdown of the BBB, extravasation of fluid and proteins, accumulation of inflammatory cells, disruption of myelin, dissolution and phagocytosis of myelin debris followed by astrocytic gliosis and subsequent loss of axons.

Experimental Demyelination

In trying to understand the contribution of various pathological factors to the MR image some investigators have isolated the various components to study their individual contributions. These studies have shown that T_1 and T_2 prolongation produced by edema is a consistent finding. There are, however, no specific findings that will help distinguish edema from other causes of increased signal. In addition, specific and isolated demyelination can be produced by toxic elements. It would be important to also study such specific demyelination, in the absence of inflammation and edema, to identify the exact MRI changes associated with specific demyelination. Isolated gliosis can be found in certain experiments of nature. Mesial temporal sclerosis is one particular such pathological change. Studies of this phenomenon in patients with epilepsy have shown that gliosis is indeed associated with increased T_1 and T_2 relaxation times, but the relative contribution of free water versus bound water is not known.

Experimental models for MS have contributed a great deal toward the understanding of the immunology of the autoimmune process (36). In addition, experimental allergic encephalomyelitis (EAE) has now begun to provide information for the interpretation of the MRI appearances in acute demyelination (34,37).

EAE can be produced in susceptible species and strains by sensitization to various myelin proteins including myelin basic protein (MBP). MBP makes up approximately 30% of the proteins of normal CNS myelin. Both acute, hyperacute, and chronic, relapsing and remitting, courses of EAE are made possible by various manipulations of the experimental conditions and proper selection of the susceptible species and strain. Chronic relapsing EAE in guinea pigs has been shown to involve both acute inflammatory lesions in the initial stages and demyelination in the chronic stage. Primate EAE is not only both inflammatory and demyelinating, but can also be necrotic with interstitial hemorrhage. The early lesions in EAE are relatively small perivascular inflammatory ones that progressively coalesce into large irregularly shaped ones. Guinea pig EAE has been studied by Karlik and Noseworthy (37). They have found that the acute lesions of EAE, particularly those that are edematous, can be associated with prolongation of T_1 and T_2 values. They also found a very interesting phenomenon, that is, the presence of inflammatory cells in an otherwise edematous lesion can be associated with normalization of the T_1 and T_2 values and therefore potentially, normalization of the MR image. Their work in the guinea pig spinal cord has pioneered attempts at interpretation of the MR image by providing knowledge of the specific tissue parameters involved in inflammatory demyelination.

Stewart et al have produced EAE in primates (34,35). Her studies have demonstrated that:
1. Acute EAE lesions can be seen before the onset of clinical signs. These MRI lesions are due to a progressive increase in T_1 and T_2 values in the lesion area.
2. MRI is an accurate indicator of the gross extent of the lesions in this model.
3. Contrary to previously held thoughts, the most longstanding lesions are the most hemorrhagic.
4. The EAE lesions in primates vary considerably in their histologic components, and the MRI characteristics of those lesions also vary. Microscopically similar lesions seem to have the same MR characteristics.
5. With progression of the disease process, the MRI characteristics change, mostly showing prolongation of T_1 and T_2 values over time.

She has now developed a model for relapsing MRI visable primate EAE. She has one animal that has had bilateral asymptomatic MRI visable lesions that resolved spontaneously only to be followed by new lesions in different areas of the CNS. This particular animal is currently under observation.

SUMMARY

The most obvious impact of MRI in MS has been in accuracy of diagnosis. More recently MRI is providing a number of insights into the study of disease activity as follows:
1. MRI can be used as a quantitative index of the extent of disease.
2. Frequent carefully repositioned MRI scans can detect a degree of disease activity in otherwise stable MS that suggests that signs of clinical activity are truly the tip of the iceberg.
3. Experimental and pathological correlation studies are beginning to show the way toward the use of MRI and NMR in tissue characterization.
These new findings mean that frequent and carefully quantitative MRI scans are going to be a necessary component in future clinical investigations in MS, including clinical trials.

ACKNOWLEDGEMENTS

Thanks go to the following for their help in all of the UBC studies: Matt Bergstrom, Ken Berry, Paul Cutler, Kathy Eisen, Rochelle Farquhar,

Ed Grochowski, Sharon Hall, Stan Hashimoto, John Hooge, Carey Isaac, Kathy Jardine, Lorne Kastrukoff, Robert Koopmans, Joel Oger, Matt Palmer, Wendy Stewart, Ernest Willoughby, and Gerry Walker.

These studies have been supported by the British Columbia Health Care Research Foundation, the Medical Research Council of Canada, the National Multiple Sclerosis Society (US), the Multiple Sclerosis Society of Canada, the Kroc Foundation and the Jacob W. Cohen Fund for Research in Multiple Sclerosis.

REFERENCES

1.   Stewart WA, Hall LD, Churg A, Oger J, Hashimoto SA, Paty DW: Magnetic resonance imaging (MRI) in multiple sclerosis (MS): Pathological correlation studies in eight cases. Neurology 36(Suppl 1):320,1986

2.   Prineas JW, Kwon EE, Sharer LR, Cho E-S: Massive early remyelination in acute multiple sclerosis. Neurology 37(Suppl 1):109, 1987

3.   Poser CM, Paty DW, Scheinberg L, McDonald WI, Davis FA, Ebers GC, Johnson KP, Sibley WA, Silberberg DH, Tourtellotte WW: New diagnostic criteria for multiple sclerosis: Guidelines for research protocols. Ann Neurol 13:227-231, 1983

4.   Ebers GC, Vinuela FV, Feasby T, Bass B: Multifocal CT enhancement in MS. Neurology 34:341-346, 1984

5.   Fog T, Linnemann F: The course of multiple sclerosis. Acta Neurol Scand 46 (Suppl 47):1-175, 1970

6.   Antunes JL, Schlesinger EB, Michelsen WJ: The abnormal brain scan in demyelinating diseases. Arch Neurol 30:269-271, 1974

7.   Seaman WB, Ter-Pogossian MM, Schwartz HG: Localization of intracranial neoplasms with radioactive isotopes. Radiology 62:30-36, 1954

8.   Brooks DJ, Leenders KL, Head G, Marshall J, Legg NJ, Jones T: Studies on regional cerebral oxygen utilization and cognitive function in multiple sclerosis. J Neurol Neurosurg Psychiat 47:1182-1191, 1984

9.   Sheremata WA, Sevush S, Knight D, Ziajka P: Altered cerebral metabolism in MS. Neurology 34(Suppl 1):118, 1984

10.   Glydensted C: Computer tomography of the cerebrum in multiple sclerosis. Neuroradiology 12:33-42, 1976

11.   Sears ES, McCammon A, Bigelow R, Hayman LA: Maximizing the harvest of contrast enhancing lesions in multiple sclerosis. Neurology 32:815-820, 1982

12.   Sears ES, Tindall RSA, Zarnow H: Active multiple sclerosis. Arch Neurol 35:426-434, 1978

13.   Vinuela FV, Fox AJ, Debrun GM, Feasby TE, Ebers GC: New perspective in computed tomography of multiple sclerosis. Amer J Radiol 139:123-127, 1982

14.   Brant-Zawadzki M, Norman D: Magnetic resonance imaging of the central nervous system. Raven Press, New York, 1987

15.   Young IR, Hall AS, Pallis CA, Bydder GM, Legg NJ, Steiner RE: Nuclear magnetic resonance imaging of the brain in multiple sclerosis. Lancet 2:1063-1066, 1981

16.   Bydder GM, Steiner RE, Young IR, Hall AS, Thomas DJ, Marshall J, Pallis CA, Legg NJ: Clinical NMR imaging of the brain: 140 cases. Amer J Radiol 139:215-236, 1982

17.   Brant-Zawadzki M, Davis PL, Crooks LE, Mills CM, Norman D, Newton TH, Sheldon P, Kaufman L: NMR demonstration of cerebral abnormalities: Comparison with CT. Amer J Radiol 140:847-854, 1983

18.   Robertson WD, Li D, Mayo J, Genton M, Paty DW: Magnetic resonance imaging in the diagnosis of multiple sclerosis. World Congress

of Neurology. In press, Can J Radiology, 1987

19. Paty DW, Asbury AK, Herndon RM, McFarland HF, McDonald WI, McIlroy WJ, Prineas JW, Scheinberg LC, Wolinsky JS: Use of Magnetic resonance imaging in the diagnosis of multiple sclerosis: Policy statement. Neurology 36:1575, 1986

20. Paty DW, Hashimoto S, Hooge J, Oger J, Kastrukoff LF, Eisen AA, Eisen K, Purves S, Brandejs V, Robertson W, Li D: Magnetic resonance imaging in the diagnosis of multiple sclerosis (MS): A prospective study of comparison with clinical evaluation, evoked potentials, and oligoclonal banding. Neurology

21. Ebers GS, Dennis MD, Bulman E, Sadovnick AD, Paty DW, Warren S, Hader W, Murray J, Seland P, Duquette P, Grey T, Nelson R, Nicolle M, Brunet D: A population-based study of multiple sclerosis in twins. N Engl J Med 315:1638-1642, 1986

22. McFarland HF, Patronas NJ, McFarlin DE, Mandler RN, Beall SS, Cross AH, Goodman A, Krebs H: Studies of multiple sclerosis in twins using magnetic resonance. Neurology 35(Suppl 1):137, 1985

23. Johnson MA, Li DKB, Bryant DJ, Payne JA: Magnetic resonance imaging: Serial observations in multiple sclerosis. Amer J Neuroradiol 5:495-499, 1984

24. Li D, Mayo J, Fache S, Robertson WD, Paty D, Genton M: Early experience in nuclear magnetic resonance imaging in multiple sclerosis. Ann NY Acad Sci 436:483-486, 1984

25. Isaac C, Li D, Genton M, Jardine C, Grochowski E, Palmer M, Kastrukoff LF, Oger J, Paty W: Multiple sclerosis: A serial study using MRI in relapsing patients. In the press, Neurology (Neurology 36(Suppl 1):177, 1986)

26. Willoughby E, Grochowski E, Li D, Oger J, Kastrukoff LF, Paty D: A prospective study of MRI in multiple sclerosis. In the press, Annals of Neurology (Neurology 37(Suppl):231, 1987)

27. Koopmans R, Li DKB, Oger JJF, Kastrukoff L, Paty DW: Serial MRI studies in chronic progressive MS - an assessment of disease activity. Manuscript in preparation, 1988

28. Oger J, Kastrukoff L, Paty D: Multiple sclerosis: Relationship between suppressor cell function, IgG secretion in vitro, and the attacks of multiple sclerosis as studied by serial clinical and MRI examinations. Ann Neurol 120:161, 1986

29. Oger J, Kastrukoff LF, Li DKB, Paty DW: Multiple Sclerosis: In relapsing patients immune functions vary with disease activity as assessed by MRI. In the press, Neurology

30. Paty DW, Bergstrom M, Palmer M, MacFadyen J, Li D: A quantitative magnetic resonance image of the multiple sclerosis brain. Neurology 35(Suppl 1):137, 1985

31. Palmer MR, Bergstrom M, Grochowski E, Apted C, Li DK, Genton M, Hashimoto SA, Paty DW: Magnetic resonance imaging (MRI) in multiple sclerosis (MS): Quantitative changes in the size of lesions over 6 months in the placebo limb of a therapeutic trial. Can J Neurol Sci 13:168, 1986

32. Koopmans RA, Li D, Grochowski E, Cutler D, Paty DW: MRI characteristics of benign and chronic progressive multiple sclerosis. Can J Neurol Sci 14:241, 1987

33. Stewart WA, Hall LD, Berry K, Paty DW: Correlation between NMR scan and brain slice: Data in multiple sclerosis. Lancet 2:412, 1984

34. Stewart WA, Alvord EC, Hruby S, Hall LD, Paty DW: Magnetic resonance imaging (MRI) of experimental allergic encephalomyelitis (EAE) in primates. J Neuropath Exp Neurol

35. Stewart WA, Alvord EC, Hruby S, Hall LD, Paty DW: Early detection of experimental allergic encephalomyelitis by magnetic resonance imaging. Lancet 2:898, 1985

36. Raine CS: Multiple sclerosis and chronic relapsing EAE: Comparative ultrastructural neuropathology. In: Hallpike JF, Adams CWM,

Tourtellotte WW (eds) Multiple sclerosis. Chapman & Hall, London , pp 413-460, 1983

37. Karlik SJ, Strejan G, Gilbert JJ, Noseworthy JH: NMR studies in experimental allergic encephalomyelitis (EAE): Normalization of T1 and T2 with parenchymal cellular infiltration. Neurology 36:1112-1114, 1986

38. Paty DW, Li DKB: Neuroimaging in multiple sclerosis. Theodore, W: Clinical Neuroimaging. Allan Liss, Inc. pp 249-278, 1988

growth factors, 29-34
  human, 1-11
  mouse, 164-167
Oligodendrocyte-type 2 astrocyte, 19
Optic neuritis, 239, 247
Orvivirus, 208

P_0 protein, 20, 229
P_2 protein, 20, 228
Parainfluenza, 130
Paramixovirus, 124
Patch clamp recording, 97, 113-119
Pelizaeus-Merzbacher disease, 20, 94
Peroxidase, 227, 230
Phorbol-dibutyrate, 33, 39
Phorbol-ester, 42, 104
Phospholipase, 229
Plasma cells, 238
Plasmin, 229
Platelet derived growth factor, 20, 30
Pokeweed mitogen, 192, 209
Positron emission tomography (PET), 260
Progressive multifocal
    leukoencephalopathy (PML),
    55, 132-133, 153
Proliferation index, 35
Prostaglandin GE2, 197
Protease, 221
Protein kinase C, 44, 104

Quaking mutant mouse, 66
Quivering mutant mouse, 77

Rabies virus, 130
Radioimmunoassay, 239
Radionuclide brain scanning, 260
Rat
  myelin-deficient, 20
Reactive oxidative intermediate, 229
Refsum's disease, 20
Remyelination, 97
Reovirus, 124
Retrograde transport, 157
Retrovirus, 10, 132, 145-151
Retroviral construct, 19
RNA virus, 145-146
Rubella, 180

Saltatory conduction, 10
Schwann cells, 17, 29, 49, 61, 157
  growth factors for, 29-34
Scrapie, 130
Semliki forest virus, 124, 134-135
Serotonin, 104
Serum-free medium, 4-5
Simian virus I and V, 130
Slow virus infection, 146
Somatic cell hybrid, 76

Subacute sclerosing panencephalitis
    (SSPE), 208, 254
Substance P, 103
Sulfatide, 2, 9
SV-40, 132

T antigen, 133
T cell, 187-198
  CD4 subset, 187
  CD8 subset, 187
  helper, 190
  receptors, 189
  suppressor, 190
Tear, 208
Tetrodoxofin, 98
Theiler's murine encephalomyelitis
    virus, 121, 125, 135, 154
Transgenic mouse, 133
Tropical spastic paraparesis (TSP),
    131, 146-148

Vaccinia, 123
Varicella zoster, 208
Virus, 121-128, 129-143, 145-152,
    153-171
Visna, 135, 145
von Recklinghausen's disease, 20

X-ray diffraction, 228

DATE DUE